Genetic and Genomic Resources of Grain Legume Improvement

Genetic and Genomic Resources of Grain Legume Improvement

Edited by

Mohar Singh
National Bureau of Plant Genetic Resources, Pusa, New Delhi, India

Hari D. Upadhyaya
International Crops Research Institute for Semi-Arid Tropics, Patancheru, Hyderabad, India

Ishwari Singh Bisht
National Bureau of Plant Genetic Resources, Pusa, New Delhi, India

ELSEVIER

AMSTERDAM • BOSTON • HEIDELBERG • LONDON • NEW YORK • OXFORD
PARIS • SAN DIEGO • SAN FRANCISCO • SINGAPORE • SYDNEY • TOKYO

Elsevier
32 Jamestown Road, London NW1 7BY, UK
225 Wyman Street, Waltham, MA 02451, USA

First edition 2013

British Library Cataloguing-in-Publication Data
A catalogue record for this book is available from the British Library

Library of Congress Cataloging-in-Publication Data
A catalog record for this book is available from the Library of Congress

ISBN: 978-0-12-397935-3

For information on all Elsevier publications
visit our website at store.elsevier.com

This book has been manufactured using Print On Demand technology. Each copy is produced to order and is limited to black ink. The online version of this book will show color figures where appropriate.

Contents

Preface

Grain legumes have great potential in alleviating protein hunger and malnutrition among resource poor peoples in the developing countries. Besides, grain legumes have symbiotic nitrogen fixing bacteria in root nodules, fix their own nitrogen, thereby reducing, in many situations, the cost of nitrogen inputs by farmers. Globally, ~1.1 million grain legume accessions are preserved in various gene banks. These genetic resources are the reservoir of several valuable genes/alleles for the present and future crop improvement programmes. In view of this, an effort has been made to collect and analyse the scattered scientific information on these resources in a book form on the current status of genetic and genomic resources of grain legume improvement.

The book entitled "*Genetic and Genomic Resources of Grain Legume Improvement*" comprises 12 chapters contributed by the eminent legume curators/researchers around the world. The first introductory chapter summarizes the landmark research on genetic and genomic resources in grain legumes. Each of the subsequent chapters (2–12) mainly deals with aspects related to genetic and genomic resources in 11 crops, namely common bean, pea, chickpea, faba bean, cowpea, lentil, pigeonpea, peanut, Asian *Vigna* species, grass pea and horsegram. Each chapter provides a comprehensive account of information on origin, distribution, diversity and taxonomy; erosion of genetic diversity from the traditional areas; status of germplasm resources conservation; germplasm evaluation and maintenance; use of germplasm in crop improvement; limitations in germplasm use; germplasm enhancement through wide crosses and integration of genetic and genomic resources in crop improvement. A complete review of the entire gamut of published work was not feasible in this single volume. However, the renowned contributors of individual chapters have tried to provide important references on significant research work published in the leading international journals/periodicals on different aspects of genetic and genomic resources. The editors are extremely grateful to all our eminent authors for their outstanding contributions in the preparation of this book. We have also been quite fortunate to know them, both academically and personally, and our communication has been very cordial and friendly during the entire process of preparation of this manuscript. We are highly indebted to Professor K.C. Bansal, Director, National Bureau of Plant Genetic Resources, Pusa, New Delhi, India for providing necessary support and guidance in the preparation of this manuscript. The editors are highly indebted to *Elsevier Insights* for shepherding the book through the editorial process with a complete academic approach. Thanks are also due to Ms. Megha Bakshi working as Project Assistant with us for her technical inputs during the course of compilation, processing and typographical work of all the chapters. Originally, the book has been intended for scientists, professionals and graduate students, whose interests centre upon genetic and genomic resources management in grain legumes. It is hoped that this book will serve as a reference for legume curators/breeders, policy makers, taxonomists, agronomists, molecular biologists and biotechnologists, teachers and students in biology and agriculture.

Editors

List of Contributors

Amri Ahmed International Center for Agricultural Research in the Dry Areas (ICARDA), Aleppo, Syria

Hamwieh Aladdin International Center for Agricultural Research in the Dry Areas (ICARDA), Aleppo, Syria

Surendra Barpete International Center for Agricultural Research in the Dry Areas (ICARDA), Tel Hadya, Aleppo, Syria

Michael Baum International Center for Agricultural Research in the Dry Areas (ICARDA), Tel Hadya, Aleppo, Syria

Ranjana Bhattacharjee International Institute of Tropical Agriculture (IITA), Ibadan, Nigeria

Ishwari Singh Bisht National Bureau of Plant Genetic Resources, Pusa, New Delhi, India

Ousmane Boukar International Institute of Tropical Agriculture (IITA), Ibadan, Nigeria

R.K. Chahota Department of Agricultural Biotechnology, CSK Himachal Pradesh Agricultural University, Palampur, Himachal Pradesh, India

Clarice Coyne United States Department of Agriculture, Agricultural Research Service, Plant Introduction and Testing Unit, Washington State University, Pullman, WA, USA

Christian Fatokun International Institute of Tropical Agriculture (IITA), Ibadan, Nigeria

Maalouf Fouad International Center for Agricultural Research in the Dry Areas (ICARDA), Aleppo, Syria

C.L. Laxmipathi Gowda International Crops Research Institute for the Semi-Arid Tropics (ICRISAT), Patancheru, Hyderabad, India

Badara Gueye International Institute of Tropical Agriculture (IITA), Ibadan, Nigeria

Priyanka Gupta International Center for Agricultural Research in the Dry Areas (ICARDA), Tel Hadya, Aleppo, Syria

Naresh Kumar Department of Agricultural Biotechnology, CSK Himachal Pradesh Agricultural University, Palampur, Himachal Pradesh, India

Shiv Kumar International Center for Agricultural Research in the Dry Areas (ICARDA), Aleppo, Syria

P. Lava Kumar International Institute of Tropical Agriculture (IITA), Ibadan, Nigeria

Lucia Lioi CNR, Istituto di Genetica Vegetale, Bari, Italy

P.N. Mathur Bioversity International, Office for South Asia, New Delhi, India

Nigel Maxted School of Biosciences, University of Birmingham, Edgbaston, Birmingham, UK

Rebecca McGee United States Department of Agriculture, Agricultural Research Service, Grain Legume Breeding and Physiology Unit, Washington State University, Pullman, WA

Nawar Mohammed International Center for Agricultural Research in the Dry Areas (ICARDA), Aleppo, Syria

Angela R. Piergiovanni CNR, Istituto di Genetica Vegetale, Bari, Italy

J.C. Rana NBPGR, Regional Research Station, Phagli, Shimla, India

Robert Redden Australian Temperate Field Crops Collection, DPI-VIDA, Horsham, VIC, Australia

K.N. Reddy International Crops Research Institute for the Semi-Arid Tropics (ICRISAT), Patancheru, Hyderabad, India

Manish Roorkiwal International Crops Research Institute for the Semi-Arid Tropics (ICRISAT), Patancheru, Hyderabad, India

A. Sarker South Asia and China Regional Program (SACRP) of ICARDA, New Delhi, India

Rachit Saxena International Crops Research Institute for the Semi-Arid Tropics (ICRISAT), Patancheru, Hyderabad, India

S.K. Sharma Department of Agricultural Biotechnology, CSK Himachal Pradesh Agricultural University, Palampur, Himachal Pradesh, India

Shivali Sharma International Crops Research Institute for the Semi-Arid Tropics (ICRISAT), Patancheru, Hyderabad, India

T.R. Sharma Department of Agricultural Biotechnology, CSK Himachal Pradesh Agricultural University, Palampur, Himachal Pradesh, India

Bao Shiying Institute of Grain Crops, Yunnan Academy of Agricultural Sciences, Kunming, China

Mohar Singh National Bureau of Plant Genetic Resources, Pusa, New Delhi, India

Petr Smýkal Department of Botany, Palacký University Olomouc, Olomouc, Czech Republic

Yang Tao Institute of Crop Science/National Key Facility for Crop Gene Resources and Genetic Improvement, Chinese Academy of Agricultural Sciences, Beijing, China

H. Thomas Stalker Department of Crop Science, North Carolina State University, Raleigh, NC

Hari D. Upadhyaya International Crops Research Institute for the Semi-Arid Tropics (ICRISAT), Patancheru, Hyderabad, India

Rajeev K. Varshney International Crops Research Institute for the Semi-Arid Tropics (ICRISAT), Patancheru, Hyderabad, India

Zong Xuxiao Institute of Crop Science/National Key Facility for Crop Gene Resources and Genetic Improvement, Chinese Academy of Agricultural Sciences, Beijing, China

1 Introduction

Mohar Singh[1], *Hari D. Upadhyaya*[2] *and*
Ishwari Singh Bisht[1]

[1]National Bureau of Plant Genetic Resources, Pusa, New Delhi, India;
[2]International Crops Research Institute for the Semi-Arid Tropics (ICRISAT), Patancheru, Hyderabad, India

Major grain legumes, including common bean, pea, chickpea, faba bean, cowpea, lentil, pigeon pea, peanut, Asian *Vigna*, grass pea and horsegram, occupy considerable area under cultivation globally and form important constituents of global diets for both vegetarian and nonvegetarian peoples. These grain legumes have the ability to fix nitrogen, which reduces fertilizer use in agriculture, besides their high protein content. Despite this significant role, global production has increased only marginally in the past 50 years. The slow production growth, along with increasing human population and improved buying capacity, has substantially reduced per capita availability of grain legumes. Further, production can be enhanced more if the loss caused by several biotic and abiotic stresses is minimized. To overcome these major constraints, there is a need to identify stable donors in genetic resources for discovering useful genes and alleles and designing crops resilient to climate change. However, excellent performance has been achieved by applying new approaches for germplasm characterization and evaluation like development of core sets, mini-core sets, reference sets and trait-specific subsets, etc. In parallel, genomic resources such as molecular markers including simple sequence repeats (SSRs), single nucleotide polymorphism (SNPs), diversity arrays technology (DArT) and transcript sequences, e.g. expressed sequence tags (ESTs) and short-read transcript sequences, have been developed for important legume crops. It is anticipated that the use of genomic resources and specialized germplasm such as mini-core collection and reference sets will facilitate identification of trait-specific germplasm, trait mapping and allele mining for resistance to various biotic and abiotic stresses and also for useful agronomic traits. Furthermore, the advent of next-generation sequencing technologies coupled with advances in bioinformatics offers the possibility of undertaking large-scale sequencing of crop germplasm accessions, so that modern breeding approaches such as genomic selection and breeding by design can be realized in the coming future for legume genetic enhancement. Here we summarize brief details on the genetic and genomic resources research on important grain legumes.

Genetic and Genomic Resources of Grain Legume Improvement. DOI: http://dx.doi.org/10.1016/B978-0-12-397935-3.00001-3

1.1 Common Bean

The common, or kidney, bean (*Phaseolus vulgaris* L.) is the centrepiece of the daily diet of more than 300 million people. It is the most important food legume, far ahead of other legumes. Nutritionists characterize the common bean as a nearly perfect food, because of its high protein content and high amounts of fibre, complex carbohydrates and other dietary elements. The common bean was domesticated more than 7000 years ago in two centres of origin – Meso-America (Mexico and Central America) and the Andean region. Over the millennia, farmers grew complex mixtures of bean types across various production systems, resulting in a vast array of genetic diversity in common beans with a wide variety of colours, textures and sizes to meet the growing conditions and taste preferences of many different regions. Given current trends in population growth and bean consumption, demand for this crop in Latin America, sub-Saharan Africa and even in Europe and other parts of the world can be expected to grow in the future. International Centre for Tropical Agriculture (CIAT) scientists are convinced that new bean cultivars with higher yields, multiple disease resistance and greater tolerance to drought and low soil fertility will enable farmers to increase bean productivity and achieve greater yield stability. New production technology, together with the bean crop's wide adaptability, will help it remain an attractive option for small-farmer cropping systems. One potent source of solutions to problems in bean production is the great genetic diversity available for research and development in the world *Phaseolus* collection maintained at CIAT's Genetic Resources Unit (GRU) in trust for the Food and Agriculture Organization (FAO). The collection includes over 36,000 entries, of which 26,500 are cultivated *Phaseolus vulgaris*, about 1300 are wild types of common bean (http://isa.ciat.cgiar.org/urg/main.do?language=en), and the rest are distant relatives of the common bean. CIAT scientists have also created more manageable core collections. The core collection of domesticated common bean contains about 1400 accessions, while the collection of wild common bean consists of about 100 accessions. In recent years bean researchers at CIAT and in national programs of Latin America and sub-Saharan Africa have been evaluating the core collection for a wide range of useful traits, such as insect and disease resistance and tolerance to low phosphorus. Useful materials have been identified and incorporated into breeding programs at CIAT and elsewhere.

While focusing mainly on dry beans, CIAT scientists are also working to improve the green snap beans. Demand for fresh snap beans for domestic consumption or export is growing in Africa, Asia and Latin America, and sales are an excellent source of cash income for small farmers. Much of the CIAT's strategic research on dry beans, especially that dealing with diseases and pests, is readily applicable to snap beans. Classical breeding within the primary gene pool of common bean has given excellent results in the last two decades, with tangible benefits to the farming community. More recently, CIAT scientists have begun to integrate various biotechnology techniques into problem-solving research on the crop. CIAT scientists have succeeded in hybridizing common bean with the distantly related species *Phaseolus acutifolius*, or tepary bean, which possesses genes for resistance to common bacterial blight (CBB), leafhoppers and drought. The resulting breeding lines have shown high levels of resistance to CBB. CIAT researchers have also developed a molecular marker–assisted approach to improving beans for resistance to bean golden mosaic virus (BGMV) that has cut breeding time and effort by about 60%. The results of recent molecular marking and selection work are highly encouraging, demonstrating not only the effectiveness of the strategy by the Standing Committee on Agricultural Research (SCAR) for selecting BGMV-resistant beans but also its efficiency.

1.2 Pea

Pea (*Pisum sativum* L.) is one of the world's oldest domesticated crops. Its area of origin and initial domestication lies in the Mediterranean, primarily in the Middle East. The range of wild representatives of *P. sativum* extends from Iran and Turkmenistan through Anterior Asia, northern Africa and southern Europe. The genus *Pisum* contains the wild species *P. fulvum* found in Jordan, Syria, Lebanon and Israel; the cultivated species *P. abyssinicum* from Yemen and Ethiopia, which was likely domesticated independently of *P. sativum*; and a large and loose aggregate of both wild (*P. sativum* subsp. *elatius*) and cultivated forms that comprise the species *P. sativum* in a broad sense.

Currently, no international organization conducts pea breeding and genetic resources conservation, and no single collection predominates in size and diversity. Important genetic diversity collections of *Pisum* with over 2000 accessions are found in national gene banks in at least 15 countries, with many other smaller collections worldwide. A high level of duplication exists between the collections, giving a misleading impression of the true level of diversity. However, the numbers of original pea landraces mainly from Europe, Asia, the Middle East and North Africa/Ethiopia have not been documented. The much smaller collections of wild relatives of pea are less widely distributed; there is more clarity when tracing these accessions to their origin. There are still important gaps in the collections, particularly of wild and locally adapted materials, that need to be addressed before these genetic resources are lost forever (Maxted, Shelagh, Ford-Lloyd, Dulloo, & Toledo, 2012). Many studies have been conducted on *Pisum* germplasm collections to investigate genetic and trait diversity. Several major world pea germplasm collections have been analysed by molecular methods and core collections were formed. The key priority is the collection and conservation of the historic landraces and varieties of each country in *ex situ* gene banks. The overall goal should be to ensure maintenance of variation for adaptation to the full range of agro-ecological environments, end uses and production systems. Wild peas have less than 3% representation in various national collections despite their wide genetic diversity. There is an urgent need to fully sample this variation, since natural habitats are being lost due to increased human population, increased grazing pressure, conversion of marginal areas to agriculture and ecological threats due to future climate change. It is urgent to implement a comprehensive collection of wild relatives of peas representing the habitat range from the Mediterranean through the Middle East and Central Asia while these resources are still available, since these are likely to contain genetic diversity for abiotic stress tolerance. Genetic diversity available in wild *Pisum* species has been poorly exploited. The most attention has been given to *P. fulvum* as a donor of bruchid resistance and source of novel powdery mildew resistance (*Er3*). Relatively few genotypes with high degree of relatedness have been used as parents in modern pea breeding programs, leading to a narrow genetic base of cultivated germplasm. There are several current efforts to make either genome-wide introgression lines or at least simple crosses with the intent of broadening the genetic base. Further investigations, particularly in the wild *P. sativum* subsp. *elatius* gene pool, are of great practical interest. Molecular approaches will allow breeders to avoid the linkage drag from wild relatives and make wide crosses more successful and practical.

1.3 Chickpea

The genus *Cicer* comprises one cultivated and 43 wild species. Chickpea probably originated from southeastern Turkey. Four centres of diversity were identified in the Mediterranean, Central Asia, the Near East and India, as well as a secondary centre of origin in Ethiopia. Further, chickpeas spread with human migration toward the west and south via the Silk Route. It is grown and consumed in large quantities from Southeast Asia to India and in the Middle East and Mediterranean countries. It ranks second in area and third in production among the pulses worldwide. Most production and consumption of chickpea takes place in developing countries. It is a true diploid and predominantly self-pollinated legume, but cross-pollination by insects sometimes occurs. Thirty five of the chickpea wild relatives are perennials and the other nine (including the cultivated species) are annuals. Based on seed size and shape, two main kinds of chickpea are recognized: the desi type, closer to the putative progenitor (*C. reticulatum*), is found predominantly in India and Ethiopia and has small, angular, coloured seeds and a rough coat. They have a bushy growth habit and blue-violet flowers. The kabuli type, predominantly grown in the Mediterranean region, has large, beige-coloured and owl-head-shaped seeds with a smooth seed coat. Their plants have a more erect growth habit and white flowers. It is estimated that more than 80,000 accessions are conserved in more than 30 gene banks worldwide (http://apps3.fao.org/wiews/germplasm_query.htm?i_l=EN). The gene bank at International Crops Research Institute for the Semi-Arid Tropics (ICRISAT), India, is one of the largest gene banks, holding greater than 20,000 accessions of chickpea from about 60 countries. Other major collections (more than 12,000 accessions) are held at the National Bureau of Plant Genetic Resources (NBPGR), New Delhi, India; International Center for Agricultural Research in the Dry Areas (ICARDA) in Aleppo, Syria; Australian Temperate Field Crops Collection, Victoria, Australia; the United States Department of Agriculture (USDA); and the Seed and Plant Improvement Institute, Iran. Currently there is a reasonable number of wild annual

Cicer species, but still limited availability of perennial species. Less than 1% of the *Cicer* accessions (conserved in about 10 gene banks worldwide) represent wild species. Priority should be given to the conservation of chickpea in primary and secondary centres of diversity. *Cicer* genetic resources could be much better utilized. A representative core collection (10% of the entire collection) and a mini-core collection (10% of the core or 1% of entire collection) are being developed at the ICRISAT and evaluated extensively for useful traits (Upadhyaya & Ortiz, 2001). However, recent advances in plant biotechnology have resulted in the development of a large number of molecular markers, genetic and physical maps, as well as the generation of expressed sequenced tags, genome sequencing and association studies showing marker–trait associations, which has facilitated the identification of quantitative trait loci (QTLs) and discovery of genes/alleles associated with resistance to several abiotic and biotic stresses, beside agronomic traits.

1.4 Faba Bean

Faba bean (*Vicia faba* L.) is a major food and feed legume, because of the high nutritional value of its seeds, which are rich in protein and starch. Seeds are consumed dry, fresh, frozen or canned. The main faba bean producer countries are China, some in Europe, Ethiopia, Egypt and Australia. Geographical distribution and objectives of the breeding programs developed for this species therefore reflect where consumption is highest. In relation to the size of the market and in comparison with soybean, the faba bean selection programs are few and small. The role of *ex situ* and on-farm collections is even stronger for this crop due to the absence of a natural reservoir of wild accessions and to the modernization of agriculture, which progressively phases out numerous landraces. Botanic and molecular data suggest that the wild ancestor of this species has not yet been discovered or has become extinct. At the world level, more than 38,000 accession entries are included in about 37 listed germplasm collections. A large genetic variability has already been identified in *V. faba* in terms of floral biology, seed size and composition, and also tolerance to major biotic and abiotic stresses. More knowledge is needed on the interactions of *V. faba* with parasitic and pollinator insects, on traits related to environmental adaptation and impacts on nitrogen fixation in interaction with soil rhizobia and on bioenergy potential, which strengthens the demand for new and large phenotyping actions. Diversity analysis through genotyping is just beginning. The use of amplified fragment length polymorphism (AFLP) or SSR markers has allowed genetic resources to be distinguished according to their geographical origin and structuring of germplasm collections. Conservation of gene sequences among legume species and the rapid discovery of genes offer new possibilities for the analysis of sequence diversity for *V. faba* genes and evaluation of their impact on phenotypic traits. Projects that combine genotyping and phenotyping must be continued on *V. faba*, so that core collections can be defined; these will help in the discovery of genes and alleles of interest for faba bean breeders (Rispail et al., 2010).

1.5 Cowpea

Cowpea (*Vigna unguiculata* (L.) Walp.) is cultivated widely in the tropics and has multipurpose uses: as food for human beings, fodder for livestock and atmospheric nitrogen fixers. Cowpea grains rich in protein are consumed in different forms in several parts of the tropics. The average grain yield of cowpea in West Africa is approximately 492 kg ha^{-1}, which is much lower than its potential yields. This low productivity is due to a host of diseases, insects, pests, parasitic weeds, drought, poor soils and low plant population density in the farmer's field. Despite a large number of cowpea accessions (about 15,000) maintained at the International Institute of Tropical Agriculture (IITA), recent studies demonstrated that genetic diversity in cultivated cowpea is low. Researchers, however, found a high level of random amplified polymorphic DNA marker diversity in landraces from Malawi. However, *ex situ* collection of cowpea and wild *Vigna* germplasm from different parts of the world were assembled in the IITA gene bank. These genetic resources have been explored to identify new traits and to develop elite cowpea varieties. Many cowpea varieties with high yield potential have been developed and adopted by the farmers. Efforts are continuing to develop better-performing varieties using conventional breeding procedures, while molecular tools are being developed to facilitate progress in cowpea breeding (Agbicodo et al., 2010).

1.6 Lentil

Lentils have been part of the human diet since Neolithic times, being one of the first crops domesticated in the Near East. Archaeological evidence reveals that they were eaten 9500–13,000 years ago. Lentil colours range from yellow to red-orange to green, brown and black. Lentils also vary in size, and are sold in many forms, with or without the skins, whole or split. Lentils are relatively tolerant to drought and are grown throughout the world. The FAO has reported that the world production of lentils primarily comes from Canada, India, Turkey and the United States. About a quarter of the worldwide production of lentils is from India, most of which is consumed in the domestic market. Canada is the largest export producer of lentils in the world.

Extensive collections of lentil germplasm now exist in various gene banks around the world. This germplasm including wild *Lens* species has been used in plant introduction strategies and in efforts to widen the potential sources of increasing genetic diversity in the breeding programmes of lentil. Improved techniques are emerging to overcome hybridization barriers between species, and as a result interspecific hybrids have been successfully obtained between species. Several interspecific recombinant inbred line populations have been developed. Selected and backcrossed lentil lines are currently in advanced yield trial stages, and desirable traits such as yield, disease resistance and agronomic traits have been incorporated into cultivated lentil especially from *Lens ervoides*, generating a wider spectrum of variability. Secondly, further expansion of the overall pool of germplasm and examination of allelic variation at the nucleotide level will benefit lentil-breeding programmes by augmenting phenotype-based variation to further advance cultivar development. Genomic resources for lentils are limited now, but this situation is changing rapidly as the cost of genotyping has declined. As a result, two successive EST projects were undertaken under the NAPGEN EST project initiative and an Agricultural Development Fund project initiative. It has been emphasized that creation of intraspecific and interspecific genetic populations, genetic maps, association maps, QTLs and marker-assisted selection technologies for implementation in the breeding programme will enhance deployment of genes responsible for traits of interest. The economical use of genomic technologies for use in germplasm resource management and genetic improvement is on the near horizon.

1.7 Pigeon Pea

Pigeon pea (*Cajanus cajan* (L.) Millspaugh) is an important grain legume of the Indian subcontinent, Southeast Asia and East Africa. More than 85% of the world pigeon pea is produced and consumed in India, where it is a key crop for food and nutritional security of the people. The centre of origin is the eastern part of peninsular India, including the state of Orissa, where the closest wild relatives occur. Though pigeon pea has a narrow genetic base, vast genetic resources are available for its genetic improvement. The ICRISAT gene bank maintains about 13,216 accessions, whereas the Indian NBPGR bank maintains a total of about 12,900 accessions. Evaluation of small-sized subsets such as core (10% of whole collection) and mini-core (about 1% of the entire collection), developed at the ICRISAT, has resulted in identification of promising diverse sources for agronomic and nutrition-related traits, as well as resistance to major biotic and abiotic stresses for use in pigeon pea improvement programs. Wild relatives of pigeon pea are the reservoir of several useful genes, including resistance to diseases, insect pests and drought, as well as good agronomic traits, and have contributed to the development of cytoplasmic male sterility systems for pigeon pea improvement. Availability of genomic resources, including the genome sequence, will facilitate greater use of germplasm to develop new cultivars with a wider genetic base.

1.8 Peanut

The domesticated peanut (*Arachis hypogaea* L.) is an amphidiploid or allotetraploid having two sets of chromosomes from two different species, thought to be *A. duranensis* and *A. ipaensis*. These likely combined in the wild to form the tetraploid species *A. monticola*, which gave rise to the domesticated peanut. This domestication may have taken place in Paraguay or Bolivia, where the wildest strains grow today. Certain cultivar groups are preferred for particular uses based on differences in flavour, oil content, size, shape and disease resistance. For many uses, the different cultivars are interchangeable. Most peanuts marketed in the shell are of the Virginia type, along with some Valencias selected for large size and the attractive appearance of the shell. Spanish peanuts are used mostly for peanut candy, salted nuts and peanut butter. Most runners are used to make peanut butter. Although India and China are the world's largest producers of peanuts, they account for a small part of international trade, because most of their production is consumed domestically as peanut oil. Exports of peanuts from India and China are equivalent to less than 4% of world trade. The major producers/exporters of peanuts are the United States, Argentina, Sudan, Senegal and Brazil. These five countries account for 71% of total world exports. In recent years, the United States has been the leading exporter of peanuts in the world.

Further, the number of accessions in the ICRISAT gene bank are about 13,500. Most of them have been characterized and evaluated for their reaction to diseases, insect pests and other desirable agro-morphological characteristics, leading to identification of 506 useful genetic stocks. Most of the germplasm is conserved as pods or seeds in the gene bank, while rhizomatous *Arachis* species are conserved as whole plants. ICRISAT serves as the world's largest repository of peanut germplasm and has distributed about 60,000 peanut germplasm samples free of cost to the international scientific community.

Despite significant progress, peanut genetic resource activities still suffer from several limitations in assembly and characterization. The establishment of a peanut genetic resources network is proposed to overcome many such limitations. However, sufficient numbers of molecular markers that reveal polymorphism in cultivated peanut are available for diversity assessments. In a study, the amount and distribution of genetic variation within and among six peanut botanical varieties, as well as its partitioning among three continents of origin (South America, Asia and Africa) was assessed at 12 SSR loci by means of 10 sequence-tagged microsatellite site primers. Discriminant function analysis reveals a high degree of accordance between variety delimitation on the basis of morphological and molecular characters. Landraces from Africa and Asia were more closely related to each other than to those from South America. Nei's unbiased estimate of gene diversity revealed very similar levels of diversity within botanical varieties. Landraces from South America had the highest diversity and possessed 90% of alleles, compared with Africa (63%) and Asia (67%).

1.9 Asian *Vigna*

Asian *Vigna* species constitute an economically important group of cultivated and wild species, and a rich diversity occurs in India and other Asian countries. Taxonomically, cultigen and conspecific wild forms are recognized in all major cultivated Asiatic pulses, mung bean (*V. radiata*), urd bean (*V. mungo*), rice bean (*V. umbellata*) and azuki bean (*V. angularis*) except for moth bean (*V. aconitifolia*), which has retained a wild-type morphology. The cultivated species, *V. radiata* and *V. mungo*, are of Indian origin. The domestication of *V. aconitifolia* is also apparently Indian, whereas that of *V. angularis* and *V. umbellata* is Far Eastern. The green gram is already a popular food throughout Asia and other parts of the world. The present level of its consumption can be expected to increase. The black gram, although very popular in India, is less likely to generate sufficient demand to stimulate production significantly outside its traditional areas. The azuki bean has generated interest as a pulse outside traditional areas of production and consumption, and consumer demand for it could increase in the near future. Perhaps the most interesting future exists for rice bean, which has a high food value and tolerance to biotic and abiotic stresses. It possibly has the highest yielding capacity of any of the Asian *Vigna* and could become a useful crop, if a sizeable consumer demand were built up. Moth bean has a future in India as a pulse crop. *V. trilobata* is probably most useful as a forage crop in semi-arid conditions. The fullest possible range of landraces and cultivars needs to be collected and conserved together with the conspecific wild-related species. The wild germplasm resources have a great potential for widening the genetic base of the *Vigna* gene pool by interspecific hybridization. The available genetic resources with valuable characters will therefore be required to make extended cultivation economically attractive.

1.10 Grass Pea

Grass pea presents a fascinating paradox; it is both a lifesaver and a destroyer. It is easily cultivated and can withstand extreme environments from drought to flooding. However, when eaten as a large part of the diet over a long enough period (which is often the case during famine), it can permanently paralyse adults from the knees down and cause brain damage in children, a disorder named lathyrism. Grass pea has a long history in agriculture. It was first domesticated some 7000–8000 years ago in the eastern Mediterranean region and has a history of cultivation in southern parts of Europe, North Africa and across Asia. Today it is mostly grown in India, Pakistan, Bangladesh and Ethiopia. More recently, grass pea has become popular as a forage crop in Kazakhstan, Uzbekistan, South Africa and Australia.

Recently ICARDA at Aleppo, Syria, together with Ethiopian breeders, has undertaken a project to develop cultivars with low neurotoxin levels. The role of diversity in breeding programmes was instantly clear: the toxins found in African and Asian grass pea plants are seven times more toxic than Middle Eastern types. The Centre for Legumes in Mediterranean Agriculture (CLIMA) in Australia has also recently produced a low-toxin grass pea variety. The use of grass pea diversity in breeding has shown how the genetic resources of a crop can be used to improve its nutritional value for human health. The ICARDA scientists used the diversity found in the world's largest collection of grass pea and its relatives, stewarded at the ICARDA seed bank in Syria, with more than 3000 accessions. Large *Lathyrus* collections are also conserved in France, NBPGR in India, Bangladesh and Chile. Despite this research, much additional work is needed in order to produce locally adapted, low-toxin varieties and to distribute these to the farming community. Furthermore, there is a need to expand the molecular research work in species identification and their proper utilization in grass pea breeding.

1.11 Horsegram

Horsegram (*Macrotyloma uniflorum*) is one of the lesser known grain legume species. The whole seeds of horsegram are generally utilized as cattle feed. However, it is consumed as a whole seed, as sprouts, or as whole meal in India. It is quite a popular legume, especially in southern Indian states such as Karnataka, Tamil Nadu, Andhra Pradesh, northwestern Himalayan states and Uttarakhand. The chemical composition is comparable with more commonly cultivated legumes. Like other legumes, horsegram is deficient in methionine and tryptophan, though it is an excellent source of iron and molybdenum. Horsegram is also known to have many therapeutic effects – not scientifically proven – though it has been recommended in ayurvedic medicine to treat renal stones, piles, oedema, etc. A total of 1721 accessions of horsegram are being conserved in different gene banks of the world. Of these collections, about 95% are conserved at NBPGR, New Delhi, India, and its regional research station, Thrissur, Kerala, is designated as an active site for the conservation and evaluation of horsegram germplasm. No worthwhile genomic resource information on horsegram is available.

References

Agbicodo, E. M., Fatokun, C. A., Bandyopadhyay, R., Wydra, K., Diop, N. N., Muchero, W., et al. (2010). Identification of markers associated with bacterial blight resistance loci in cowpea [*Vigna unguiculata* (L.) Walp.]. *Euphytica*, *175*, 215–226.

Maxted, N., Shelagh, K., Ford-Lloyd, B., Dulloo, E., & Toledo, A. (2012). Toward the systematic conservation of global crop wild relative diversity. *Crop Science*, *52*(2), 774–785.

Rispail, N., Kalo, P., Kiss, G. B., Ellis, T. H. N., Gallardo, K., Thompson, R. D., et al. (2010). Model of legumes to contribute to faba bean breeding. *Field Crops Research*, *115*, 253–269.

Upadhyaya, H. D., & Ortiz, R. (2001). A mini core collection for capturing diversity and promoting utilization of chickpea genetic resources in crop improvement. *Theoretical and Applied Genetics*, *102*, 1292–1298.

2 European Common Bean

Lucia Lioi and Angela R. Piergiovanni

CNR, Istituto di Genetica Vegetale, Bari, Italy

Ex Hispaniis accepta cum hac inscriptione *Alubias de Indias,* id est

Clusius (1583)

2.1 Introduction

Common bean (*Phaseolus vulgaris* L.), a legume native to America, is now one of the most important crops worldwide. The rich nutritive composition, the different forms (fresh, canned, frozen pods or seeds, precooked/ dehydrated seeds, packaged dry seeds) and the versatility in cooking make it an interesting and valuable crop. Consumption patterns vary dramatically by geographic regions and among cultures. As a matter of fact, it is cultivated extensively in the five continents and spans from 52°N to 32°S latitude, and from near sea level in the continental USA and Europe to elevations of more than 3000 m above sea level (asl) in Andean South America. According to the Food and Agriculture Organization (FAO), in 2010 the total world production of all the cultivated common bean species was about 23 million tons. American countries produce nearly half of the world's supply of dry beans: Brazil, USA, Mexico and Central America are the major producers. India, China and Myanmar are the major Asian producers. In Europe, cultivation is concentrated in the regions bordering the Mediterranean basin, such as the Iberian Peninsula, Italy and the Balkan states, though the production is not sufficient to cover the whole demand.

Although far less important than cereals, common bean is a cheap source of vegetable proteins, calories and micronutrients. Like other legumes, the major limitations are the low content of sulphur-amino acids and the presence of antinutritional compounds. The main form of consumption is represented by dry seeds, however varieties suitable for other consumption forms, such as snap or shell beans, have been developed. Snap bean cultivars possess a thick succulent mesocarp and reduced or no fibre in green pod walls and sutures, while shell beans are immature seed harvested before complete desiccation in the pod. The economic relevance of common bean justifies the efforts currently in progress for the release of new varieties suitable to mechanical harvest, characterized by resistance to pests and diseases, and of high nutritional quality. In this context the European common bean germplasm can play a key role, making available to European breeders significant genetic variation useful for further improvement of the crop.

Genetic and Genomic Resources of Grain Legume Improvement. DOI: http://dx.doi.org/10.1016/B978-0-12-397935-3.00002-5

2.2 Taxonomy, Origin, Distribution and Diversity of Cultivated *Phaseolus vulgaris*

By 1753 the common bean was so common in the Old World that Linnaeus chose the name *Phaseolus* for this species, naming it *P. vulgaris* L., and proposed that it originated from India. However, the true Old World bean species are *Vigna unguiculata* (L.) Walp. (cowpea), *Vicia faba* L. (faba bean) and *Lablab purpureus* (L.) Sweet (hyacinth bean). Of these species, especially *V. unguiculata* is very similar to common bean. Linnaeus considered *V. unguiculata* as introduced from the New World. These confusions led over time to many wrong conclusions about the origin, history and classification of beans. Over the past two centuries over 400 species have been named, often with poor descriptions or lacking good type specimens. Formerly, differentiations were made between Old World beans, mostly from the genus *Vigna*, and New World beans, from the genus *Phaseolus*. So though the genus *Phaseolus* has a complex taxonomic and nomenclature history, this strictly New World genus is diagnosed by foliage bearing hooked hairs, keel petals that are laterally and tightly coiled, and inflorescence nodes that lack extra floral nectaries. The majority of species, having a Neotropical origin, are distributed in the tropics and subtropics of the New World. On the basis of current floristic knowledge, there are no *Phaseolus* species growing wild naturally in other parts of the world. The majority of species are concentrated in the western mountainous ranges of Mexico (Central America), and in the northern and central Andes (South America) between 37° North and 28° South (Debouck, 2000).

Based on evolutionary rate, the genus *Phaseolus* is approximately six million years old, suggesting that this extremely successful group of plants is relatively young. The 70 or more *Phaseolus* species are divided into 15 sections by Freytag and Debouck (2002). This classification is partly inconsistent with that of Delgado-Salinas, Bibler, and Lavin (2006), who recognized two main groups or classes. Clade A groups species distributed mostly in Mexico, but also in the southwestern United States and Central America, generally growing over 1000 m asl. Clade B species are distributed throughout the American continent from the southeast of Canada to the Andean region of South America, in lower altitude areas. The five main domesticated species, *P. vulgaris* L. (common bean), *Phaseolus coccineus* L. (runner bean), *Phaseolus polyanthus* Grenm., synonym of *Phaseolus dumosus* Macfad. (year bean), *Phaseolus acutifolius* A. Gray (tepary) and *Phaseolus lunatus* L. (Lima bean), occur among the clade B species, with the first four more closely related.

P. vulgaris ($2n = 2x = 22$), belonging to the *Phaseolus* genus (subtribe Phaseolinae, tribe Phaseoleae, family Fabaceae), is a member of a plant family that produces pods that carry a nutrient-dense high protein seed. Over a time of at least 7000 years, the common bean has evolved into a major leguminous crop. Several remains have been discovered in the Andes, but also in Mesoamerica (Kaplan & Lynch, 1999). Historical and linguistic data support the existence of specific words designating the common bean in several native Indian languages (Brown, 2006).

Before domestication, wild *P. vulgaris*, widely distributed from northern Mexico to northwestern Argentina, had already diverged into two major ecogeographical gene pools, each with its own distribution. Moreover, some wild populations from Columbia are considered to belong to a transect area (Papa & Gepts, 2003). Wild forms of the common bean comprise an additional third gene pool that is located in a restricted area between Northern Peru and Ecuador (Debouck, Toro, Paredes, Johnson, & Gepts, 1993). These populations are usually considered as the putative ancestor from which the species *P. vulgaris* originated, showing the specific type I phaseolin, the major seed storage protein considered an evolutionary marker (Kami, Velasquez, Debouck, & Gepts, 1995). From this area, wild materials were presumably dispersed towards north and south, giving rise to the Mesoamerican and Andean gene pools, respectively. An alternative hypothesis of the Mesoamerican origin of common bean, most likely located in Mexico and already supported by data obtained with multilocus molecular markers by Rossi et al. (2009) and Kwak and Gepts (2009), was confirmed on the basis of sequence data by Mamidi et al. (2011). More recently, Bitocchi, Nanni, et al. (2012) sequenced five loci of a large collection that included wild common bean accessions from Mesoamerican and Andean gene pools as well as genotypes from Northern Peru–Ecuador, characterized by the ancestral type I phaseolin. Results present clear evidence, either from phylogeny analysis or from the structure of populations, for a Mesoamerican origin of *P. vulgaris* that was most likely located in Mexico. Moreover, these last studies strongly support the occurrence of a bottleneck during the formation of the Andean gene pool that predated the domestication, as previously proposed by Rossi et al. (2009) on the basis of amplified fragment length polymorphism (AFLP) data on wild and domesticated common bean accessions. In the paper by Bitocchi, Nanni, et al. (2012) a new scenario is suggested for wild populations from Northern Peru; they could represent a relict of wild materials migrating from Central Mexico in ancient times.

Domestications from wild beans occurred independently in Mesoamerica and Andean South America and gave rise to two major distinct gene pools also within the cultivated forms. The occurrence of separate domestication events has been well established using multiple approaches, based on morphological and agronomic traits, other than biochemical and molecular markers (Chacón, Pickersgill, & Debouck, 2005; Gepts, 1988; Koenig & Gepts, 1989; Papa, Nanni, Sicard, Rau, & Attene, 2006). These two gene pools are characterized by partial reproductive isolation, thus suggesting a process of incipient speciation (Koinange & Gepts, 1992).

The number of domestication events within each gene pool is still debated. Generally a single domestication event is thought to have occurred in the Mexican state of Jalisco (Kwak, Kami, & Gepts, 2009). A similar conclusion, although hypothesized, could not be drawn for the Andean counterpart, due to the lower diversity of this material compared to Mesoamerican accessions (Nanni et al., 2011; Rossi et al., 2009). More recently, Bitocchi, Bellucci, et al. (2013) investigated the effect of domestication on genetic diversity in both gene pools, using nucleotide data from five fragment genes. This study highlighted a single domestication event within each gene pool and indicated the Oaxaca valley in Mesoamerica and southern Bolivia and northern Argentina as geographical areas of common bean domestication.

Variation within domesticated gene pools arose in part from the already diverged wild gene pools and partly from further selection under domestication. As a consequence, ecogeographical races in each of the two gene pools appeared, according to morphological traits, agro-ecological adaptation and biochemical markers (Singh, Gepts, & Debouck, 1991), and are generally congruent with the population structure identified by microsatellite markers (Kwak & Gepts, 2009). Cultivars from Mesoamerica, consisting of the races Durango, Jalisco and Mesoamerica, usually are small- or medium-seeded (>25 g or 25–40 g/100 seed weight, respectively) and have S phaseolin type (Figure 2.1). The small-seeded navy and black beans belong to the Mesoamerica race, Pinto and Great Northern beans belong to the Durango race, and small red and pink beans belong to the Jalisco race.

In a more recent study on race structure within the Mesoamerican gene pool as determined by microsatellite markers, the Jalisco and Durango races were found more closely related, an expected result due to the similar geographical range from which they have originated in Central Mexico (Diaz & Blair, 2006). Based on morphological and ecological criteria, the races Nueva Granada, Peru and Chile have been identified in the South America counterparts. They have large seeds (>40 g/100 seed weight) with T, C, H and A phaseolin patterns (Diaz & Blair, 2006; Singh et al., 1991). Among them, the race Nueva Granada is the most widely

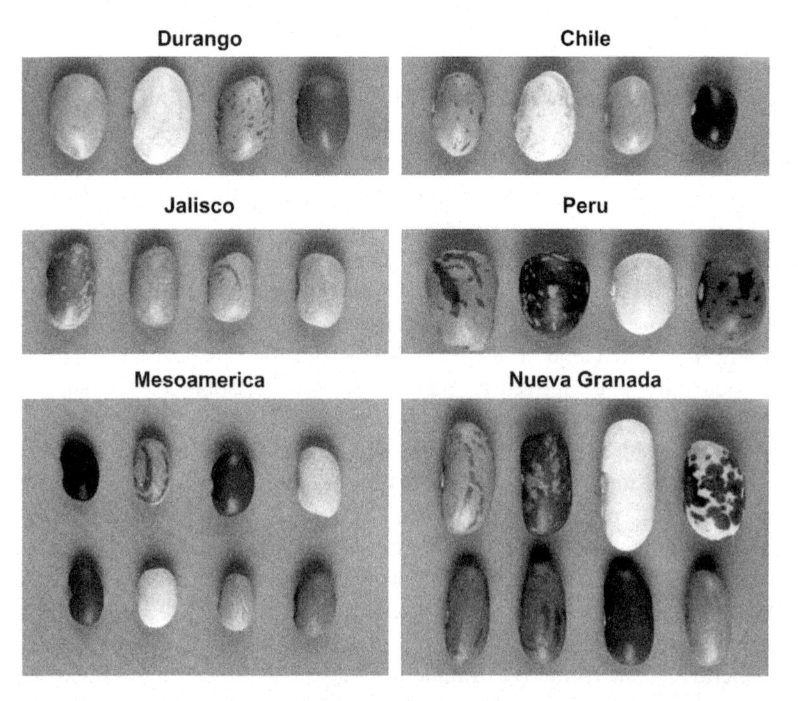

Figure 2.1 Characteristics of dry seeds from different races of cultivated common bean. Right: Middle American races; left: Andean South American races.
Source: Photo courtesy of S. P. Singh, University of Idaho, ID, USA.

cultivated, including the majority of commercial large-seeded kidney cultivars and most snap beans. Race Peru includes the yellow beans, race Chile mainly the vine cranberry beans.

2.3 Introduction and Dissemination in Europe

At European contact, Amerindian agriculture was based on a group of three major crops: maize, squash and beans. Two bean plants climbing on a living stake can be seen in the drawing of an indigenous American planting seeds with the aid of a digging stick, in the manuscript titled *Histoire naturelle des Indes*, known as the Drake Manuscript (dated about 1586). The illustration, titled 'The manner and style of gardening and planting of the Indians', also shows multieared maize, a cucurbit vine bearing many large round fruits, capsicum pepper and a pineapple. Beans were sown in the same hole with maize, and the two crops complemented each other both as crops and as food. Maize acts as support of the climbing beans and is nitrogen demanding, while beans are nitrogen fixing as a result of *Rhizobium* symbiosis. Furthermore, maize and beans complement each other nutritionally, since maize seeds are deficient in the essential amino acid lysine; conversely, bean seed is deficient in the sulphur-containing amino acids (cysteine and methionine). The mixture of beans and tortillas (maize pancakes) provided a complete protein food that was the basis of Aztec and Mayan diets (Janick, 2011).

The knowledge of the ways through which the common bean was introduced in Europe is fragmentary, but it is likely that after the discovery of the Americas many introductions were made from many places. It is well known that the two common bean gene pools arrived in Europe at different times. If the Mesoamerican common beans arrived in Europe just after the discovery of America, the Andean counterpart reached Spain in 1528, after the exploration of Peru. Common bean spread into Europe in a very short time, probably as a consequence of the high similarity of seeds with those of cowpea, *V. unguiculata*, a legume grown in Europe for millennia. Already in about 1508 the common bean was depicted in France in the prayer book of Anne de Bretagne, Queen of France and Duchess of Brittany (Figure 2.2). The image of a bean plant was identified by Jussieu (1772) as *Phaseolus flore luteo* and successively by Camus (1894) as the taxon entity *P. vulgaris* L. (Paris, Daunay, Pitrat, & Janick, 2006). The New World plant appears in the festoons of fruits, vegetables and flowers including over 170 species of plants, which surround the gorgeous frescoes painted between 1515 and 1517 by Giovanni Martini da Udine at Villa Farnesina in Rome (Caneva, 1992).

The first description of common bean in European herbal references was done by Leonhard Fuchs, who reported in *De historia stirpium* (Fuchs, 1542) that the common bean had a climbing habit, white or red flowers, and red, white, yellow, skin-coloured or liver-coloured seeds with or without spots (Figure 2.3). However, it cannot be excluded that Fuchs reported a combination of traits belonging to both *P. vulgaris* and *P. coccineus*. Subsequent descriptions were done by Roesslin in 1550, by Oellinger in 1553 and by Dodonaeus in 1554 (Zeven, 1997). A brief

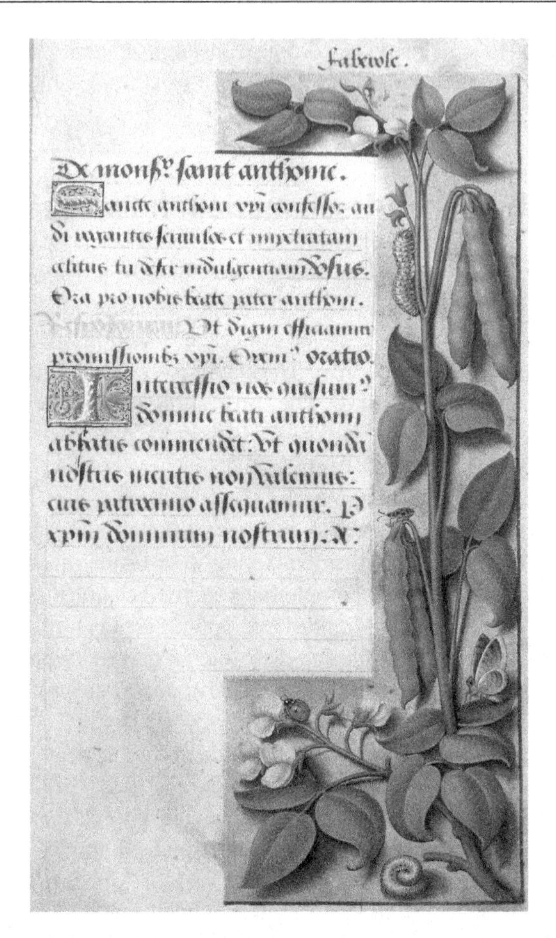

Figure 2.2 A common bean plant depicted in France in the prayer book of Anne de Bretagne, Queen of France and Duchess of Brittany (1508).
Source: Photo courtesy of Bibliothèque Nationale de France (BnF, Paris, France).

selection of old manuscripts (1493–1774) mentioning *P. vulgaris* or its synonyms is reported by Krell and Hammer (2008).

The beginning of cultivation in Italy is supported by documents that fixed 1532 as the year in which the humanist and literate Pierio Valeriano received a bag of bean seeds as compensation for his work at the Pope Clemente VII court. The Pope had obtained the seeds from the Spanish Emperor Charles V, who ruled some Italian possessions at that time. After sowing the common bean seeds in his fields located in Belluno province (northeastern Italy), Valeriano described the cultivation technique, the plant and seed morphology, and the supposed therapeutic properties of seeds in his poem 'De Milacis Cultura'. During the fifteenth and sixteenth centuries, common bean was introduced from Spain into Portugal, as a consequence of the flourishing commerce of this country with the Spanish region of Galicia (Rodiño, Santalla,

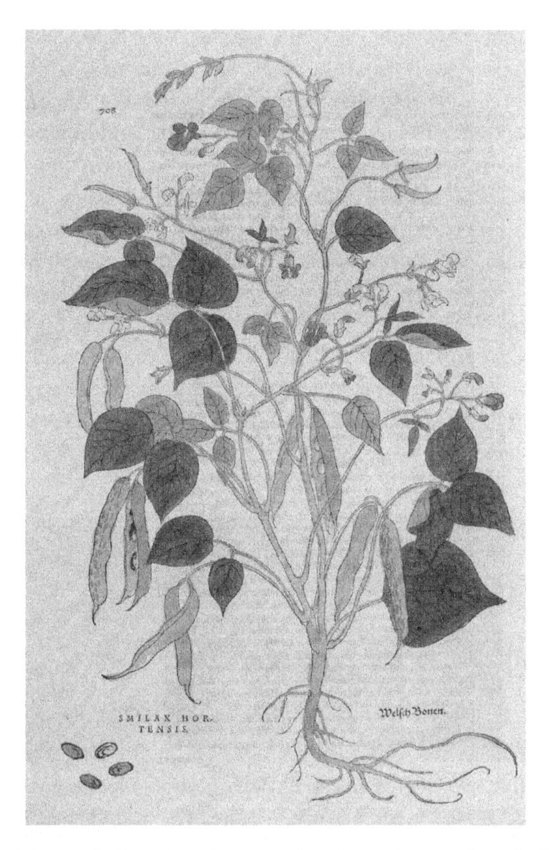

Figure 2.3 One of the early European images of common bean called *Smilax hortensis* from L. Fuchs's herbal reference *De historia stirpium* (Fuchs, 1542).
Source: Photo courtesy of Biblioteca Riccardiana, Florence, Italy.

Montero, Casquero, & De Ron, 2001). Historical documents support the introduction of *Phaseolus* seeds from Italy and Spain to the present Hungary, part of the exchange of botanical species and scientific information among naturalists (Barona, 2007). Fine illustrations and botanical descriptions of *Phaseolus* plants are present in the *Stirpium per Pannonia, Austriam etc.* (Clusius, 1583) under the names of *Phaseolus purkircherianus* and *Phaseolus africanus*, tentatively identified as *P. lunatus* and *P. coccineus* by K. Hammer (pers. commun.). In 1669 common bean was cultivated on a large scale in the Dutch province of Zeeland (Van der Groen, 1669), and after 20 years Valvasor (1689) reported the presence of the pulse in Slovenia. Over time, the dissemination across Europe surely occurred through seed exchanges among farmers being facilitated by territorial contiguity and similarity of environments.

In the early decades of the sixteenth century, the common beans introduced into Europe were surely subjected to selective pressures that gave rise to the loss of part of the germplasm carried from America. The driving forces of the genetic erosion that occurred in the early times were nature and farmers. Particularly, the ability to survive in the new environments, the tolerance to long days and the resistance to pests and diseases represented important selecting factors. In addition, farmers took good care of their precious beans by sowing those having the most desirable features such as seed colour and size, resistance to biotic and abiotic stress, and good culinary quality. This process produced over the time a myriad of landraces well adapted to restricted areas of cultivation distributed in Europe. As a consequence, each country selected its own set of landraces able to fulfil the expectations of local populations. An example of morphological variation present in Italian common bean germplasm is shown in Figure 2.4. In the countries characterized by a high diversification of growing environments, the process of differentiation was more pronounced, so that each region had its own set of landraces. However, only in relatively recent times and for some European countries have detailed lists of the cultivated landraces been compiled. Authors of the eighteenth and nineteenth centuries mentioned the great variation found in Spain (Moreno, Martinez, & Cubero, 1983), and

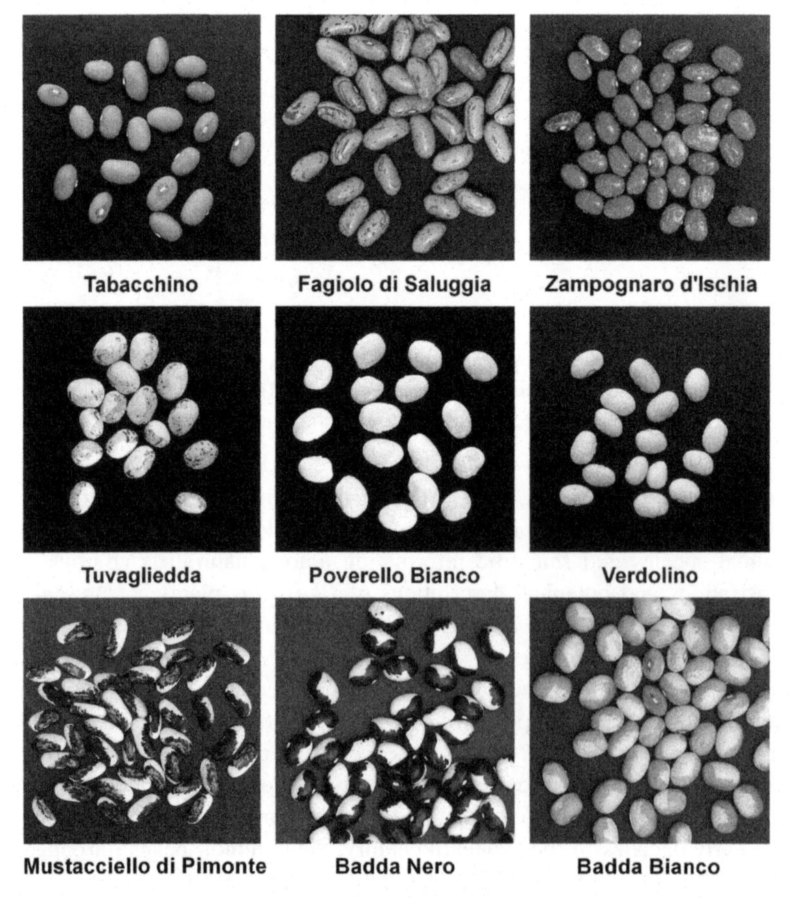

Figure 2.4 Seed morphological variation in Italian common bean landraces.

Puerta Romero (1961) classified the different cultivars used as traditional food by the Spanish on the basis of morpho-agronomic characters. A book describing 472 common bean landraces cultivated in Italy was published by Comes (1910), while investigations on the phenotypic variation within 1500 landraces grown in the Netherlands were performed by Nijdam (1947).

Starting from 1990s, systematic studies on the European common bean landraces have been carried out by recording morphological and agronomical traits, seed quality traits and phaseolin pattern. This last biochemical marker allows the attribution of the landraces to one of the two major gene pools of the crop. The prevalence of the Andean types (76%) was first described by Gepts and Bliss (1988) and was confirmed by subsequent studies at national (Lioi, 1989; Logozzo et al., 2007; Ocampo, Martin, Sanchez-Yelamo, Ortiz, & Toro, 2005; Rodiño et al., 2001) and regional (Escribano, Santalla, Casquero, & De Ron, 1998; Limongelli, Laghetti, Perrino, & Piergiovanni, 1996; Lioi, Nuzzi, Campion, & Piergiovanni, 2012; Piergiovanni, Taranto, Losavio, & Pignone, 2006) levels. Within the European germplasm, the distribution of phaseolin types parallels that observed for American genotypes. Types C and T are clearly predominant within the Andean gene pool, while type S is prevalent within the Mesoamerican one. Evaluations carried out by using DNA-based markers have evidenced a very high variation present within the Iberian germplasm. Based on these evidences, Santalla, Rodiño, and De Ron (2002) suggested Spain as a secondary diversification centre for the common bean.

It is well known that due to the environmental changes produced by human activities over time populations of plant and animal species have become small, fragmented and isolated. This trend also pertains to the common bean, but a detailed analysis of the studies published in the last decade evidences that, though the cultivation of common bean landraces is fragmented and confined to marginal areas, a significant number of landraces still survive on farm, mainly in the Iberian Peninsula (Moreno et al., 1983) and Italy (Piergiovanni & Lioi, 2010). This means that a significant fraction of the common bean variation present at the beginning of the twentieth century has been conserved up to present times. Generally, the perpetuation of landrace cultivation is not homogeneous within the countries. For example, Galicia appears to be the Spanish region still showing a wide common bean variation (Escribano et al., 1998). On the other hand, it is worthy to note that only 60% of the landraces grown in Catalonia (Spain) belong to the Andean gene pool, while in the rest of Spain 80% of landraces are of Andean origin (Rodrigo, 2000).

As concerns Italy, common bean landraces are still cultivated mainly in hilly areas along the Apennine ridge of the central and southern regions, such as Basilicata, Lazio and Abruzzo (Limongelli et al., 1996; Piergiovanni et al., 2006). Geographical isolation, as well as a lack of good roads until recent times, could explain the persistence of landraces in these areas. Unfortunately, it must be noticed that frequently landraces are mainly grown by elders for private use and only occasionally are sold in local markets. This, in addition to the diffusion of intensive agricultural systems based on commercial varieties, exposes the landraces to a high risk of loss in the coming years.

2.4 Status of Germplasm Resources Conservation (*Ex-Situ, In-Situ, On-Farm*)

It is generally accepted that significant amounts of genetic erosion have occurred and are still occurring mainly as consequence of the destruction of ecosystems and habitats by several pressure factors. Multiple strategies have been adopted to prevent the loss of genetic variation of plant species. One of them is *ex situ* conservation, which consists in the maintenance of germplasm accessions in gene bank facilities to avoid changes of genetic structure as well as extinction. Gene banks should not be considered as seed museums but as a source of genetic resources available to the user community. For a crop like common bean, as well as for its wild relatives, *ex situ* conservation can be carried out by storing seeds for long periods at low temperature and moisture. However, some hindrances associated with *ex situ* conservation can affect the genetic integrity of the conserved accessions. For materials preserved as seeds, periodic rejuvenation is required to counterbalance the declining of seed viability. Protocols adopted worldwide are designed to minimize the possibility that the genetic structure of stored samples could be modified by mutations, selection, random drift or accidental contamination. Large *Phaseolus* germplasm collections were developed to acquire, maintain, evaluate, document and distribute germplasm, in order to aid scientists in improving the quality and productivity of this crop. These collections stored all over the world include genotypes of both domesticated and wild species of *Phaseolus*. Seed samples are generally available on request for research or breeding purposes, with the addition of a paper trail for material transfer agreements. In the germplasm bank of the Genetic Resources Program of the International Center for Tropical Agriculture (CIAT; Cali, Colombia), the largest and most diverse bean collection in the world is preserved. This gene bank belongs to the Consultative Group for International Agricultural Research (CGIAR) and stores about 36,000 accessions of *Phaseolus* spp., corresponding to 44 taxa from 109 countries (http://isa.ciat.cgiar.org/urg/main.do?language=en). The largest segment of this collection corresponds to the primary centres of origin in the Neotropics, especially Mexico, Peru, Colombia and Guatemala, but there are also important segments from Europe and Africa, and to a lesser extent from Asia. A collection of about 15,000 accessions is housed at the Western Regional Plant Introduction Station, Pullman, Washington, USA (http://www.ars.usda.gov/Main/site_main.htm?docid=9065). The main collections of *Phaseolus* germplasm in Europe are those of the Institut für Pflanzengenetik und Kulturpflanzenforschung (IPK; Gatersleben, Germany), with about 9,000 accessions (http://gbis.ipk-gatersleben.de/gbis_i/), and of the N.I. Vavilov Research Institute of Plant Industry (VIR; Russia), with about 6,000 accessions (http://www.vir.nw.ru/data/dbf.htm). Of noteworthy interest is the collection of wild Phaseoleae – Phaseolinae species held at the National Botanic Garden of Meise, Belgium. This collection covers a very wide genetic diversity and currently includes 1886 accessions representing 225 taxa of the Phaseoleae tribe, chiefly centred on the Phaseolinae subtribe. *Phaseolus* and *Vigna* are the best represented genera with 41 species (712 accessions) and 67 species (978 accessions), respectively (http://www.br.fgov.be/research/collections/living/phaseolus/). Additionally, smaller collections are scattered all over the world. All together, these collections represent a substantial source of genetic diversity that is generally freely available for plant genetics and breeding research. An overview on the status of the smaller European germplasm collections was reported in the *Catalogue of Bean Genetic Resources* compiled in 2001 as an initiative of the European Union PHASELIEU project partners (Amurrio, Santalla, & De Ron, 2001). Bioversity International, a member of the CGIAR Consortium and a partner of FAO of the UN, currently coordinates the European Cooperative Programme for Plant Genetic Resources (ECPGR), which helps to rationally and effectively conserve the plant genetic resources. The platform implemented by ECPGR allows access to passport data of common bean accessions stored at more than 20 worldwide gene banks (http://www.ecpgr.cgiar.org/germplasm_databases.html).

In recent decades there has been increasing interest in the use of *in situ* conservation for wild relatives of crop species and for crop species themselves. This approach is based on the maintenance of the ecosystem as a whole and is the elective strategy for preservation of crop wild relatives. *In situ* conservation of crop plants, specifically designed as on-farm conservation, is based on the genetic resources maintenance by custodian farmers who continue to grow and use traditional varieties or landraces, allowing their evolution to be continued in the environment where they are traditionally cultivated. In this way a source of adapted germplasm is available for plant breeding and other users. The results of a study on the effectiveness of on-farm conservation of common bean landraces showed that this type of conservation is really the most effective to maintain the diversity present in the original populations (Negri & Tiranti, 2010). However, it should be kept in mind that on-farm conservation is a complementary rather than an alternative strategy to gene banks.

Small-scale farming systems such as home garden conservation should also be included as a further potential reservoir of agricultural biodiversity. Even in Europe some studies document its role in securing crop genetic diversity, shaping the landscape and maintaining the cultural heritage of a community (Galluzzi, Eyzaguirre, & Negri, 2010). Recently Szabó (2009) proposed the common bean as a model taxon for monitoring trends in European home garden diversity. In fact, we still do not know adequately the home garden–based diversity for the most important crops.

2.5 Germplasm Evaluation and Use

Knowledge about genetic variation within germplasm collections plays a key role for their utilization. Although this task is a complex, expensive and time-consuming exercise, it is one research area that benefits crop improvement, since it supports decisions concerning breeding methodology and management of genetic resources. The evaluation of genetic diversity also supports the resource allocation decisions that affect the long-term maintenance of germplasm collections. Starting from these considerations, it is evident that the collection management requires robust, rapid and cheap methods to perform detailed characterizations of stored accessions. It is accepted that passport data are not sufficient predictors for evaluating diversity within germplasm collections, because geographical associations could become less clear due to human migrations or to seed exchange among farmers that caused genetic material to be carried from one to another region.

An important question in germplasm evaluation is the number of markers to be used. To be significant, any diversity analysis should be based on the estimation of a high number of traits covering the maximum range of phenotypic and genotypic variation at the same time. Some decades ago, the estimation of genetic variation of stored collections was based only on some morpho-agronomic traits and electrophoretic protein profiles. In recent years advanced DNA-based methodologies to characterize germplasm collections have started to be widely applied. However, the evaluation of very large collections, such as that of common bean maintained at CIAT, is feasible only for easily scored traits. For these reasons gene bank managers are always seeking methodological approaches that would allow analysing the stored collections in a relatively short time and at acceptable costs.

The creation of a core collection, a subset of accessions incorporating a representative sample of the variation within the whole collection with a minimum of redundancy, allows an increase in the number of traits taken into consideration, especially the most investigated ones, such as resistance to pests and diseases, tolerance to specific pedo-climatic conditions, nutritional value of grains and so on. Since two gene pools exist in common bean, studies on the germplasm collections should include samples from the Middle American and Andean regions or not, according to final aims. The effectiveness of this strategy has been evidenced by Skroch, Nienhuis, Beebe, Tohme, and Pedraza (1998), who compared the genetic variation present in a core to that of the whole collection of Mexican beans at CIAT by using random amplified polymorphic DNA (RAPD) markers. Recently, a core collection from a total of 544 European accessions was developed by using sampling approaches based on both information available in the gene bank databases and phaseolin patterns. This first attempt at the development of a European core collection will help assess the contribution of the two American gene pools to the European germplasm and their relative usefulness for breeding purposes (Logozzo et al., 2007). Information derived from studies on whole or core collections could serve more efficiently the breeders working at the selection of improved common bean varieties (Pérez-Vega, Campa, De la Rosa, Giraldez, & Ferreira, 2009). In fact, the value of genetic material rests in the characteristics it possesses, in the worth of the product obtained as a result of its utilization and in the contribution it makes to land management and production processes.

An example of the use of stored accessions is the identification in the germplasm at CIAT of a wild common bean accession carrying a mutation that prevents the accumulation of all components of lectins, a family of closely related seed storage proteins considered to be antinutritional factors. From this material, bean lines producing seeds without lectins were developed with the aim of improving the nutritional characteristics of bean seeds used for both food and feed (Campion, Perrone, Galasso, & Bollini, 2009). In lectin-free lines, mutations for reduced phytic acid accumulation in the seeds were induced successively. These new lines have more digestible proteins, a higher level of free phosphorus and increased bioavailability of bivalent cations (Campion, Sparvoli, et al., 2009).

Another relevant case is the finding within the CIAT germplasm of some wild Mexican bean accessions resistant to weevils. The resistance has been associated to the presence of a particular class of proteins, the arcelins, components of the multigene family of lectins. So far, seven arcelin variants have been identified, all in wild accessions. Different attempts have been made to breed resistance traits of wild common bean into cultivars, and some successful examples have been described (Cardona, Kornegay, Posso, Morales, & Ramirez, 1990). Moreover, recently a wild accession resistant to both bruchid beetles, the weevil *Acanthoscelides obtectus* Say and the Mexican weevil *Zabrotes subfasciatus* Bohemian, has been collected in Mexico (Zuagg et al., 2013). Thus, common bean wild materials have been confirmed to be a useful source of desirable traits for future breeding purposes.

Other than wild materials, landraces are universally considered a good source of precious variation. They constitute an important resource for breeders because of their considerable genotypic variation and high adaptation to particular environmental conditions. As a consequence, the wide genetic diversity present in southwestern European landraces could be an excellent source for bean breeding, this material being unimproved adapted germplasm. The screening of stored landraces can be a multitask exercise such as that carried out by Rodiño, Monteagudo, De Ron, and Santalla (2009), who studied the variability among common bean lines selected from ancestral landraces maintained at the MBG-CSIC gene bank (Pontevedra, Spain) to identify groups of lines with superior traits. They found accessions having good expression of some pod and seed quality traits that would be appreciated by both consumers and producers, lines having notable performances that could be used to improve yield and lines showing some tolerance or resistance to pathogens, which would be essential for the development of resistant cultivars. In some cases the screening is focused on a well-defined objective, such as the search for potential resistance sources to anthracnose caused by *Colletotrichum lindemuthianum* (Sacc & Magnus), one of the most devastating diseases of common bean in mild and wet areas of northern Spain. Although the screening of the bean collection maintained at Villaviciosa (Asturias, Spain) did not allow the identification of resistant accessions, some materials showing moderate resistance were found (Ferreira, Campa, Pérez-Vega, & Giraldez, 2008).

2.6 A Glimpse at Crop Improvement

The genetic basis of the commercial common bean classes is narrow as compared to worldwide germplasm, which in contrast shows a wide diversity of seed and pod traits, plant growth habit, phenological traits, flowering time, photoperiod sensitivity, adaptation to different soil types, wide range of resistance to diseases and stress, and different nutritional seed quality. The genetic variation of common bean germplasm has been widely used by breeders to further enhance the crop since the late nineteenth century and early twentieth century. However, so far a large part of the variation observed in gene pools, races and wild relatives has not been used in breeding. The major limitation to its utilization can be attributed to the lack of adaptation of germplasm to new ecological niches, the presence of undesired traits such as seed shattering and the time-consuming analysis of progenies.

Due to the economic value of common bean, several breeding programmes are presently in progress throughout the world. Breeders freely crossed between Mesoamerican and Andean gene pools, as well as among races, although intergene pool crosses have had only limited success, suggesting an ongoing process of speciation (González, Rodiño, Santalla, & De Ron, 2009; Koinange & Gepts, 1992). Breeding can also involve gene introgression from additional gene pools. Indeed, the secondary and the tertiary gene pools of common bean, covering a range of environments from cool moist highlands to hot semi-arid regions, could be an important resource for the genetic improvement of common bean, which will increasingly suffer from the increase of temperatures and moisture, and from drought periods, as a consequence of climatic changes (Beebe, Rao, Mukankusi, & Buruchara, 2012).

Several species of *Phaseolus* can be hybridized to common bean. The species belonging to its secondary gene pool, such as *P. coccineus*, *P. polyanthus* and *P. costaricensis* Freytag & Debouck, can freely be crossed with each other without embryo rescue, particularly when common bean is used as the female parent. *P. coccineus* has been more commonly used in wide crosses with *P. vulgaris*, especially for traits such as cold temperature tolerance, root rot and bean yellow mosaic virus resistance. However, hybrid progenies may be partially sterile, preventing the recovery of desired stable traits. The tertiary gene pool of common bean comprises *P. acutifolius* and *P. parvifolius* Freytag; crosses of common bean with these two species are successful, but require embryo rescue, and backcrosses to the recurrent common bean parent are often required to restore hybrid fertility. Genes for disease resistance have been successful moved from *P. acutifolius* to common bean. Crosses with other species, such as *P. lunatus*, *P. filiformis* Benth. and *P. angustissimus* A. Gray have been attempted without producing viable hybrid progenies, so these species could be considered the quaternary gene pool of common bean (Singh, 2001).

Early maturity, adaptation to higher altitude, upright plant type, high pod quality and seed yield, and some resistances to diseases such as viruses and rust, insect pests, and drought and abiotic constraints such as deficiency of nitrogen, phosphorus and zinc or tolerance to aluminium and manganese toxicity have been bred into common bean cultivars. Most, if not all, commonly used crop breeding methods have been employed with common bean (Beaver & Osorno, 2009). Differences in genetic distance among gene pools, races and species dictate specific breeding methods and strategies. The results and the efficiency of the different methods applied have been the object of some detailed reviews (Graham & Ranalli, 1997; Kelly, 2010; Singh, 2001). Challenges such as drought, root rot, heat, depleted soils, excessive rainfall and new and old pests and diseases pose new breeding targets and require increased efforts to address them. To overcome some of the inherent difficulties faced by conventional plant breeding, new biotechnology tools have been developed and are growing in importance and use. Molecular approaches, such as marker-assisted selection (MAS), can support breeders facilitating and accelerating the transfer of desired traits. A detailed report on implementation and adoption of MAS in common bean breeding is provided by Miklas, Kelly, Beebe, and Blair (2006), who reported highlighted examples of MAS success in gene pyramiding, rapid and simpler detection, and selection of resistance genes. Slower progress has been obtained in the improvement of nitrogen fixation, insect resistance and tolerance to abiotic stresses. Moreover, progress in increasing seed yield potential has been only moderately successful, because multiple constraints limit bean productivity (Beaver & Osorno, 2009).

In terms of consumer preferences, the most desirable traits are those related to the technical and nutritional quality of dry seeds, such as ease of cooking, soft coat texture, good taste and protein content. Cooking time is certainly one of the factors that limit the home consumption of dry bean. Some studies showed that it is an oligogenic trait with high genetic variation but also significantly affected by the growing location. The recent identification of quantitative trait loci (QTLs), which define the location of genes governing this target trait, is the first step in future breeding programmes (Garcia et al., 2012).

It should be keep in mind that common bean is most produced and consumed in developing countries, where yield is often affected by deficiencies and toxicities of minerals in soil. This is the case of aluminium toxicity that negatively affects the yield in acid soils of tropic regions. Studies conducted at CIAT have shown that some accessions of *P. coccineus* are more resistant than common bean to aluminium toxicity. Butare et al. (2012) crossed an Al-sensitive common bean with an Al-resistant *P. coccineus* accession, obtaining recombinant inbred lines, among which were promising resistant common bean genotypes.

Finally, since each region has different agro-techniques, pedo-climatic conditions, biotic and abiotic constraints, and consumer preferences, breeding programmes must be tailored to the needs of farmers and consumers who will use the new cultivars.

2.7 Biochemical and Molecular Diversity

Electrophoretic analysis of seed storage proteins has proven to be a valuable tool in tracing the evolution of crop plants, especially for identification of the wild progenitors and gathering additional information on the evolutionary and domestication patterns.

The structural and functional features of phaseolin, the major seed protein of common bean, make it a useful marker. This protein, accounting for 50% of total protein stored in the cotyledons and 35–46% of the total seed nitrogen, is coded by a cluster of closely related genes that may arise by successive duplication and diversification from an ancestral gene. The divergence processes include insertions, nucleotide substitutions, duplications or deletions of repeats (Kami & Gepts, 1994). In addition co- and post-translational modifications, including cleavage of the signal peptide, different glycosylation of polypeptides (Lioi & Bollini, 1984) and charge variation due to amino acid substitution resulted in the formation of slightly heterogeneous phaseolin polypeptides in the Mr 54–44 kDa, reflecting genotype divergence.

In a pioneering work by Gepts (1988), phaseolin was used as a marker in describing the domestication patterns and worldwide dissemination of common bean. Phaseolin electrophoretic analysis of wild and domesticated materials supported the hypothesis of multiple domestication events, thought to be the cause of parallel geographic phaseolin variation between wild and cultivated forms. The Mesoamerican domestication gave rise to small-seeded S phaseolin cultivated materials, while large-seeded T, C, H and A phaseolin were observed in the southern Andes (Koenig, Singh, & Gepts, 1990). Moreover, it has been shown that phaseolin is a useful biochemical marker to follow the dispersal pathway of common bean from domestication areas into Europe. This revealed that the European common beans arose from the introduction of domesticated beans from both of the American gene pools. A higher frequency of Andean phaseolin types (76%) with respect to Mesoamerican ones (24%) was first recorded in European germplasm by Gepts and Bliss (1988). This was successively confirmed by Lioi (1989), analysing a large collection of accessions mainly from Italy, Greece and Cyprus. The prevalence of Andean types within the European common bean germplasm stored in some international gene banks has been recently confirmed by Logozzo et al. (2007), who analysed a collection of 544 accessions all from European regions, showing that the Andean phaseolin types T (45.6%) and C (30.7%) prevailed over the Mesoamerican S type (23.7%). A summary of the results from different studies are reported in Figure 2.5. Despite a large variation in sample sizes and sampling strategies among these investigations, the presence of all three major phaseolin types (C, T and S) was observed in all the areas considered, suggesting a large seed exchange among the European countries. Over a total of 1309 European accessions considered, a prevalence of Andean phaseolin types at a single-country level was confirmed, with a global 79.6% versus 20.4% of Mesoamerican types. Differences in the frequencies of each Andean phaseolin type have also been observed. In the countries along the Mediterranean arc such as on the Iberian Peninsula, in Italy and the Balkan area, phaseolin C was the most common type. Conversely, in accessions from France, Central Europe and Sweden, the T type was the prevailing one. A relatively high frequency of Mesoamerican types was observed in Central Europe (27%) and France (30%) compared to Mediterranean countries, where the frequency is lower, reaching a mean value of 18%. European S types showed a larger seed size than those from the centre of domestication. Logozzo et al. (2007) suggested two hypotheses to explain this finding: a preferential introduction of Durango and Jalisco races that, among Mesoamerican races, possess larger seeds, or a selection towards larger seeds within S types after introduction in Europe.

It has been suggested that crop expansion from America to Europe resulted in a reduction of diversity because a strong founder effects due to adaptation to new environments and consumer preferences, followed by evolution probably involving hybridization and recombination between the Andean and Mesoamerican gene pools (Gepts, 1999). Papa et al. (2006) estimated a loss of diversity around 30% and a low differentiation between the gene pools in Europe, when compared with the differences in the Americas, suggesting a combination of greater gene flow or convergent evolution for adaptation to European environments. More recently Angioi et al. (2010) using six chloroplast microsatellite (cpSSR) markers, confirmed that European common beans arose from both gene pools, but the bottleneck effect of the introduction into Europe might not have been so strong. Moreover, they estimated that hybrids between the two gene pools occurred at higher frequencies in Central Europe and lower frequencies in Italy and Spain. Moreover they suggest that not only some of the countries therein, but the entire European continent can be regarded as a secondary diversification centre for *P. vulgaris*.

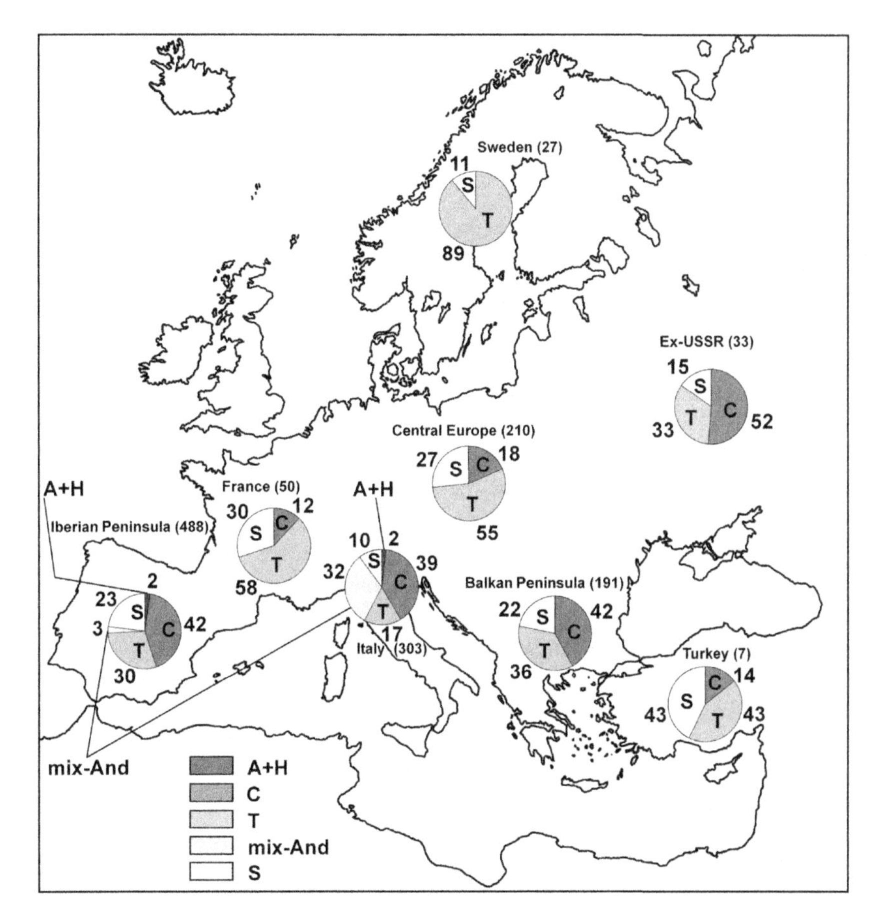

Figure 2.5 Distribution (%) of phaseolin type frequencies across Europe. Number in parentheses next to the geographical region name refers to sample size.
Source: Data from Gepts and Bliss (1988), Rodiño et al. (2001), Šustar-Vozlič et al. (2006), Logozzo et al. (2007), Pérez-Vega et al. (2009) and Piergiovanni and Lioi (2010).

Molecular markers have been shown to be effective indicators for genetic variation underlying agronomic traits with some advantages over morphological traits, such as the ability to distinguish among accessions with similar morphology and discriminate polymorphism over far more loci than isozymes or seed storage proteins. Molecular markers span broader genomic areas and present different types of inheritance, so they have also been used to better estimate the levels of diversity and to understand the effects of migration and selection on the maintenance of polymorphism in the European beans. There are several papers on the characterization of European germplasm of *P. vulgaris* using different molecular markers. Some studies were based on random PCR markers, such as RAPDs (Mavromatis et al., 2010), inter-simple sequence repeats (ISSRs) and AFLP (Svetleva et al., 2006; Šustar-Vozlič, Maras, Kavornik, & Meglič, 2006). Other molecular markers such as SSR, which are more specific in target, were used to assess diversity among landraces (Lioi et al., 2005). Moreover, recently some studies were carried out to fingerprint specific landraces using different molecular markers (Lioi et al., 2012; Paniconi, Gianfilippi, Mosconi, & Mazzuccato, 2010).

2.8 The Germplasm Safeguarded Through the Attribution of Quality Marks

The crop landraces managed by local communities as part of their farming systems have been maintained on farm until today. The persistence of these landraces can be associated with the presence of elder farmers, a cultural value for the local communities, economic and/or geographic isolation of cultivation areas, and utilization in the preparation of traditional local dishes, medicinal practices or religious ceremonies. Despite the lack of coordinated efforts, these farmers have practiced de facto the on-farm conservation of genetic resources, adopting a cost-efficient approach as compared to the *ex situ* method. The protection of the autochthonous germplasm in regions where agriculture still maintains traditional practices is considered a priority, even though the de facto on-farm maintenance cannot guarantee the survival of landraces over time. In 1997 an International Plant Genetic Resources Institute (IPGRI; Rome, Italy) project started to promote the on-farm conservation of locally selected varieties in 10 pilot countries (IPGRI, 1997). Successively, a first inventory of on-farm conservation and management activities in Europe was compiled by the '*in situ*/on-farm task force' of ECPGR promoted by Bioversity International, formerly IPGRI (Negri et al., 2000). More recently, the European Community (EC) provided new financial resources to support the on-farm conservation (commission Directive 2008/62/ECoj 20 June 2008) in relation to *agricultural landraces and varieties which are naturally adapted to the local and regional conditions and threatened by genetic erosion.*

Starting in the 1990s the EC set down the rules (EC Reg. n. 2081/92 and 2082/92, recently substituted by EC Reg. 510/2006) for the attribution of origin and quality marks to local typical products for human consumption (i.e. vegetables, fruits, cereals and meat) of the European countries. In this way these products can be easily distinguished from the commodities belonging to the same category. Three marks were introduced: protected designation of origin (PDO); protected geographic indication (PGI) and traditional specialities guaranteed. The main difference among them is related to how closely the quality specificities of the products are linked to the geographical area of which they bear the name. Contrary to individual brands, these quality marks have a collective dimension involving a group of producers that may be identified with a geographical reference.

The aim of the EC marks is the creation of a legal framework for the protection and promotion of brand names of Europe's traditional agricultural products and foods. In this way, the work of thousands of farmers and artisanal food producers is safeguarded, the European Union's rural heritage is preserved, and the quality and performances of a food product carrying the mark are recalled by consumers. As concerns vegetables, fruits and cereals, the attribution of the EC marks to elite ecotypes could sustain their on-farm conservation over time, encouraging the farmers to continue their growing. In fact, the production of certified products generally assures similar or higher incomes compared to modern varieties. The EC marks are attributed on the basis of instances describing deeply the history of each ecotype; the connection with a recognizable geographical area; the agronomic, nutritional, organoleptic and other peculiarities; and the discrimination of these products from the similar commercial ones. The achievement of these objectives requires collaboration between researcher institutions with different competences and the local communities. The different steps required to obtain the attribution of the European quality marks are schematized in Figure 2.6. An example of this road applied to common

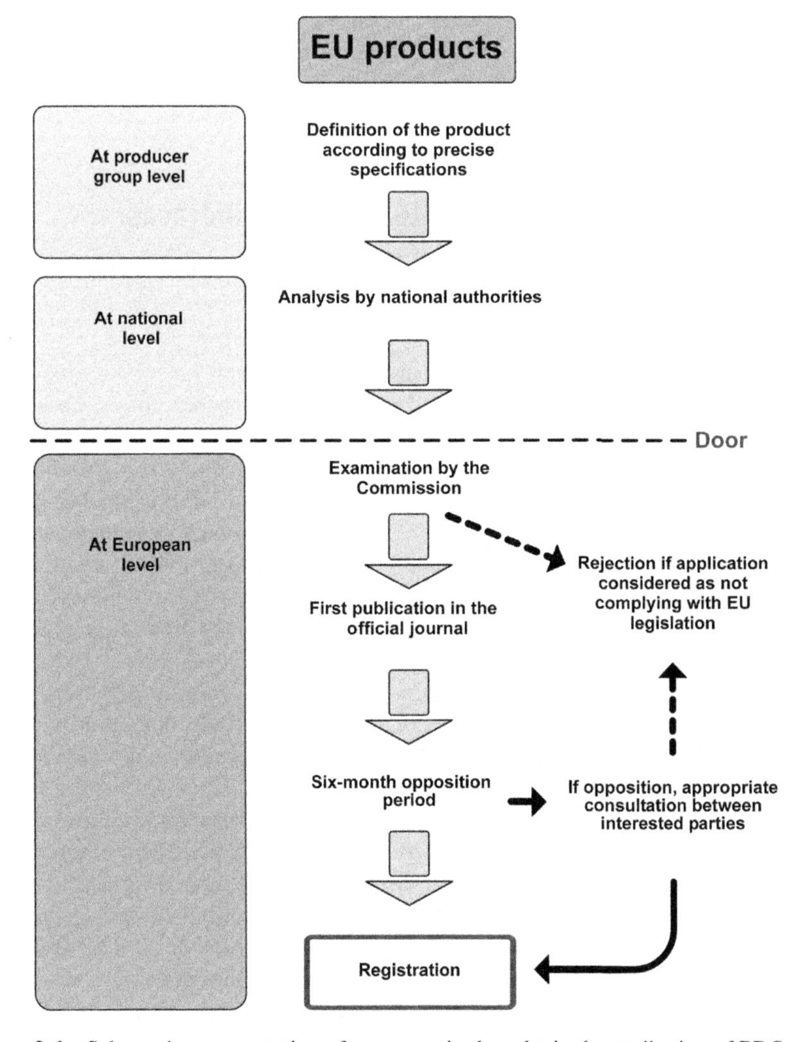

Figure 2.6 Schematic representation of steps required to obtain the attribution of PDO or PGI marks.

bean ecotypes cultivated in the Basilicata region (southern Italy) was described by Piergiovanni and Laghetti (1999). Still today the EC quality marks have been attributed to *Phaseolus* spp. ecotypes grown in five countries (Table 2.1). It is worthy to note that some marks have been attributed to a single ecotype (i.e. Fagiolo di Cuneo or Fagiolo di Sorana), while others regard a group of them grown in the same geographical area (i.e. Judías de el Barco de Ávila, as well as Fagioli di Lamon).

In some European countries other quality marks have been implemented by local authorities such as regions, provinces and municipalities, or associations such as Slow Food. Some Italian common bean landraces have obtained similar brands such as the municipal denomination of origin (De.CO.) assigned to Fagioli di Cortale (Calabria region), or the creation of a Slow Food mark for Fagiolo Gialet (Veneto region). The benefits of these further initiatives, though focused at the local level, can be described in terms of increased income to farmers, safeguarding of precious germplasm and maintenance of whole agro-ecosystems, which can be considered an advantage for the entire community (Negri, 2011).

2.9 Characterization and Evaluation of Landraces: Some Case Studies

2.9.1 Ganxet Bean

The Ganxet bean is a landrace cultivated in Catalonia (Spain) for a long time and is the most prestigious among those cultivated in the region (Sánchez, Sifres, Casañas, & Nuez, 2007). Originating in Mesoamerica, it probably reached the Catalan coast in the nineteenth century. Ganxet is a white-seeded type, very easily recognizable by a marked hooked shape (*ganxet* means 'little hook' in Catalan). The organoleptic properties, highly appreciated by consumers, can explain the persistence of its cultivation up to now (Casañas et al., 1999). However, the original germplasm has suffered from the introgression of other common bean varieties, including new improved varieties introduced in recent times in the territory traditionally devoted to the cultivation of Ganxet. Many transitional forms between Ganxet and non-Ganxet beans are presently under cultivation in Catalonia, as testified by a very high variation recorded within the germplasm currently used by Catalan farmers (Casañas et al., 1999). Variation is mainly related to the degree of hook and flatness of seed, while memories describe a much more homogeneous germplasm. Understanding the evolutionary history of Ganxet represents a model to elucidate the evolution of a landrace sharing the cultivation area with other common beans. AFLP and RAPD analyses of Ganxet germplasm carried out by Sánchez, Sifres, Casañas, and Nuez (2008) detected a limited variability among the lines representing the Ganxet prototype, while the variability increases as the studied material moves farther away from the typical seed morphology. The molecular markers used by these authors proved that the source of the introgression is mainly the Great Northern market class. Populations belonging to this market class are more productive than true Ganxet-type lines, so crosses between them tend to be more productive and for this reason more attractive for farmers (Casañas et al., 1999).

Table 2.1 List of European Common Bean Landraces that Have Obtained a Quality Mark

Country	Type of Mark	Local Name
Bulgaria	Slow Food Presidium	Smilyan beans[a]
France	EC PGI	Haricot Tarbais (reg. 06.06.2000)
Greece	EC PGI	Fasolia Gigantes Elefantes Kato Nevrokopiou (reg. 21.01.1998)
		Fasolia kina Messosperma Kato Nevrokopiou (reg. 21.01.1998)
		Fasolia Gigantes Elefantes Prespon Florinas (reg. 18.07.1998)
		Fasolia (Plake Megalosperma) Prespon Florinas (reg. 18.07.1998)
		Fasolia Gigantes Elefantes Kastorias (reg. 12.08.2003)
		Fasolia Vanilies Feneou (reg. 24.05.2012)
Italy	EC PGI	Fagiolo di Lamon della vallata bellunese (reg. 02.07.1996)[a]
		Fagiolo di Sarconi (reg. 02.07.1996)[a]
		Fagiolo di Sorana (reg. 14.06.2002)
		Fagiolo di Cuneo (reg. 20.05.2011)
	EC PDO	Fagiolo Cannellino di Atina (reg. 05.08.2010)
		Fagioli Bianchi di Rotonda (reg. 12.03.2011)
	Slow Food Presidium	Fagiolo Dente di Morto di Acerra
		Fagiolo di Controne
		Fagioli Badalucco, Conio e Pigna[a]
		Piattella Canavesana di Cortereggio
		Fagiolo Badda di Polizzi[a]
		Fagiolo di Sorana
		Fagiolo Rosso di Lucca
		Fagiolo Gialet della Val Belluna
Poland	EC PDO	Fasola Piękny Jaś z Doliny Dunajca (reg. 25.10.2011)
		Fasola Wrzawska (reg. 13.01.2012)
	EC PGI	Fasola Korczynska (reg. 13.07.2010)
Spain	EC PGI	Judias de el Barco de Avila (reg. 21.06.1996)[a]
	EC PDO	Mongeta del Ganxet (reg. 23.12.2011)
	Slow Food presidium	Ganxet bean
Sweden	Slow Food presidium	Öland Island brown beans[a]
Switzerland	Slow Food presidium	Swiss dried green beans[a]

[a]More than one type.

This example may be representative of the transformation that other common bean landraces have undergone over time. For Ganxet bean, the extremely hooked shape of seed can help to ensure the survival of the true landrace, because farmers and consumers can easily recognize and reject materials that differ greatly from the standard shape. Conversely, for those landraces that do not show easily distinctive agro-morphological traits, a similar approach cannot be applied and the discrimination of the traditional landrace prototype from possible hybrids appears to be not as easy.

2.9.2 Prespon Florinas and Kastorias Beans

In Greece, common bean is an important crop, cultivated areas being located in the northern and central parts of the country, Macedonia, Thrace and Thessaly regions (Mavromatis et al., 2010). As for other countries, the autochthonous material has been progressively replaced with modern cultivars; only in some areas do farmers continue to maintain local landraces. Since presently the demand for organic food is mainly oriented toward products of plant origin, the performances of commercial cultivars and Greek landraces grown under organic farming have been evaluated in detail with the aim of identifying niche markets able to sustain the on-farm conservation of local common beans. When organic agro-techniques are applied, landraces and cultivars mainly differ in yield component traits, such as seed size and weight, number of pods per plant and number of seeds per pod (Mavromatis et al., 2007, 2010). Although the highest values could be expected to be recorded in commercial cultivars, this was not the case. In particular, the landraces Kastoria and Xanthi, both from northern Greece, displayed very good performances. These results could be issued either on the promising genetic traits of these landraces or on the adaptation in organic farming, since landraces are traditionally cultivated in family farms without the use of agrochemicals. In addition, it has been shown that Kastoria beans have a protein content significantly greater than the mean value reported by Escribano, Santalla, and De Ron (1997) for Spanish landraces (28.58% vs 22.6%) (Mavromatis et al., 2007). With regard to grain quality, also some landraces traditionally grown near Prespes lake (Macedonia region) deserve particular attention, since their nutritional traits were comparable or better than those of commercial cultivars cultivated in the same environment (Ganopoulos, Bosmali, Madesis, & Tsaftari, 2012).

PGI quality marks have been awarded to Fassolia Gigantes Prespon Florinas as well as to Kastoria beans (Table 2.1), with the aim to sustain the on-farm survival of these elite landraces, together with the rural areas where they are grown. In these cases, the attribution of European protected designations derived from the high quality of grains, a successful combination of genetic characteristics, adaptation to both local microclimate conditions and traditional agro-techniques. It is worth underlining that these examples could represent a partial answer, without public investments, to the unsolved problems related to farmers' rights and genetic resources management. If it is true that landraces are the result of indigenous farmers' work and, in a sense, belong to a region, the attribution of European quality marks should mainly direct economic benefits towards the local communities.

2.9.3 Fagiolo del Purgatorio di Gradoli

Where socioeconomic conditions are weak, modern agricultural methods cannot be applied and agriculture retains traditional farming traits. The old bean populations owned by the farmers often show a high genetic variability with undesirable

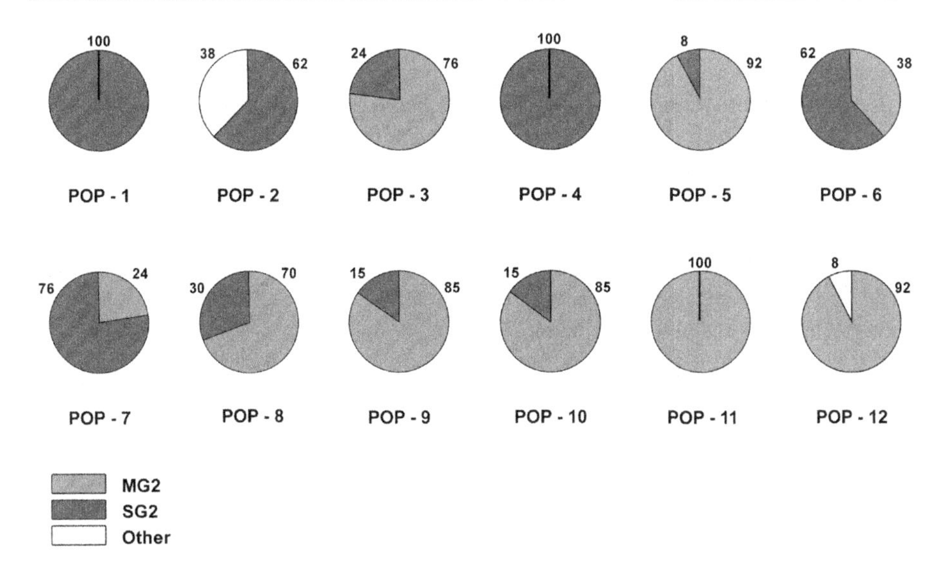

Figure 2.7 Frequencies (%) of PHA electrophoretic variants (MG2, SG2, and others) observed in 12 accessions of the common bean landrace Fagiolo del Purgatorio (Italy).

characters like low yields or susceptibility to some pests and diseases. This reduces their overall quality, posing serious constraints to their use. Producers feel the need for improving the landraces, which, in turn, could affect their genetic structure.

In central Italy a white, small-seeded (100 seed weight <25 g) common bean, belonging to the Mesoamerican gene pool, named Fagiolo del Purgatorio, has been cultivated since the eighteenth century in Gradoli and Acquapendente, Lazio region (Lioi et al., 2005). The perpetuation of the cultivation of this landrace over time is attributable to the ritual consumption of dishes prepared with dry seeds in a lunch for poor people, organized every year during Lent, by the brotherhood 'Confraternità del Purgatorio' (Piergiovanni & Lioi, 2010). Recently, Fagiolo del Purgatorio has been the object of a multidisciplinary study. This analysis represents a preliminary action necessary for drawing up disciplinary rules for a conservation consortium, as well as for the request of a European quality mark. Twenty-three samples representative of the germplasm currently used by farmers growing the Fagiolo del Purgatorio were analysed for morpho-agronomic traits, biochemical markers (phaseolin and phytohaemagglutinin electrophoretic profiles), molecular markers (AFLP and SSR), seed nutritional quality and resistance to pest and diseases (Lioi et al., 2007). Data collected in this study showed the existence within the tested germplasm of two nuclei showing differences detectable using different methodologies. The characteristic traits of the two nuclei were:

a. determinate growth habit, low number of nodes per plant, low yield, high susceptibility to bean common mosaic virus (BCMV), phytohaemagglutinin type SG2;

b. semi-determinate growth habit, high number of nodes per plant, high yield, low susceptibility to BCMV, phytohaemagglutinin type MG2.

On the basis of SSR and AFLP marker profiles, the Fagiolo del Purgatorio populations were grouped in two subclusters confirming biochemical and agronomic data and suggesting that more than one constitutive nucleus has contributed to the genetic background of this landrace. Different frequencies of the phytohaemagglutinin electrophoretic variants (MG2, SG2, and others) observed in 12 Fagiolo del Purgatorio accessions are reported in Figure 2.7.

The safeguarding of a landrace characterized by a complex genetic structure, such as Fagiolo del Purgatorio, poses some problems. First of all the on-farm conservation should be based on a sufficient number of populations to assure the same chances of co-evolution to both nuclei in the traditional areas of cultivation. On the other hand, the market requirements as well as the constraints to obtain one of the European quality marks could encourage the selection of one of the two nuclei, irremediably modifying the genetic structure of the landrace.

2.10 Conclusions

The worldwide common bean germplasm is characterized by a high degree of genetic diversity. The entire European continent can be regarded as a secondary diversification centre, as consequence of five centuries of uninterrupted cultivation and unconscious selection, coupled to a capillary diffusion of this crop. Given the wide diversification of European common bean germplasm, the overall number of accessions stored *ex situ* and landraces still surviving on farms is remarkable. Taking into account the abovementioned studies, only a multidisciplinary approach can be fully effective to characterize this precious material and to help plan adequate safeguard actions. The creation of an inventory of European landraces could be an important goal for the improved safeguarding of landraces and future uses in breeding programmes.

Acknowledgement

The authors thank Salvatore Cifarelli for technical assistance and Gabriella Sonnante for critical reading of the manuscript.

References

Amurrio, M., Santalla, M., & De Ron, A. M. (Eds.), (2001). *Catalogue of Bean genetic resources. PHASELIEU – FAIR-PL97-3463*. Spain: Misión Biológica de Galicia (CSIC).

Angioi, S. A., Rau, D., Attene, G., Nanni, L., Bellucci, E., Logozzo, G., et al. (2010). Beans in Europe: Origin and structure of the European landraces of *Phaseolus vulgaris* L. *Theoretical and Applied Genetics*, *121*, 829–843.

Barona, J. L. (2007). Clusius's exchange of botanical information with Spanish scholars. In F. Egmond, P. Hoftijzer, & R. P. W Visser (Eds.), *Carolus Clusius. Towards a cultural history of a Renaissance naturalist. History of Science and Scholarship in the Netherlands, 8* (pp. 99–116). Amsterdam: Royal Netherlands.

Beaver, J. S., & Osorno, J. M. (2009). Achievements and limitations of contemporary common bean breeding using conventional and molecular approaches. *Euphytica, 168*, 145–175.

Beebe, S., Rao, I. R., Mukankusi, C., & Buruchara, R. (2012). Improving resource use efficiency and reducing risk of common bean production in Africa, Latin America, and the Caribbean. In C. Hershey (Ed.), *Issues in tropical agriculture. I. Eco-efficiency: From vision to reality*. Cali, Colombia: CIAT.

Bitocchi, E., Bellucci, E., Giardini, A., Rau, D., Rodriguez, M., Biagetti, E., et al. (2013). Molecular analysis of the parallel domestication of the common bean (*Phaseolus vulgaris*) in Mesoamerica and the Andes. *New Phytologist, 197*(1), 300–313.

Bitocchi, E., Nanni, L., Bellucci, E., Rossi, M., Giardini, A., Spagnoletti Zeuli, P., et al. (2012). Mesoamerican origin of the common bean (*Phaseolus vulgaris* L.) is revealed by sequence data. *Proceedings of the National Academy of Sciences of the United States of America, 109*(14), E788–E796.

Brown, C. H. (2006). Prehistoric chronology of the common bean in the New World: The linguistic evidence. *American Anthropologist, 108*(3), 507–516.

Butare, L., Rao, I., Lepoivre, P., Cajiao, C., Polania, J., Cuasquer, J., et al. (2012). Phenotypic evaluation of interspecific recombinant inbreed lines (RILs) of *Phaseolus* species for aluminium resistance and shoot and root growth response to aluminium-toxic acid soil. *Euphytica, 186*(3), 715–730.

Campion, B., Perrone, D., Galasso, I., & Bollini, R. (2009). Common bean (*Phaseolus vulgaris* L.) lines devoid of major lectin proteins. *Plant Breeding, 128*, 199–204.

Campion, B., Sparvoli, F., Doria, E., Tagliabue, G., Galasso, I., Fileppi, M., et al. (2009). Isolation and characterisation of an lpa (low phytic acid) mutant in common bean (*Phaseolus vulgaris* L.). *Theoretical and Applied Genetics, 118*, 1211–1221.

Camus, J. (1894). Les noms des plantes du Livre d'Heures d'Anne de Bretagne. *Journal de Botanique, 8*, 325–335; 345–352; 366–375; 396–401.

Caneva, G. (1992). La Loggia di Psiche: Una delle prime testimonianze dell'introduzione di piante americane in Europa. *Rendiconti Accademia dei Lincei, 3*(2), 163–172.

Cardona, C., Kornegay, J., Posso, C., Morales, F., & Ramirez, H. (1990). Comparative value of four arcelin variants in the development of dry bean lines resistant to the Mexican bean weevil. *Entomologia Experimentalis et Applicata, 56*, 197–206.

Casañas, F., Bosch, L., Pujolà, M., Sanchez, E., Sorribas, X., Baldi, M., et al. (1999). Characteristics of a common bean landrace (*Phaseolus vulgaris* L.) of a great culinary value and selection of a commercial inbred line. *Journal of the Science and Food Agriculture, 79*, 693–698.

Chacón, S, Pickersgill, B., & Debouck, D. G. (2005). Domestication patterns in common bean (*Phaseolus vulgaris* L.) and the origin of the Mesoamerican and Andean cultivated races. *Theoretical and Applied Genetics, 110*, 432–444.

Clusius, C. (1583). *Rariorum aliquot stirpium per Pannoniam, Austriam, & vicinas quasdam provincias observatarum*. Antwerp, Belgium: Ex officina Christophori Plantini.

Comes, O. (1910). Del fagiolo comune: Storia, filogenesi, qualità sospettata tossicità e sistemazione delle sue razze ovunque coltivate. *Atti Istituto Incoraggiamento di Napoli, 61*, 75–145.

Debouck, D. G. (2000). Biodiversity, ecology and genetic resources of *Phaseolus* beans – Seven answered and unanswered questions. *Proceedings of the 7th MAFF International Workshop on Genetic Resources Part 1 Wild Legumes AFFRC & NIAR* (pp. 95–123). Japan.

Debouck, D. G., Toro, O., Paredes, O. M., Johnson, W. C., & Gepts, P. (1993). Genetic diversity and ecological distribution of *Phaseolus vulgaris* (Fabaceae) in northwestern South America. *Economic Botany, 47*, 408–423.

Delgado-Salinas, A., Bibler, R., & Lavin, M. (2006). Phylogeny of the genus *Phaseolus* (Leguminosae): A recent diversification in an ancient landscape. *Systematic Botany, 31*, 779–791.

Diaz, L. M., & Blair, M. W. (2006). Race structure within the Mesoamerican gene pool of common bean (*Phaseolus vulgaris* L.) as determined by microsatellite markers. *Theoretical and Applied Genetics, 114*, 143–154.

Escribano, M. R., Santalla, M., Casquero, P. A., & De Ron, A. R. (1998). Patterns of genetic diversity in landraces of common bean (*Phaseolus vulgaris* L.) from Galicia. *Plant Breeding, 117*, 49–56.

Escribano, M. R., Santalla, M., & De Ron, A. M. (1997). Genetic diversity in pod and seed quality traits of common bean populations from Northwestern Spain. *Euphytica, 93*, 71–81.

Ferreira, J. J., Campa, A., Pérez-Vega, E, & Giraldez, R. (2008). Reaction of a bean germplasm collection against five races of *Colletotrichum lindemuthianum* identified in Northern Spain and implications for breeding. *Plant Disease, 92*, 705–708.

Freytag, G. F., & Debouck, D. G. (2002). Taxonomy, distribution, and ecology of the genus *Phaseolus* (Leguminosae-Papilionoideae) in North America, Mexico and Central America. *SIDA, Botanical Miscellany, 23*, 1–300.

Fuchs, L. (1542). *De Historia stirpium*. Basel: In officina Isingriniana.

Galluzzi, G., Eyzaguirre, P., & Negri, V. (2010). Home gardens: Neglected hotspots of agro-biodiversity and cultural diversity. *Biodiversity and Conservation, 19*, 3635–3654.

Ganopoulos, I., Bosmali, I., Madesis, P., & Tsaftari, A. (2012). Microsatellite genotyping with HRM (high resolution melting) analysis for identification of the PGI common bean variety Plake Megalosperma Prespon. *European Food Research and Technology, 234*, 501–508.

Garcia, R. A. V., Rangel, P. N., Bassinello, P. Z., Brondani, C., Melo, L. C., Sibov, S. T., et al. (2012). QTL mapping for the cooking time of common beans. *Euphytica, 186*, 779–792.

Gepts, P. (1988). Phaseolin as an evolutionary marker. In P. Gepts (Ed.), *Genetic resources of Phaseolus beans: their maintenance, domestication, evolution, and utilization* (pp. 215–241). Dordrecht, Netherlands: Kluwer Academic Publisher.

Gepts, P. (1999). Development of an integrated genetic linkage map in common bean *(Phaseolus vulgaris* L.) and its use. In S. Singh (Ed.), *Bean breeding for the 21st century* (pp. 53–91). Dordrecht, Netherlands: Kluwer Academic Publisher. ; 389–400.

Gepts, P., & Bliss, F. A. (1988). Dissemination pathways of common bean (*Phaseolus vulgaris*, Fabaceae) deduced from phaseolin electrophoretic variability. II Europe and Africa. *Economic Botany, 42*, 86–104.

González, A. M., Rodiño, A. P., Santalla, M., & De Ron, A. M. (2009). Genetics of intra-gene pool and inter-gene pool hybridization for seed traits in common bean (*Phaseolus vulgaris* L.) germplasm from Europe. *Field Crops Research, 112*, 66–76.

Graham, P. H., & Ranalli, P. (1997). Common bean (*Phaseolus vulgaris* L.). *Field Crops Research, 53*, 131–146.

IPGRI, (1997). *Newsletter for Europe n. 11*. Rome, Italy: IPGRI.

Janick, J. (2011). New World crops: Iconography and history. *ISHS Acta Horticulturae, 916*, 14–93. http://www.actahort.org/books/916/916_9.htm.

Jussieu, A. (1722). *Réflexions sur diverses dénominations françoises de plantes qui sont dépeintes dans un manuscrit ancien du cabinet du roy (including a copy of Catalogus stirpium elegantissime)*. Paris: Bibliothèque Centrale, Museum National d'Histoire Naturelle.

Kami, J. A., & Gepts, P. (1994). Phaseolin nucleotide sequence diversity in *Phaseolus*. I. Intraspecific diversity in *Phaseolus vulgaris*. *Genome, 37*, 751–757.

Kami, J., Velasquez, V. B., Debouck, D. G., & Gepts, P. (1995). Identification of presumed ancestral DNA sequences of phaseolin in *Phaseolus vulgaris*. *Proceedings of the National Academy of Sciences of the United States of America, 92*, 1101–1104.

Kaplan, L., & Lynch, T. F. (1999). *Phaseolus* (Fabaceae) in archaeology: AMS radiocarbon dates and their significance for pre-Colombian agriculture. *Economic Botany, 53*(3), 261–272.

Kelly, J. D. (2010). *The story of bean breeding. White paper prepared for Bean CAP & PBG works on the topic of dry bean production and breeding research in the US*. Michigan State University. http://www.beancap.org/_pdf/Story_of_Bean_Breeding_in_the_US.pdf.

Koenig, R. L., & Gepts, P. (1989). Allozyme diversity in wild *Phaseolus vulgaris:* Further evidence for two major centers of genetic diversity. *Theoretical and Applied Genetics, 78*, 809–817.

Koenig, R. L., Singh, S. P., & Gepts, P. (1990). Novel phaseolin types in wild and cultivated common bean (*Phaseolus vulgaris* Fabaceae). *Economic Botany, 44*, 50–60.

Koinange, E. M. K., & Gepts, P. (1992). Hybrid weakness in wild *Phaseolus vulgaris* L. *Journal of Heredity, 83*, 135–139.

Krell, K., & Hammer, K. (2008). 500 jahre gartenbohne (*Phaseolus vulgaris* L.) in Europa: *Botanik, einführungsgeschichte und genetische ressourcen. Schriften des Vereins zur Nutzpflanzenvielfal*. Schandelah, Germany: VEN.

Kwak, M., & Gepts, P. (2009). Structure of genetic diversity in the two major gene pools of common bean (*Phaseolus vulgaris* L., Fabaceae). *Theoretical and Applied Genetics, 118*, 979–992.

Kwak, M., Kami, J. A., & Gepts, P. (2009). The putative Mesoamerican domestication center of *Phaseolus vulgaris* is located in the Lerma-Santiago basin of Mexico. *Crop Science, 49*, 554–563.

Limongelli, G., Laghetti, G., Perrino, P., & Piergiovanni, A. R. (1996). Variation of seed storage proteins in landraces of common bean (*Phaseolus vulgaris* L.) from Basilicata, Southern Italy. *Euphytica, 92*, 393–399.

Lioi, L. (1989). Geographical variation of phaseolin pattern in an old world collection of *Phaseolus vulgaris*. *Seed Science and Technology, 17*, 317–324.

Lioi, L., & Bollini, R. (1984). Contribution of processing events to the molecular heterogeneity of four banding types of phaseolin, the major storage protein of *Phaseolus vulgaris* L. *Plant Molecular Biology, 3*, 345–353.

Lioi, L., Nuzzi, A., Campion, B., & Piergiovanni, A. R. (2012). Assessment of genetic variation in common bean (*Phaseolus vulgaris* L.) from Nebrodi mountains (Sicily, Italy). *Genetic Resources and Crop Evolution, 59*, 455–464.

Lioi, L., Piergiovanni, A. R., Pignone, D., Puglisi, S., Santantonio, M., & Sonnante, G. (2005). Genetic diversity of some surviving on-farm Italian common bean (*Phaseolus vulgaris* L.) landraces. *Plant Breeding, 124*, 576–581.

Lioi, L., Piergiovanni, A. R., Soressi, G. P., Nigro, C., Tamietti, G., Turina, M., et al. (2007). Caratterizzazione, selezione, risanamento e valutazione di cultivar locali di fagiolo comune (*Phaseolus vulgaris* L.). *Italus Hortus, 14*, 31–40.

Logozzo, G., Donnoli, R., Macaluso, L, Papa, R., Knupffer, H., & Spagnoletti Zeuli, P. L. (2007). Analysis of the contribution of Mesoamerican and Andean gene pools to European common bean *(Phaseolus vulgaris* L.) germplasm and strategies to establish a core collection. *Genetic Resources and Crop Evolution, 54*, 1763–1779.

Mamidi, S., Rossi, M., Annam, D., Moghaddam, S., Lee, R., Papa, R., et al. (2011). Investigation of the domestication of common bean (*Phaseolus vulgaris*) using multilocus sequence data. *Functional Plant Biology*, *38*, 953–967.

Mavromatis, A. G., Arvanitoyannis, I. S., Chatzitheodorou, V. A., Khah, E. M., Korkovelos, A. E., & Goulas, C. K. (2007). Landraces versus commercial common bean cultivars under organic growing conditions: A comparative study based on agronomic performance and physicochemical traits. *European Journal of Horticultural Science*, *72*(5), 214–219.

Mavromatis, A. G., Arvanitoyannis, I. S., Korkovelos, A. E., Giakountis, A., Chatzitheodorou, V. A., & Goulas, C. K. (2010). Genetic diversity among common bean (*Phaseolus vulgaris* L.) Greek landraces and commercial cultivars: Nutritional components, RAPD and morphological markers. *Spanish Journal of Agricultural Research*, *8*(4), 986–994.

Miklas, P. N., Kelly, J. D., Beebe, S. E., & Blair, M. W. (2006). Common bean breeding for resistance against biotic and abiotic stresses: From classical to MAS breeding. *Euphytica*, *147*, 105–131.

Moreno, M. T., Martinez, A., & Cubero, J. I. (1983). Bean production in Spain. In *Potential for field beans (*Phaseolus vulgaris *L.) in West Asia and North Africa*. Proceedings of Regional Workshop (pp. 70–85). Aleppo, Syria.

Nanni, L., Bitocchi, E., Bellucci, E., Rossi, M., Rau, D., Attene, G., et al. (2011). Nucleotide diversity of a genomic sequence similar to SHATTERPROOF (PvSHP1) in domesticated and wild common bean (*Phaseolus vulgaris* L.). *Theoretical and Applied Genetics*, *123*, 1341–1357.

Negri, V. (2011). Policies supportive of on-farm conservation and their impact on custodian farmers in Italy. In S. Padulosi, N. Bergamini, & T. Lawrence (Eds.), *On farm conservation of neglected and underutilised species: status, trends and novel approaches to cope with climate change*. Rome: Bioversity International. Proceedings of an International Conference, Frankfurt, 14–16 June 2011.

Negri, V., Becker, H., Onnela, J., Sartori, A., Strajeru, S., & Laliberté, B. (2000). A first inventory of on farm conservation and management activities in Europe including some examples of cooperation between the formal and informal sector. In B. Laliberté, L. Maggioni, N. Maxted, & V. Negri (Eds.), In situ*/on farm conservation network task forces meeting*. ECP/GR, Isola Polvese, Perugia, 19–21 May 2000 (pp. 14–30). Rome: IPGRI.

Negri, V., & Tiranti, B. (2010). Effectiveness of *in situ* and *ex situ* conservation of crop diversity. What a *Phaseolus vulgaris* L. landrace case study can tell us. *Genetica*, *138*, 985–998.

Nijdam, F. E. (1947). Rassen en vormen van landbouwstanbonen, die ten behoeve van het eigen gebruik in Nederland worden verbound. *Technische Berichten Peulvrachten Studie Combinatie*, *42*, 10.

Ocampo, C. H., Martin, J. P., Sanchez-Yelamo, M. D., Ortiz, J. M., & Toro, O. (2005). Tracing the origin of Spanish common bean cultivars using biochemical and molecular markers. *Genetic Resources and Crop Evolution*, *52*, 33–40.

Paniconi, G., Gianfilippi, F., Mosconi, P., & Mazzuccato, A. (2010). Distinctiveness of bean landraces in Italy: The case study of the '*Badda*' bean. *Diversity*, *2*, 701–716.

Papa, R., & Gepts, P. (2003). Asymmetry of gene flow and differential geographical structure of molecular diversity in wild and domesticated common bean (*Phaseolus vulgaris* L.) from Mesoamerica. *Theoretical and Applied Genetics*, *106*, 239–250.

Papa, R., Nanni, L., Sicard, D., Rau, D., & Attene, G. (2006). The evolution of genetic diversity in *Phaseolus vulgaris* L. In T. J. Motley, N. Zerega, & H. Cross (Eds.), *Darwin's harvest: New approaches to the origins, evolution, and conservation of crops*. Columbia University Press.

Paris, H. S., Daunay, M. C., Pitrat, M., & Janick, J. (2006). First known image of *Cucurbita* in Europe, 1503–1508. *Annals of Botany*, *98*, 41–47.

Pérez-Vega, E., Campa, A., De la Rosa, L., Giraldez, R., & Ferreira, J. J. (2009). Genetic diversity in a core collection established from the main genebank in Spain. *Crop Science*, *49*, 1377–1386.

Piergiovanni, A. R., & Laghetti, G. (1999). The common bean landraces from Basilicata (Southern Italy): An example of integrated approach applied to genetic resources management. *Genetic Resources and Crop Evolution*, *46*, 47–52.

Piergiovanni, A. R., & Lioi, L. (2010). Italian common bean landraces: History, genetic diversity and seed quality. *Diversity*, *2*, 837–862.

Piergiovanni, A. R., Taranto, G., Losavio, F. P., & Pignone, D. (2006). Common bean (*Phaseolus vulgaris* L) landraces from Abruzzo and Lazio regions (Central Italy). *Genetic Resources and Crop Evolution*, *53*, 313–322.

Puerta Romero, J. (1961). *Variedades de judias cultivadas en España*. Madrid: Publicaciones del Ministerio de Agricultura.

Rodiño, A. P., Monteagudo, A. B., De Ron, A. M., & Santalla, M. (2009). Ancestral landraces of common bean from the south of Europe and their agronomic value for breeding programs. *Crop Science*, *49*, 2087–2099.

Rodiño, A. P., Santalla, M., Montero, I., Casquero, P. A., & De Ron, A. M. (2001). Diversity of common bean (*Phaseolus vulgaris* L.) germplasm from Portugal. *Genetic Resources and Crop Evolution*, *48*, 409–417.

Rodrigo, A. P. (2000). Caracterización morfoagronómica y bioquímica del germoplasma de la judía común (*Phaseolus vulgaris* L.) de España. Ph.D. Thesis, Universidad de Santiago de Compostela, Santiago, Spain.

Rossi, M., Bitocchi, E., Bellucci, E., Nanni, L., Rau, D., Attene, G., et al. (2009). Linkage disequilibrium and population structure in wild and domesticated populations of *Phaseolus vulgaris* L.. *Evolutionary Applications*, *2*(4), 504–522.

Sánchez, E., Sifres, A., Casañas, F., & Nuez, F. (2007). Common bean (*Phaseolus vulgaris* L.) landraces in Catalonia, a Mesoamerican germplasm hotspot to be preserved. *Journal of Horticultural Science and Biotechnology*, *82*(4), 529–534.

Sánchez, E., Sifres, A., Casañas, F., & Nuez, F. (2008). The endangered future of organoleptic prestigious European landraces: Ganxet bean (*Phaseolus vulgaris* L.) as an example of a crop originating in the Americas. *Genetic Resources and Crop Evolution*, *55*, 45–52.

Santalla, M., Rodiño, A. P., & De Ron, A. M. (2002). Allozyme evidence supporting southwestern Europe as a secondary center of genetic diversity for the common bean. *Theoretical and Applied Genetics*, *104*, 934–944.

Singh, S. P. (2001). Broadening the genetic base of common bean cultivars: A review. *Crop Science*, *41*, 1659–1675.

Singh, S. P., Gepts, P., & Debouck, D. G. (1991). Races of common bean (*Phaseolus vulgaris*, Fabaceae). *Economic Botany*, *45*(3), 379–396.

Skroch, P. W., Nienhuis, J., Beebe, S., Tohme, J., & Pedraza, F. (1998). Comparison of Mexican common bean (*Phaseolus vulgaris* L) core and reserve germplasm collections. *Crop Science*, *38*, 488–496.

Šustar-Vozlič, J., Maras, M., Kavornik, B., & Meglič, M. (2006). Genetic diversity and origin of Slovene common bean (*Phaseolus vulgaris* L.) germplasm as revealed by AFLP markers and phaseolin analysis. *Journal of American Horticultural Science*, *131*(2), 242–249.

Svetleva, D., Pereira, G., Carlier, J., Cabrita, L., Leitão, J., & Genchev, D. (2006). Molecular characterization of *Phaseolus vulgaris* L. genotypes included in Bulgarian collection by ISSR and AFLP™ analyses. *Scientia Horticulturae*, *109*, 198–206.

Szabó, A. T. (2009). *Phaseolus* as a model taxon for monitoring trends in European home garden diversity: a methodological approach, and a proposal. In A. Bailey, P. Eyzaguirre, & L. Maggioni (Eds.), *Crop genetic resources in European home gardens. Proceedings of ECP/GR Biodiversity International Related Workshop*, Ljubliana, Slovenia, 2007 (pp. 37–54). Rome: Biodiversity International.

Valvasor, J. V. (1689). *Die ehre des hertzogthums grain. Reprint in 1978*. Ljubljana, Slovenija: Mladinska Knjiga.

Van der Groen, J. (1669). Den Nederlandtsen hovenier. Reprint 1988, Utrecht, The Netherlands.

Zeven, A. C. (1997). The introduction of the common bean (*Phaseolus vulgaris* L.) into Western Europe and the phenotypic variation of dry beans collected in the Netherlands in 1946. *Euphytica, 94*, 319–328.

Zuagg, I., Magni, C., Panzeri, D., Daminati, M. G., Bollini, R., Benrey, B., et al. (2013). QUES, a new *Phaseolus vulgaris* genotype resistant to common bean weevils, contains the Arcelin-8 allele coding for new lectin-related variants. *Theoretical and Applied Genetics, 126*(3), 647–661.

3 Peas

Petr Smýkal[1], Clarice Coyne[2], Robert Redden[3] and Nigel Maxted[4]

[1]Department of Botany, Palacký University Olomouc, Olomouc, Czech Republic, [2]United States Department of Agriculture, Agricultural Research Service, Plant Introduction and Testing Unit, Washington State University, Pullman, WA, USA, [3]Australian Temperate Field Crops Collection, DPI-VIDA, Horsham, VIC, Australia, [4]School of Biosciences, University of Birmingham, Edgbaston, Birmingham, UK

3.1 Introduction

Of all the legumes, pea has its prominent place in plant biology and particularly in genetics, owing to work of J.G. Mendel (1866). Although not fully recognized and supported internationally, pea remains today one of the most important temperate pulses, fodder and vegetable crops and currently ranks second only to common bean as the most widely grown grain legume in the world, with primary production in temperate regions and global production of 10.4 million tons in 2011 (Food and Agriculture Organization, FAO, 2011). Pea seeds are rich in protein (23–25%), slowly digestible starch (50%), soluble sugars (5%), fibre, minerals and vitamins (Bastianelli, Grosjean, Peyronnet, Duparque, & Régnier, 1998). On a worldwide basis, legumes contribute about one-third of humankind's direct protein intake, while also serving as an important source of fodder and forage for animals and of edible and industrial oils. Peas have a wide variety of end uses with leaves, green pods, unripe seed and dry mature seed used as food and feed uses include direct grazing, hay and silage. One of the most important attributes of legumes is their capacity for symbiotic nitrogen fixation, underscoring their importance as a source of nitrogen in both natural and agricultural ecosystems (Phillips, 1980). Pea, as with other legumes, also accumulates natural products (secondary metabolites) such as isoflavonoids that are considered beneficial to human health through anticancer and other health-promoting activities (Dixon & Sumner, 2003).

3.2 Origin, Distribution, Diversity and Systematics

Pea (*Pisum sativum* L.) is one of the world's oldest domesticated crops. Archaeological evidence dates the existence of pea back to 8000 BC (Baldev, 1988) in the Near East, in Europe it has been found since the Stone and Bronze Ages, and in India since 200 BC. (De Candolle, 1882). Domesticated about 10,000 years ago (Abbo, Lev-Yadun, & Gopher, 2010; Ambrose, 1995; De Candolle, 1882; Kislev & Bar-Yosef, 1988; Smartt,

Genetic and Genomic Resources of Grain Legume Improvement. DOI: http://dx.doi.org/10.1016/B978-0-12-397935-3.00003-7

1990; Vavilov, 1949; Zohary & Hopf, 2000), pea, among other grain legumes, accompanied cereals and formed important dietary components of early civilizations in the Middle East and Mediterranean. These regions are also the area of origin and initial domestication. *Pisum sativum* subsp. *elatius* and subsp. *sativum* are found naturally in Europe, northwestern Asia and south to temperate Africa, while *P. fulvum* is restricted only to the Middle East. *Pisum abyssinicum* is found in Ethiopia and Yemen (Maxted & Ambrose, 2001). Cultivation of pea spread from the Fertile Crescent to today's Russia, and westwards through the Danube valley into Europe and to ancient Greece and Rome, which further facilitated its spread to northern and western Europe. In parallel, pea was moved eastward to Persia, India and China (Chimwamurombe & Khulbe, 2011; Makasheva, 1973). In pea, explosive pod indehiscence and seed dormancy (hard seededness) were probably the greatest barriers to domestication (Smartt, 1990) that had to be overcome. Other traits selected during domestication and development of modern cultivated forms include a number of characters that are determined by one or a few genes, such as *a* (lack of anthocyanin production) and *r* (wrinkled seed in garden types), which improved palatability, and *p* and *v* for the absence of sclerenchymatic tissue in pods. Domestication has also resulted in increased seed and pod size in pea (although not as markedly as in other crops) with a correlated increase in leaf size and stem strength (Swiecicki & Timmerman-Vaughan, 2005; Weeden, 2007).

There are several records of garden peas in the writing of the old Greeks and Romans, as well as in the herbal references of several centuries ago. There is discussion on cultivation of pea in ancient India and Egypt (De Candolle, 1882), indicated by both linguistic and archaeological evidence. Theofrastus of Greece (died 287 BC) records the use of *orobos* for the vetch, *erebinthos*, for the chickpea and *pisos* for the pea. Subsequently the transfer of Greeks *pisos* to Rome, become *Pisum*, a name passed to the English as *peason*, then *pease* or *peasse*, which after the drop of *s* became the universal name among English-speaking people (Mikić, 2012). This interesting paleolinguistics study shows roots directly related to traditional Eurasian pulse crops. Pea had entered China via India by the first century BC (Makasheva, 1973). We are not certain when pea cultivation was taken up by Romans, as neither Cato (149 BC) nor Varro (27 BC) name *pisum*, but use more general terms such as pulses or legumes, which are known to include lentils and chickpea (Cubero, Perez de la Varga, & Fratini, 2009). In the first century BC pea was mentioned by the Romans Collumela, Pliny and Virgil. Hybridization studies were done with pea well before Mendel. Knight (1799) began his work on hybridization using pea in 1787, publishing the results in 1799. Later Goss (1824) noted the phenomena that later Mendel formulated into the principles. The domestication of pea has been experimentally tested, both to determine the genetic basis, which led to cultivated crop from wild plant (Weeden, 2007) and to research wild pea harvesting (Abbo et al., 2010). The so-called domestication syndrome in the case of pulses applies to increases in seed size, reduction or elimination of pod shattering, loss of germination inhibition, shoot basal branching, seed toxins and antimetabolites (Plitman & Kislev, 1989; Smartt, 1990; Zohary & Hopf, 2000). All together, at least 11 loci involved in domestication traits have been identified (Weeden, 2007). In addition self-pollination reinforced fertility barriers between wild and cultivated populations, facilitating the fixation of desired genotype (Zohary & Hopf, 2000). We know that mutation of *A* gene (flower colour, seed testa pigmentation) also improved seed quality and reduced seed dormancy. Loss of *Np* increased seed size but reduced tolerance to bruchid beetle attack. The recessive *r* gene allele improved seed quality (sweetness from higher free sugar at the expense of starch) but reduced seed size. Photoperiod response genes *Sn* and *Hr* influence or are loosely linked to genes influencing root/shoot ratio (Weeden, 2007).

3.2.1 Phenotypic and Molecular Characterization of Diversity

There are several user-defined classifications of cultivated pea diversity. Four simply inherited characters determine the main use types of peas within subsp. *sativum*: the presence or absence of pod parchment, flower anthocyanin, leaflet occurrence and whether the starch grains in the dry seed are simple or compound (Green, 2008). This classification is similar to that proposed by Lehman (1954), except for the *afila* type, which was unknown at that time. There are two other characters used to establish groups based on their prevalence in cultivated material: the presence/absence of tare leaves and marrowfat seeds. Marrowfat seed type has simple starch grains and irregularly compressed seeds, often misinterpreted as wrinkled seeds. Early data from electrophoretic patterns of albumin and globulin proteins (Waines, 1975), allozymes (Hoey, Crowe, Jones, & Polans, 1996) and chloroplast DNA polymorphism (Palmer, Jorgensen, & Thompson, 1985) separated *P. fulvum* as a distinct species and *P. sativum* as an aggregate of '*humile*', *P. sativum* subsp. *elatius* and *P. sativum*. Interesting chemosystematic studies were made by Harborne (1971) and Pate (1975). Due to their widespread occurrence and chemical stability, flavonoids are well accepted as chemical markers in plant taxonomy (Gottlieb, 1982). Studies of leaf flavonol glycosides showed that *P. fulvum* contains quercetine 3-glucoside, primitive cultivars from Nepal and *P. abyssinicum* contain kaempferol and quercetine 3-sophoroside, while modern pea cultivars contain kaempferol and quercetine 3 (coumaroyl-sophorotrioside). Moreover, Harborne (1971) reported that petals of wild peas contain delphinidin, petunidin and malvidin 3-rhamnoside-5-glucosides, while petals of garden pea contain in addition pelargonidin, cyanidin and peonidin 3-rhamnoside-5-glucosides. Unfortunately, the yellow *P. fulvum* petals were not studied. Importantly the *Pisum* genus contains the flavonoid phytoalexin pisatin, which is shared with genera in *Lathyrus* but not found in *Vicia* species (Bisby, Buckingham, & Harborne, 1994). Serological reactions of *Pisum* taxa were done by Kloz (1971) indicating close relationship of all taxa except *P. fulvum* and *P. abyssinicum*. He was possibly the first to indicate that *P. abyssinicum* might originated from hybridization between *P. sativum* subsp. *elatius* and *P. fulvum*.

Recent phylogenetic studies based on retrotransposon insertion markers support the model of *P. sativum* subsp. *elatius* as a paraphyletic group, within which all *P. sativum* are nested (Jing et al., 2005; Jing et al., 2010; Nasiri, Haghnazari, & Saba, 2010; Vershinin, Allnutt, Knox, Ambrose, & Ellis, 2003). The study done by Hoey et al. (1996) using morphological, allozyme and RAPD characteristics on a set of Ben-Ze'ev and Zohary (1973) accessions resulted in separation of *P. fulvum* and 'southern *humile*', while cultivated peas were among *P. sativum* subsp. *elatius* accessions. The position of 'northern *humile*' varied between sister groups to cultivated peas and *P. sativum* subsp. *elatius*. More recently, studies of ITS sequence variation (Polans & Saar, 2002; Saar & Polans, 2000) and histone H1 subtype 5 gene (Zaytseva, Bogdanova, & Kosterin, 2012) have supported this. Extensive phylogenetic relationship of pea diversity was reconstructed using both amplified fragment length polymorphism (Ellis, Poyser, Knox, Vershinin, & Ambrose, 1998), its derived retrotransposon insertion-based marker method, sequence-specific amplification polymorphisms (SSAP) (Majeed et al., 2012; Pearce, Knox, Ellis, Flavell, & Kumar, 2000; Vershinin et al., 2003), and gene sequences (Jing et al., 2007; Zaytseva et al., 2012). *P. fulvum* and *P. abyssinicum* formed neighbouring but separate branches, a subset of *P. sativum* subsp. *elatius* was positioned between *P. fulvum* and *P. abyssinicum*, and further branches were found within the cultivated pea. The most recent studies of *P. abyssinicum* placed it between *P. fulvum* and a subset of *P. sativum* subsp. *elatius* (Ellis, 2011; Jing et al., 2010; Smýkal et al., 2011; Vershinin et al., 2003) and revealed its very low genetic diversity, which could be explained by passage through a genetic bottleneck.

A general feature of molecular phylogenetic analysis of *Pisum* has been the impact of introgression on pea diversity and evolution (Jing et al., 2007). Moreover, high conservation between SSAP (Vershinin et al., 2003), retrotransposon insertions (Jing et al., 2005) and gene-based derived (Jing et al., 2007) trees was observed, in spite of the fact that they derived from different genomic compartments. Another study on relationships among wild *Pisum* used a combination of mitochondrial, chloroplast and nuclear genome markers (Kosterin & Bogdanova, 2008; Kosterin, Zaytseva, Bogdanova, & Ambrose, 2010), separating *P. fulvum* and *P. abyssinicum* accessions and about half of those of wild *P. sativum* from the rest of the wild and all cultivated *P. sativum*. The distinction within *P. sativum* coincided with the cytogenetic classes of Ben-Ze'ev and Zohary (1973).

3.2.2 Biosystematics and Taxonomy

Pea belongs to the Leguminosae plant family, the third-largest flowering plant family, with 800 genera and over 18,000 species (Lewis, Schrirer, Mackinder, & Lock, 2005). Papilionoideae is the largest subfamily, with 476 genera and about 14,000 species. It is estimated that all papilionoids shared a common ancestor around 50 mya, which experienced a 50 kb inversion in its chloroplast genome (Doyle et al., 1997; Lavin, Herendeen, & Wojciechowski, 2005). The largest group of papilionoids is Hologalegina, with nearly 4000 species in 75 genera. This group includes the large galegoid tribes (Galegeae, Vicieae, Trifolieae, etc.), united by the loss of one copy of the chloroplast inverted repeat (IR). Tribe Fabeae (syn. Vicieae) contains five genera: *Lathyrus* (grasspea/sweet pea) (about 160 species); *Lens* (lentils) (4 species); *Pisum* (peas) (3 species); *Vicia* (vetches) (about 160 species) and monotypic genus *Vavilovia formosa*. Recent comprehensive phylogenetic analysis of 262 species (70%) of Fabeae tribe has shown that *Pisum* and *Vavilovia* are nested in *Lathyrus*, the genus Lens is nested in *Vicia* (Schaefer et al., 2012) and as consequence current generic and infrageneric circumscriptions do not reflect monophyletic groups and should be revised.

Further, the classification of *Pisum* based on morphology and karyology has changed over time from being considered a genus with five species (Govorov, 1937) to a monotypic genus (Lamprecht, 1966; Marx, 1977). Kupicha (1981) and Davis (1970) recognized only two species, *P. fulvum* and *P. sativum*, and did not consider the third putative species *P. abyssinicum*. Numerous names have been proposed for wild representatives of *P. sativum*. All *Pisum* species are true diploid with $2n=2x=14$. In the review of Yarnell (1962) *P. humile* and *P. sativum* were considered conspecific, even though they might differ by inversions and translocations. Importantly, the other 'species' such as *P. abyssinicum*, *P. jomardi* and *P. arvense* were also considered conspecific. Frustratingly, *P. abyssinicum*, native to Ethiopia and Yemen, has few seed accessions available and has been excluded from many *Pisum* studies, and as a result its true taxonomic status is still a matter of debate (Maxted & Ambrose, 2001). Based on morphological characteristics, Govorov (1937) labelled it as a separate cultivated species, while Makasheva (1979) regarded it as a subspecies. A serious karyological barrier for crossing to *P. sativum* (Ben-Ze'ev & Zohary, 1973) and clear-cut phenotypic differences support the view of its species status (Lamprecht, 1963). High genetic homogeneity and distinction of *P. abyssinicum* was revealed by numerous morphological, allozyme (Weeden & Wolko, 2001) as well as molecular analyses (Jing et al., 2005, 2010; Pearce et al., 2000; Vershinin et al., 2003). Although its origin is not fully understood, it has been proposed that it was domesticated independently some 5000 years ago (Jing et al., 2010; Vershinin et al., 2003). The centre of pea genetic diversity is the broad area of the Fertile Crescent through Turkey, Syria, Iraq, Israel and Lebanon. It extends further east to Central Asia (Iran, Afghanistan, Pakistan and Turkmenistan) (Smýkal et al., 2011). Vavilov (1949) has considered Ethiopia together with the Mediterranean countries and Central Asia as primary centres, with Near East secondary. Phylogenetically, there are two wild populations variously described as subspecies of *P. sativum* or as species, *P. sativum* subsp. *elatius* Bieb. and *P. humile* Boiss and Noe (syn. *P. syriacum* (Berger) Lehm.) (Ben-Ze'ev & Zohary, 1973). These two wild groups are morphologically, ecologically and also genetically distinct. Crossing experiments undertaken by Ben-Ze'ev and Zohary (1973) included genotypes of *P. sativum* subsp. *elatius*, *P. humile*, *P. fulvum* and *P. sativum* to define the primary gene pool as *P. sativum* aggregate including wild *P. sativum* subp. *elatius*, a secondary gene pool composed of *P. fulvum* and a tertiary gene pool consisting only of *Vavilovia formosa*. The domestication of cultivated pea from northern populations of '*humile*' was proposed by Ben-Ze'ev and Zohary (1973), but the source could equally be the 'northern elatius' (Kosterin et al., 2010; Smýkal et al., 2011). The most used classification is of Maxted and Ambrose (2001), to which *Vavilovia formosa* is added to classify four species; namely, *P. sativum* L., subsp. *sativum* (includes var. *sativum* and var. *arvense*); subsp. *elatius* (Bieb.) Aschers. & Graebn (includes var. *elatius*, var. *brevipedunculatum* and var. *Pumilio*), *P. fulvum* Sibth. & Sm.; *P. abyssinicum* A. Br.; *Vavilovia formosa* (*P. formosum*) (Stev.) Fed. This classification, which amends the classification used in review paper of Smýkal et al. (2011), will now be used in this chapter.

Since the botanical description of recognized *Pisum* species and subspecies is often lacking or fragmentary, we would like to provide it here in detail.

a. ***P. sativum* subsp. *elatius*** grows as a tall climber (up to 3 m) in humid forested valleys from the Caspian coast through the Caucasus to the Mediterranean region, including its islands and northern African coast, extending north to the Black Sea coast and Hungarian plains.

It is found at altitudes from 0 to 1700 m above sea level (asl) (Maxted & Ambrose, 2001). It has large (20–30 mm), often bicolour flowers and long peduncles (2–4× longer than stipules) most often with two flowers (1–3), producing large pods (50–80×10–12 mm). Leaflets are two to four paired, ovate-elliptic, entire or subdentate. This subspecies has a chromosomal translocation difference from cultivated *P. sativum*, but it is interfertile, although some nucleo-cytoplasmatic conflict has been reported in specific crosses (Bogdanova, Galieva, & Kosterin, 2009). Former subspecies *pumilio* (now as *P. sativum* subsp. *elatius* var. *pumilio*) or synonymous *P. humile*, has shorter internodes (20–40 cm stem length), shorter peduncles, smaller (40–45×7–10 mm) often pigmented pods and small flowers (15–18 mm). It is distributed from the Mediterranean through Turkey, Syria and Israel to Iran in steppe habitats. Compare to *Pisum* subsp. *elatius* found in higher altitudes, from 700–1800 m at least in Syria (Maxted & Ambrose, 2001). Comparison of data from the expeditions to Syria and previous herbarium passport data from Turkey reveals differences in circumstances. For example, in Syria discrete variation exists in altitude, rainfall and parent rock or soil type, correlated with an allopatric association between subsp. *elatius* and var. *pumilio*. However, in Turkey, where these varieties have been found sympatric, mild and overlapping climatic conditions have been reported (Mumtaz, Shehadeh, Ellis, Ambrose, & Maxted, 2002).

b. ***Pisum fulvum*** is distinguished by its weak slender stems (10–45 cm), one to two paired dentate leaflets, peduncle as long as the incised-dentate stipules, usually with single small (10–15 mm), yellow to orange flowers. Pods are small (30–40×5–10 mm) and pigmented, seeds are dark brown to velvet black with subpapillose testa. Some *P. fulvum* accessions possess amphicarpic character, with basal pods growing into the ground. It grows on open arid (300–450 mm annual rainfall) rocky limestone slopes (30–1500 m asl).

c. ***P. abyssinicum*** is 30–60 cm tall, with ovate, obtuse, irregularly dentate 4–5 cm long stipules up to the top and also along the inner margin, with semicordate acute basal lobes. The stipules are as long as internodes. Peduncles are shorter (1/3 to 1/2) than the stipules at the time of flowering, but prolonged thereafter, one-flowered with small flowers. Flowers are pale, calyx lobes narrow lanceolate, standard only half open, whitish, wings shorter bright or pale purple-red, keel shorter than wings and narrow. Pods 40–50 mm long, with four to six seeds. Seeds globular-cubic, brownish red, violet, brown or grayish green. Most with one pair of leaflets and branched tendrils. Leaflets ovate, elliptical or obovate, obtuse, mucronulate, sharply or incisely dentate except of lower third, 3–4 cm long. Entire plants often have a bluish green colour. *P. abyssinicum* has been described from Ethiopia and Yemen.

d. ***Vavilovia formosa (P. formosum)*** is a perennial herbaceous species. It has long roots and underground rhizomes that form an important part of the plant's biology and are possibly crucial to its conservation strategy, as they may enable established plants to survive (Akopian et al., 2010). The anatomical investigations of stem structure showed that stems have lateral wings each with cortical vascular bundle, which are not prominent and hard to observe by morphological examination (Zorić et al., 2010). The leaf is compound, with small, semisagittate, foliaceous stipules, one pair broad, cuneate-obovate to suborbicular. The leaf does not terminate with tendrils, but with mucrolike rachis, similar to that in faba bean. The flowers are often solitary, axillary and pedunculate, with small and/or inconspicuous bracts, lacking bracteoles, and having a campanulate calyx, pink or purple in colour and likely insect pollinated, albeit with no detailed data available (Atlagić et al., 2010). Pods are linear-oblong and dehiscent, 20–35 mm long and bearing from three to five seeds (Davis, 1970). Seeds are globose or oval and smooth, usually with dark blotches on the surface. The geographical distribution of *Vavilovia* is widespread, but rather limited by its ecology. The centre of its range is the central and eastern Caucasus, with a distribution across neighbouring montane areas of Iran, Iraq, Lebanon, Syria and Turkey (Akopian et al., 2010; Mikič et al., 2009). Vavilovia is typically found at altitudes of 1500–3500 m in high mountainous areas, on shale or rocky substrates such as loose limestone scree. This enigmatic species has received considerable attention recently, both for conservation and diversity as well as phylogeny studies (Mikič et al., 2009, 2013; Oskoueiyan, Osaloo, Maassoumi, Nejadsattari, & Mozaffarian, 2010).

3.3 Status of Germplasm Resources Conservation

3.3.1 Conservation of Cultivated Gene Pool

About 98,000 pea accessions are preserved worldwide. The total germplasm collection is much smaller owing to substantial overlap (on an average 20%, but some particularly smaller collections are duplicates up to 90%). There are 25 larger collections preserving pea diversity, holding together around 72,000 accessions. The remaining 27,000 accessions are distributed over 146 collections worldwide. As shown in Table 3.1, only 1876 (2%) of these are wild pea relatives, approximately one-quarter (24,000) each are commercial varieties, 8500 landraces, while 600 and 6000 represent breeding and recombinant inbred lines or mutant stocks, respectively (Figure 3.1). In the case of true wild *Pisum* species, there are only 706 *P. fulvum*, 624 *P.* subsp. *elatius*, 1562 *P. sativum* subsp. *sativum* (syn. *P. humile/syriacum*) and 540 *P. abyssinicum* accessions (Figure 3.1) preserved *ex situ* in collections. Moreover, when passport data on geographical origin are summarized, there is a large bias (17%) towards Western and Central European accessions, as these regions represent modern pea breeding activities. Substantially less well represented are Mediterranean (2.5%), Balkan (2%) regions, Caucasus (0.8%) and Central Asia (2%) centres of pea crop domestication and diversity (Table 3.1; Figure 3.2), where higher variation can be anticipated. Currently, no international centre conducts pea breeding, since International Center for Agricultural Research in the Dry Areas (ICARDA) in Syria relinquished the international mandate for genetic conservation of peas, and worldwide no single collection predominates in size and diversity (Table 3.1). Important genetic diversity collections of *Pisum* with over 1000 accessions are found in national gene banks of at least 15 countries (Table 3.1), with many other smaller collections worldwide (Smýkal, Coyne, et al., 2008; Smýkal et al., 2012). A high level of duplication (an estimated 20% on average) exists between the collections, thus reducing the actual level of diversity. In spite of this overlap, each represents a unique assembly. These are dominated by cultivated forms (Table 3.1; Figure 3.1), and although wild forms in these collections are highly diverse, they are comparably few and inadequately sampled (Ellis, 2011; Smýkal et al., 2011). The much smaller collections of wild relatives of pea are less widely distributed and there is more clarity when tracing these accessions to their origin, although precise collection sites are often unknown. Furthermore allelic diversity in wild material is unknown. There are still important gaps in the *ex situ* collections, particularly of wild and locally adapted materials, which need to be addressed before these genetic resources are lost forever

Table 3.1 Major Gene Banks Holding Pea Germplasm

Code	Country	Institute	Number of Accessions	Web Site	Online Catalogue	Genotyped	Phenotyped	Core
VIR	Russia	N.I. Vavilov Research Institute of Plant Industry, St. Petersburg	6790	http://www.vir.nw.ru	No	No	No	
USDA	USA	Plant Germplasm Introduction and Testing Research Station, Pullman	6827	http://www.ars-grin.gov	Yes	Partly	Yes	Formed
BAR	Italy	CNR-Istituto Di Genetica Vegetale, Bari	4558	http://www.igv.cnr.it	Yes	No	No	
SAD	Bulgaria	Institute of Plant Introduction and Genetic Resources, Sadovo	2100	http://www.genebank.hit.bg	No	No	Partly	
NGB	Sweden	NordGen, Nordic Genetic Resource Centre, Alnarp	2849	http://www.nordgen.org/sesto	Yes	Partly	Partly	
CGN	The Netherlands	Centre for Genetic Resources, Wageningen	1002	http://www.cgn.wur.nl/pgr/	Yes	No	No	
ATFC	Australia	Australian Temperate Field Crop Collection, Horsham	7432	http://www2.dpi.qld.gov.au	No	Yes	Yes	Formed
ICARDA	Syria	International Center for Agricultural Research in the Dry Areas	6105	http://www.icarda.cgiar.org	No	No	No	
GAT	Germany	Leibniz Institute of Plant Genetics and Crop Plant Research	5343	http://www.ipk-gatersleben.de	Yes	No	Yes	
ICAR	China	Institute of Crop Sciences, CAAS China	3837	http://icgr.caas.net.cn/cgris	No	Partly	No	

JIC	UK	John Innes Centre, Norwich	3567	http://www.jic.ac.uk	Yes	Yes	Yes	Formed
WTD	Poland	Plant Breeding and Acclimatization Institute Blonie, Radzikow	2896	http://www.igr.poznan.pl	Yes	No	No	
INRA	France	INRA CRG Légumineuse à grosses graines, Dijon	8839	http://195.220.91.17/legumbase	Yes	Partly	Yes	Formed
INIA	Spain	Instituto Nacional de Investigación y Tecnología Agraria	1648	http://www.inia.es	Yes	Partly	Partly	
ITACyL	Spain	Instituto Tecnológico Agrario de Castilla y León	1772	http://www.itacyl.es	No	Partly	Partly	Formed
UKR	Ukraine	Yurjev Institute of Plant Breeding, Kharkov	1671	http://www.bionet.nsc.ru	No	No	No	
CZE	Czech	AGRITEC, Research, Breeding and Services Ltd., Sumperk	1326	http://genbank.vurv.cz	Yes	Yes	Yes	Formed
CZE	Czech	Centre for Research of Vegetables and Special Crops, Olomouc	1414	http://genbank.vurv.cz/genetic/resources	Yes	No	Yes	
HUN	Hungary	Research Centre for Agrobiodiversity, Tápiószele	1205	http://www.rcat.hu	No	No	No	
CAN	Canada	Plant Gene Resources of Canada, Saskatchewan, Canada	616	http://www.agr.gc.ca/pgrc-rpc	No	Yes	Yes	Formed
SRB	Serbia	IFVCNS, Novi Sad	991	http://www.nsseme.com/en/	No	No	No	

(*Continued*)

Table 3.1 (Continued)

Code	Country	Institute	Number of Accessions	Web Site	Online Catalogue	Genotyped	Phenotyped	Core
ISR	Israel	Israel Plant Gene Bank, ARO Volcani Center	343	http://igb.agri.gov.il	Yes	Partly	Partly	
TUR	Turkey	Aegean Agricultural Research Institute, Menemen/IZMIR	236	http://www.etae.gov.tr/eng/	No	Partly	Partly	
ARM	Armenia	Institute of Botany NAS RA, Yerevan	19	http://www.sci.am/	No	No	No	
ETH	Ethiopia	Institute of Biodiversity Conservation, Addis Ababa	1768	http://www.ibc.gov.et/	No	No	No	
NBPGR	India	National Bureau of Plant Genetic Resources, New Delhi	3609	http://www.nbpgr.ernet.in	No	No	Yes	
BRA	Brazil	National Center for Vegetable Crops Research (CNPH)/EMBRAPA	1958	http://www.cnph.embrapa.br	No	No	No	
	Others (149)	FAO report on germplasm collections	28,831	http://www.fao.org	No	No	No	
		Svalbard Global Seed Vault	9670	http://www.croptrust.org	Yes	No	No	
TOTAL			98,947					

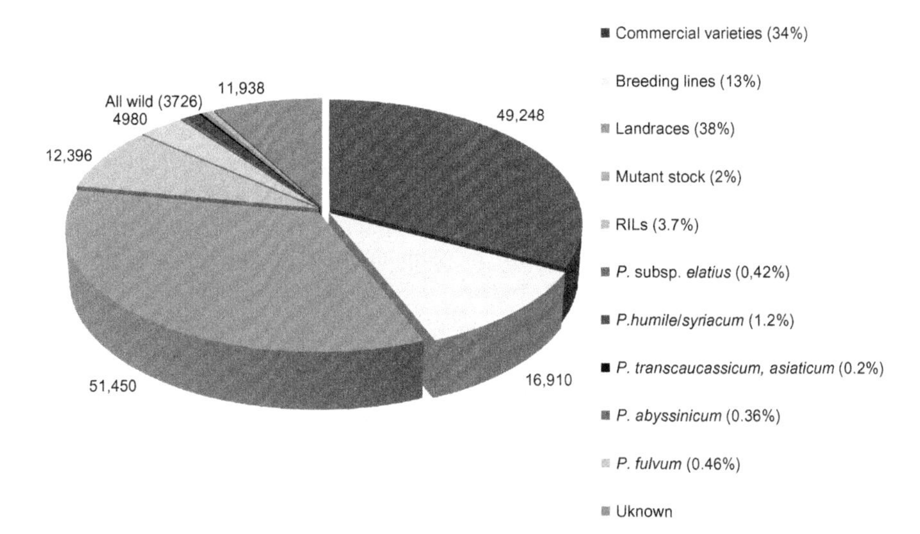

Figure 3.1 Stratification of pea germplasm collections listed in Table 3.1 by species, subspecies and breeding status, with indicated numbers and percentage of total. RILs, Recombinant Inbred Lines.

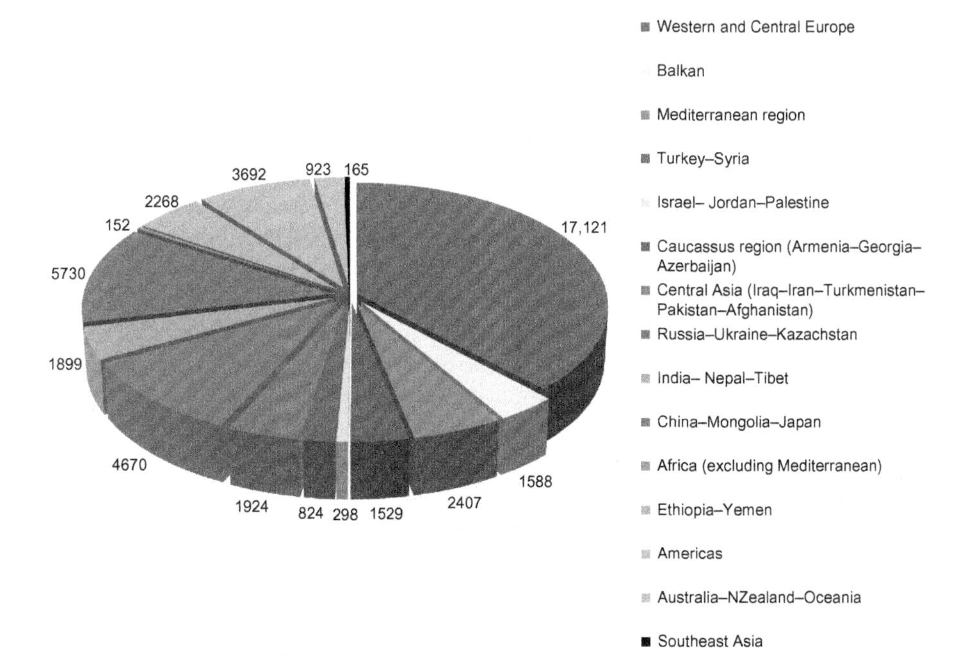

Figure 3.2 Stratification of pea germplasm collections listed in Table 3.1 (except ETH, BRA, UKR due to lack of data) by geographical regions, with indicated numbers of accessions.

due to native habitats destruction (Maxted et al., 2010). Several attempts have been made at *ex situ* conservation of *Vavilovia*, the closest *Pisum* relative, especially in the former USSR, with all of them being unsuccessful likely due to inadequate cultivation (Makasheva, 1973; Zhukovskyi, 1971). Some success was achieved in the United Kingdom (Cooper & Cadger, 1990), but these did not result in the production of new seeds or in multiplication of the plants. More promising results were produced in the Vavilov Institute during 1974–1981. Some plants survived for years, bloomed and even formed fruits with seeds (reviewed in Akopian et al., 2010). *Vavilovia* has periodically been grown in the Yerevan Botanic Garden since 1940, as well as is being recently cultivated *in vitro* (Akopian et al., 2010; Mikič et al., 2013); nevertheless, this particular species in currently vulnerable to habitat destruction and climate change, and no seeds have been preserved *ex situ* to ensure its longer term conservation. There is an urgent need to systematically sample the genetic diversity in wild relatives that was only partially captured in the domestication of pea (Ellis, 2011; Smykal et al., 2011), since natural habitats are being lost due to increased human population, increased grazing pressure, conversion of marginal areas to agriculture and ecological threats due to future climate change (Keiša, Maxted, & Ford-Lloyd, 2007; Maxted & Kell, 2012). The target areas for comprehensive collection of wild relatives of peas include the habitat from the Mediterranean through the Middle East and Central Asia, as these are likely to contain genetic diversity for abiotic stress tolerances (Coyne et al., 2011). The storing of pea seeds in gene banks (*ex situ*) is relatively inexpensive and effective, consequently it is the most common way to preserve crop diversity. In addition to gene banks, botanical gardens offer an *ex situ* alternative to seed conservation. Gardens have usually held a broad taxonomic range and consequently often a limited number of accessions of each species, limiting their effectiveness in the genetic conservation. However, major world botanical gardens manage large seed banks (e.g. the Millennium Gene Bank managed by the Royal Botanic Gardens at Kew, UK), have well managed herbarium collections, are involved in re-introduction programmes and have DNA storage facilities (known as DNA banks). The recently funded Svalbard Global Seed Vault (Table 3.1) currently preserves 9670 pea accessions, selected from several main collections as germplasm backup. Although herbarium and DNA banks are relatively of little practical use to conserve diversity, both provide very valuable sources to study genetic diversity of crop wild relatives (CWR). Digitization and public access of herbarium vouchers allows for the study of morphological traits remotely. In the case of wild *Pisum* as well as its closest allies *Vavilovia formosa*, such digitized specimen resources exist in the Royal Botanic Gardens at Kew and Edinburgh, UK. Both also have good representation of the Eastern Mediterranean and Near East (Turkey, Syria, Palestine, Israel) floristic regions. In addition, some valuable collections of *Pisum* herbarium vouchers are held at Vavilov Institute and Komarov Botanical Institute, St. Petersburg, Russia, covering largely the Caucasus and Central Asia regions (Smýkal, pers. communication). In addition to botanical gardens, several universities, particularly in the Mediterranean region, have useful herbariums. These institutions often have the most direct knowledge and access to existing genetic diversity preserved *ex situ*. Unfortunately there is often an information gap between gene banks, botanical gardens and universities.

3.3.2 Conservation of the Wild Gene Pool

In light of the growing concern over the predicted devastating impact of climate change on global biodiversity and food security, coupled with a growing world population, taking action to conserve CWR has become an urgent priority. CWR are species with a close genetic similarity to crops and many of them have the potential or actual ability to contribute beneficial traits to crops, such as resistance to biotic and abiotic stresses, besides yielding related characters (Maxted, Shelagh, Ford-Lloyd, Dulloo, & Toledo, 2012). There has been no systematic effort to conserve temperate crop species *in situ* either through genetic reserves or on farms. Passive conservation of legume species, including pea, exists in several currently protected areas for landscape ecosystems in the Mediterranean and Near East regions, which are not intended specifically to conserve wild crop relatives. Consequently native legume populations are susceptible to genetic erosion or even extinction (Maxted, Shelagh, Ford-Lloyd, Dulloo, & Toledo, 2012). Maxted, van Slageren, and Rihan (1995) was the first to proposed establishment of genetic reserves to conserve Vicieae species *in situ* in Syria and Turkey. Three reserves were established within the Global Environment Facility project in Turkey (Kaya, Kün, & Güner, 1998). Recently international initiatives include the Global Environment Facility projects ('*In situ* Conservation of CWR Through Enhanced Information Management and Field Application' and 'Design, Testing and Evaluation of Best Practices for *in situ* Conservation of Economically Important Wild Species'), the European Community–funded project 'European CWR Diversity Assessment and Conservation Forum (PGR Forum)', the FAO commissioned 'Establishment of a Global Network for the *in situ* Conservation of CWR: Status and Needs', the International Union for Conservation of Nature (IUCN) Species Survival Commission CWR Specialist Group and the European '*In Situ* and On-Farm Conservation Network'. The need to address CWR conservation is also highlighted in international and regional policy instruments, such as the Convention on Biological Diversity (CBD), the FAO Global Plan of Action for the Conservation and Sustainable Utilization of Plant Genetic Resources for Food and Agriculture (PGRFA) (FAO, 1996), the CBD Global Strategy for Plant Conservation, the International Treaty on PGRFA, the European Plant Conservation Strategy (Planta Europa, 2001), the Global Strategy for CWR Conservation and Use (Heywood, Kell, & Maxted, 2008) and most recently the European Strategy for Plant Conservation (Planta Europa, 2008). The latter strategy specifically recommends the establishment of 25 CWR genetic reserves in Europe and the undertaking of gap analysis of current *ex situ* CWR holdings, followed by filling of diversity gaps. There are a number of potential approaches to systematic CWR conservation, but each requires the precise targeting of CWR diversity that can then be sampled for gene bank storage or designation and management as a genetic reserve (Maxted & Kell, 2009). There is an extensive literature on gap analysis, which is used to identify areas in which selected elements of biodiversity are underrepresented. Maxted, Dulloo, Ford-Lloyd, Iriondo, and Jarvis (2008) have adapted the existing methodologies and proposed a specific methodology for CWR genetic gap analysis that involves four steps: (a) identify priority taxa, (b) identify ecogeographic breadth and complementary hot spots using distribution and environmental data, (c) match current *in situ* and *ex situ* conservation actions with the ecogeographical data and complementary hot spots to identify the gaps and (d) formulate a revised *in situ* and *ex situ* conservation strategy. This methodology has been applied by Maxted and Kell (2009) for 14 globally important food crop groups including pea. A combined gap analysis was undertaken for six legume genera using over 2000 unique georeferenced records; the regression analysis undertaken illustrated that none of the countries rich in *Pisum* species can be considered oversampled, with Turkey, the former Soviet Union (particularly the countries of the Caucasus), Syria, Spain and Greece warranting further *ex situ* collection, as there is a potential for finding additional diversity. In legumes, there is considerable evidence for environmental selection pressure on phenological traits. Habitats that impose high terminal drought stress select for early flowering and short life cycles as a drought escape mechanism, whereas cool, high rainfall habitats select for delayed phenology, allowing more biomass production and supporting a higher reproductive effort. This has been demonstrated in a variety of wild and domesticated Mediterranean annuals, including legumes (reviewed in Upadhyaya et al., 2011), and confirms that habitat characterization is an essential and useful ecophysiological tool to explore the mechanisms underlying specific adaptations (Berger et al., 2012). A recent Global Environment Facility funded project, 'Conservation and Sustainable Use of Dryland Agrobiodiversity in West Asia' established two genetic reserves in northeast Lebanon at Arsal and Balbak to conserve genetic diversity of wild forage legumes, fruit trees, vegetables and cereals. Both sites contain significant *Cicer*, *Lathyrus*, *Lens*, *Medicago*, *Pisum* and *Vicia* priority crop species diversity, including both *P. sativum* subspecies and *P. fulvum*.

3.3.3 Pea Mutant Collections

Pea has a large number of mutant lines, either spontaneous or induced. It has been used as a model plant species for experimental morphology and physiology in mutagenic studies. Numerous morphologically well-described mutants exist, many of them being used in genetic mapping. The earliest collection lists 21 pairs of cultivated pea lines for contrasting characters covering plant form, foliage, flowers, pods and seeds, which were the subject of genetic investigation, held within a collection of 550 cultivars (Vilmorin, 1913). Later, Blixt (1972) made a list and linkage group positions of 169 genes (loci) which occurred spontaneously or were induced. Induced mutagenesis has become widespread for the creation of novel genetic variation for selection and genetic studies (Blixt, 1972; Lamm, 1951; Lamprecht, 1964) with mutants in traits for physiology, chlorophyll, seed, root, shoot, foliage, inflorescence, flowers and pods. These genetic analyses contributed to *Pisum* genus classification. The mutant collections have been largely preserved in John Innes Centre (JIC) (585 accessions) and Nord Genebank (Blixt & Williams, 1982). Partial duplicates exist at Polish (297) and Bulgarian (150 accessions) gene banks (Table 3.1). In addition Murfet and Reid (1993) have developed and maintain developmental mutants in Tasmania. There is a pea population of 4817 lines newly established by the technique of targeting induced local lesions in genomes (TILLING) at Institut National de la Recherche Agronomique

(INRA) (Table 3.1). In addition, fast neutron-generated deletion mutant resources (around 3000 lines) are available for pea, which have been useful in identifying several developmental genes (Hellens et al., 2010; Hofer et al., 2009; Wang et al., 2008).

3.4 Germplasm Characterization and Evaluation

Traditionally germplasm diversity has been assessed by morphological descriptors, which remain the only legitimate marker type accepted by the International Union for the Protection of New Varieties of Plants. Although morphological traits represent the action of numerous genes and thus contain high information value, they can be unreliable owing to strong environmental influence on traits with low heritability. Several studies using morphological descriptors and agronomic traits have been published (Ali, Qureshi, Ali, Gulzar, & Nisar, 2007; Azmat, Ali Khan, Asif, Muhammad, & Shahid, 2012; Cupic et al., 2009; Sardana, Mahajan, Gautam, & Ram, 2007; Sarikamiş, Yanmaz, Ermiş, Bakır, & Yüksel, 2010; Smýkal, Hýbl, et al., 2008). As expected a number of traits were found to be strongly correlated, and as a result fewer traits were sufficient for evaluating morphological diversity. Principal component analysis is used to select characteristics to capture the most variability using the lowest number of descriptors. Finally, the morphological characteristics are loaded into dummy variables and clustered using various coefficients to reveal germplasm structures. In contrast, molecular markers accurately represent the underlying genetic variation and now dominate the genetic diversity.

Development of new genomic technologies has increased during the last decade, providing previously unforeseen strategies for crop breeding. Countless DNA polymorphisms are present among a set of varied genotypes, which can then be customized into user-friendly molecular markers. Different techniques exploit nucleotide polymorphisms that arise from different classes of mutation, such as substitution (point mutations), rearrangement (insertions or deletions) or error in replications of tandem-repeat DNA. Adaptation to breeder-friendly markers has relied on polymerase chain reaction (PCR)-based microsatellites or single-nucleotide polymorphism (SNP) markers because they can be easily employed in cost-effective genotyping of large segregating populations and germplasm collections (reviewed in Smýkal et al., 2012). For the analysis of pea diversity, simple sequence repeats (SSRs or microsatellites) have been popular because of their high polymorphism and information content, codominance and reproducibility (Baranger et al., 2004; Loridon et al., 2005; Smýkal, Hýbl, et al., 2008; Zong et al., 2009). More recently, expressed sequence tag (EST)-derived simple sequence repeat (eSSR) markers have become an important resource for gene discovery and comparative mapping studies (DeCaire, Coyne, Brumett, & Schultz, 2012; Mishra, Gangadhar, Nookaraju, Kumar, & Park, 2012). Alternately, highly abundant retrotransposon repeats have been used to reveal diversity, first applied in fingerprinting format of SSAP (Ellis et al., 1998; Vershinin et al., 2003) and developed into a high-throughput locus-specific genotyping technology based on insertion/deletion of Ty1-*copia* PDR1 element and used for phylogeny and genetic relationship studies, providing a highly specific, reproducible and easily scorable method (Jing et al., 2007,

2010; Smýkal, Hýbl, et al., 2008; Smýkal et al., 2011). Another class of highly abundant Angela family was identified and used for inter retrotransposon amplified polymorphism fingerprinting (Smýkal, 2006; Smýkal, Kalendar, Ford, Macas, & Griga, 2009). Using these markers, several major world pea germplasm collections have been analysed (Cupic et al., 2009; Jing et al., 2005, 2007, 2010; Majeed et al., 2012; Martin-Sanz, Caminero, Jing, Flavell, & Perez de la Vega, 2011; Nasiri et al., 2010; Sarikamis et al., 2010; Smýkal, Hýbl, 2008, 2011; Zong et al., 2009). The use of retrotransposon insertions for large-scale pea diversity analysis showed good agreement with SNPs in 49 genes and SSAP studies (Jing et al., 2007). It was further shown that both SSRs and retrotransposon-based insertion polymorphisms (RBIPs) have similarly high polymorphism information content and offer comparable diversity measurements in diversity surveys at the species level (Smýkal, Hýbl, et al., 2008). This was an important finding, as SSRs, in spite of multiple alleles detection, are more difficult to transfer between labs, while essentially binary RBIPs are simpler. Moreover, microsatellites (SSRs) display a much higher mutation rate than the nucleotide substitution rate (Cieslarová, Hanáček, Fialová, Hýbl, & Smýkal, 2011) and therefore suffer from homoplasy (the state when identical alleles have arisen by two or more different pathways of descent) in widely diverse material (Ellis, 2011; Smýkal et al., 2011). Although SSR and RBIP marker types are widespread now, their potential is at its limits. With advances in model legume sequencing and genomic knowledge, there is a switch to gene-based markers in pea (Aubert et al., 2006; Jing et al., 2007). This trend can be expected to further proliferate in line with rapid advances in high-throughput SNP generation and detection assays (Bordat et al., 2011; Deulvot et al., 2010). Functionally associated markers (i.e. cDNA/EST) have been developed to uncover and tag candidate genes and gene pathways underpinning desirable traits. This has most recently been expanded to include whole genome transcriptome analysis. With the advent of next-generation sequencing technologies, it will be possible to transfer this technology to species with relatively large genomes such as pea. The initial set of pea ESTs was developed (Gilpin, McCallum, Frew, & Timmerman-Vaughan, 1997; Künne et al., 2005; Liang et al., 2009) and recently a comprehensive transcriptome of pea was published (Franssen, Shrestha, Bräutigam, Bornberg-Bauer, & Weber, 2011). Several high-throughput pea transcriptome sequencing projects are underway and should provide a complete set of pea genes. Based on this, a custom 384-SNP array was developed and used in pea genotypic diversity surveys and mapping (Deulvot et al., 2010). In comparison to retrotransposon and microsatellite markers, the rate of SNP marker discovery is almost unlimited as sequence data from 80 gene amplicons totalling about 63.2 kb of sequence in five pea genotypes identified a total of 669 SNP and 84 indels (Aubert et al., 2006; Deulvot et al., 2010). On average, one SNP per 94 bp was detected (i.e. one in 165 bp in coding regions and one in 60 bp in noncoding regions) (Jing et al., 2007). The set of SNP markers using Illumina Veracode genotyping technology was used to construct a consensus map which includes 244 SNP markers and placed 5460 pea unigenes on the consensus map (Bordat et al., 2011). In summary most of this knowledge has been applied to characterize the distribution of genetic diversity in *Pisum* (Baranger et al., 2004; Ellis et al., 1998; Jing et al., 2005, 2007, 2010; Majeed et al., 2012; Martin-Sanz et al., 2011; Pearce et al., 2000; Sarikamiş et al., 2010;

Smýkal, Hýbl, et al., 2008; Smýkal et al., 2011; Tar'an, Zhang, Warkentin, Tullu, & Vandenberg, 2005; Vershinin et al., 2003; Zong et al., 2008, 2009) and these give a consistent view.

In spite of being a rather small genus with two or three species, *Pisum* is very diverse and its diversity is structured, showing a range of degrees of relatedness that reflect taxonomic identifiers, ecogeography and breeding gene pools (Ellis, 2011; Smýkal et al., 2011). Upon diversity analysis several core collections were formed, as well as trait-focussed cores (Upadhyaya et al., 2011). Recently, joint analysis of several large collections by RBIP markers was undertaken (Smýkal et al., 2011; Jing et al., 2010). However, Bayesian Analysis of Population Structure (BAPS) provided posterior assignments for K=2–14. Notably, all wild peas (*P. fulvum*, *P. sativum* subsp. *elatius* and *P. abyssinicum*) separated in one cluster, together with accessions of Afghan origin (Figure 3.3). Another cluster contained a large proportion of *P. sativum* subsp. *sativum* accessions of Ethiopian origin. One hundred and forty accessions of Chinese origin were distributed more broadly into 7–8 clusters.

It was proposed that the distinct differentiation of the Chinese *P. sativum* genotypes may in part reflect the historic isolation of agriculture in eastern Asia from that in southern Asia, Europe and northern Africa (Zong et al., 2009). Three relatively distinct gene pools of Chinese pea landraces have been differentiated and formed under natural and artificial selections. Gene Pool I is typically represented with resources in Inner Mongolia and Shaanxi in the north central cropping area boundary of China. Gene Pool II comprises landraces from Henan, which is the most northerly and coldest irrigated area of winter sowing. Gene Pool III includes the majority of Chinese landraces. Resources in this gene pool distribute widely in the large neighbouring cultivated

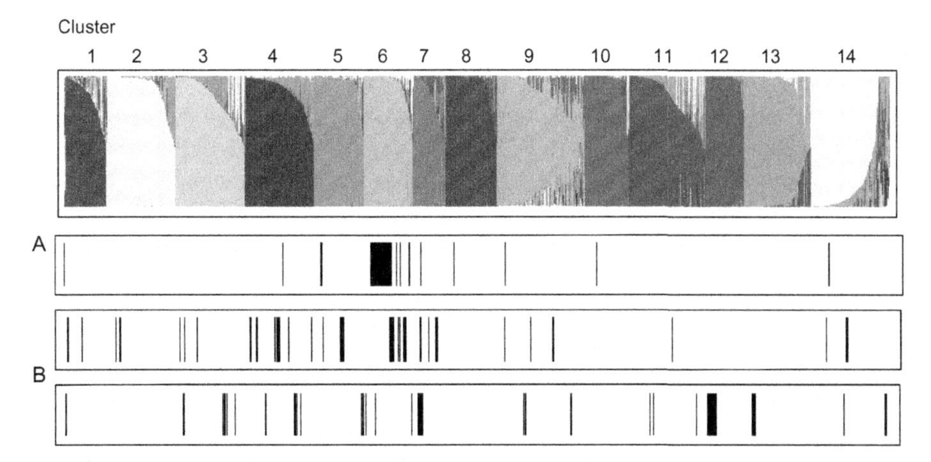

Figure 3.3 Bayesian analysis of population structure partitioning of 5641 pea accessions analysed by RBIP loci: (A) 346 wild forms (*P. fulvum*, *P. sativum* subsp. *elatius*, *P. abyssinicum*), (B) 231 accessions of Afghan–Pakistan origin (cultigen) and (C) 165 accessions of Chinese origin (cultigen).
Source: Re-analysed from Smýkal et al. (2011).

areas, especially in the west and south of China. The distinct differentiation of the three gene pools within the Chinese *P. sativum* genotypes may in part reflect a historic isolation of agriculture between northern and southern China, especially in rain-fed agriculture systems in mountain areas (Zong et al., 2009). The remaining clusters contained all cultivated material plus a set of mutant lines.

Recently, the analysis was complemented with addition of further 1518 *Pisum* accessions selected from other major European collections leading to identification of further diversity and formulation of the core collection. These results showed that wide diversity is captured in cultivated material (Figure 3.3); however, it is possible to broaden diversity using wild genotypes, which are often a source for various resistances and exotic traits. Multivariate analysis revealed close genetic relationships within cultivated materials, especially modern varieties and breeding lines, while wild material provides much of the *Pisum* genus diversity (Smýkal et al., 2011). Heterogeneity is often found within landrace accessions at individual collection sites, which is vulnerable to genetic erosion due to the small population size per accession and genetic drift during regeneration (Cieslarová et al., 2010). Taken together, as in many other inbreeding crops, relatively few genotypes with a high degree of relatedness have been used as parents in modern pea breeding programmes, leading to a narrow genetic base of cultivated germplasm (Ellis, 2011; Jing et al., 2010; Smýkal et al., 2011).

There are several current efforts to make either genome-wide introgression lines or at least simple crosses with the intent of broadening the genetic base. Further investigations, particularly in the wild *Pisum sativum* subs. *elatius* gene pool, are of great practical interest. Available molecular DNA methods will allow breeders to avoid the linkage drag from wild relatives and make wide crosses more practical and successful.

3.4.1 Sources of Resistance to Abiotic and Biotic Stresses

3.4.1.1 Abiotic Stress

One of the most important abiotic stresses is drought, which can be partly overcome by manipulation of flowering time, for example to escape the dry period which is associated with summer. As mentioned in the developmental genetics section, flowering time has been long studied in pea (Murfet & Reid, 1993). In contrast, a longer growing season or prolonged rainfall require a longer flowering time to ensure proper response. *P. sativum* subsp. *elatius* and a subset of pea landraces and winter cultivars do not flower at all under short photoperiods, but there is genetic diversity for photoperiod requirement in cultivated lines. Up to 10 loci contribute to variation related to flowering in pea, with cultivated alleles generally conferring earlier flowering and a reduction in photoperiod response. For practical purposes, the genotype *Lf Sn hr* has been adopted arbitrarily as the 'wild-type' genotype (Hecht et al., 2007; Murfet & Reid, 1993; Weller et al., 2009). *Lf* was the first pea flowering locus to be cloned and identified as a homolog of the *Arabidopsis* inflorescence identity gene *TFL1*. However, identification of functional changes in naturally occurring variants at *Lf* across *Pisum* germplasm has not been documented (Foucher et al., 2003). A 'functional candidate' approach has also been used to clone the photoperiod response locus *Hr*, a major locus controlling flowering time, with recessive *hr* alleles causing reduction, but not complete loss, of the response to photoperiod (Murfet, 1973). A single functional variant is widespread in pea germplasm and likely to underlie many of the flowering time QTL identified in this region of LG III. Naturally occurring recessive alleles at *Sn* locus confer early flowering and completely eliminate the photoperiod response, but have a restricted distribution within cultivated pea germplasm and may have arisen within a spring (*hr*) background. The dominant allele of *Hr* locus was found in a set of forage cultivars, which remain vegetative until a threshold day length of 13 h 30 min is reached. Moreover, the flowering allele *Hr* enhances the capacity of pea photoperiodic lines to produce basal laterals, which is often found in primitive accessions of *Pisum sativum*, *P. sativum* subsp. *elatius* and *P. fulvum*. Cold tolerance has been an important trait in many countries with climate suitable for autumn-sown (winter types) pea, such as Western Europe. Cold (frost) resistance has been shown to be associated with long photoperiodic requirement, in order to delay the switch of vegetative to reproductive meristem status until after winter (Lejeune-Henaut et al., 2008). Li, Redden, Zong, Berger, and Bennett (2012) used ecogeographical climatic characterization of 240 collection sites for 529 pea landraces in China to identify locations with long-term abiotic stresses, especially during the reproductive growth phase. This enabled 61 candidate accessions from these stress sites to be prioritized for phenotypic validation to confirm tolerances to frost, drought and to high temperatures. ICARDA has also had collection missions, which included high elevation sites in Kyrgyzstan, Tajikistan, Georgia and Armenia, where peas and other crop landraces were collected. Accessions from these locations could also be usefully investigated for potential frost tolerance. An ecogeographical analysis of collection sites could lead to efficiency in targeting of regions for collection of stress-tolerant landraces and of stress-tolerant CWR. Eleven populations of pea from the ICARDA Pea Germplasm Collection, comprising landraces from Mongolia, Poland, Haiti, Uganda, Spain, Eritrea, Colombia, Turkey, Denmark, Canada and Estonia were assessed for lethal temperatures (42 to 44°C) and resulted in highly significant differences among groups (Mourão, Freitas, Brito, Queiroz, & Ferreira, 2010). However, two of the tolerant varieties, Rondo and Progress, did not have heat tolerance under sustained exposure in the field in the spring of southern Australia. Genetic variation to soil constraints, such as salinity and alkaline/acidity, has been tested in pea (Leonforte, Forster, Redden, Nicolas, & Salisbury, 2012; B. Redden, Leonforte, Ford, Croser, & Slattery, 2005), and salinity-tolerant accessions have been identified from Greece and Sha'anxi province in China. These two regions were hot spots for the occurrence of salinity tolerance. Sha'anxi was one of the first Chinese provinces to develop irrigation systems over 2000 years ago, possibly leading to areas of soil salinity, but it is not clear why high levels of salinity tolerance are associated with Greece in contrast to other regions with irrigated agriculture. As an important but largely neglected factor influencing tolerance to suboptimal soil conditions, including drought, root architecture has been carefully studied on 330 pea accessions of the USDA core collection, showing large variation (McPhee, 2005). PI 261631, an accession from Spain, produced the greatest total root length and volume, as well as highest root: shoot weight ratio.

3.4.1.2 Biotic Stresses

Pea is also adversely affected by a number of fungal, viral and bacterial diseases and pests. Although some germplasm collections have been analysed by disease and pest occurrence, few examples of systematic testing and further use of resistant/tolerant genotypes have been reported. Such information would be very valuable for pea breeding and as such would be best provided in online database descriptors. There are several examples of germplasm-wide evaluation for various diseases. A set of 474 pea accessions in the Vavilov Institute originating from 28 countries was evaluated for morphological and agronomic traits at ICARDA, Syria. To screen pea cultivars for resistance to *Mycosphaerella* blight diseases under field conditions, the harvested pea seeds were transferred to Ethiopia, where the disease is endemic. Out of 581 lines evaluated, 56 lines were recorded as being promising: 16 possessed good agronomic merit, 40 lines with moderate infection levels (scored less than 2.5) were recorded as resistant and 17 of these also displayed good agronomic potential, originating from 10 countries (Priliouk et al., 1999). Another set of 242 *Pisum* accessions largely of Spanish origin were screened for resistance to *Pseudomonas syringae pv. pisi* under controlled conditions. Resistance was found to all races, including race 6 and the recently described race 8. Fifty-eight accessions were further tested for resistance to *P. syringae pv. syringae* under controlled conditions, with some highly resistant accessions identified (Martín-Sanz, Pérez de la Vega, & Caminero, 2012). Three hundred seventeen accessions largely Pakistanian and Afghan origin have been screened for resistance to *Erysiphe polygoni* or *E. pisi*, and six genotypes were found highly resistant (Ali et al., 2007; Azmat et al., 2012). In case of pea powdery mildew resistance, current cultivars rely on the presence of recessive gene *er1*, which was first reported through screening of germplasm collected in the town of Huancabamba, in the northern Peruvian Andes. The *er1* locus has been mapped and to aid selection in breeding programmes, several molecular markers linked to the *er1* locus were developed (reviewed in Smýkal et al., 2012). Recently, the underlying gene has been identified (Pavan et al., 2011) using a mutant screen. It would be interesting to conduct allele mining in a wider collection of pea germplasm to examine the natural allele diversity of this gene. New sources of partial resistance to *Fusarium* root rot have been identified in *Pisum sativum* subsp. *elatius* var. *pumilio* (Hance, Grey, & Weeden, 2004) and in three out of 44 accessions from the *Pisum* core collection (Porter, 2010) originating from Iran (PI 227258), Ethiopia (PI 226561) and India (PI 175226). Australian cultivars and breeding lines were screened for resistance to downy mildew (*Perenospora viciae*) and powdery mildew; of 88 lines tested, 14 displayed good resistance to both pathogens (Davidson, Krysinska-Kaczmarek, Kimber, & Ramsey, 2004). One hundred sixty-nine diverse pea germplasm accessions were characterized for agronomic performance, *Mycosphaerella* blight resistance and nutritional profile (Jha, Arganosa, Tar'an, Diederichsen, & Warkentin, 2012). Field screening of 165 accessions for resistance to major insect pests, i.e. pea stem fly (*Melanagromyza phaseoli*), pea leaf miner (*Chromatomyia horticola*) and pod borer (*Helicoverpa armigera*), was carried out in India and 18 accessions were identified with higher resistance to given pests (Mittal & Ujagir, 2005). Resistance to viruses has been studied in wider germplasm and sources were found in primitive landraces originating from Iran, Afghanistan, India or Ethiopia. Similarly, recessive resistance to pea seed-borne mosaic virus (PSbMV) has been identified by Hagedorn and Gritton (1973) in two Ethiopian lines (PI 193586 and PI 193835), in several accessions from India and was subsequently introduced into modern cultivars. The respective gene *eIF4E* has been identified (Smýkal, Šafářová, Navrátil, & Dostálová, 2010) and several allelic variants found while screening germplasm. A study is underway to test germplasm with virological tests as well as to find further variation in the broader gene pool (Smýkal, unpublished results).

3.4.1.3 Resistance to Biotic and Abiotic Stresses in the Wild Gene Pool

After the resistance to pea weevil was identified in *P. fulvum* (Hardie, Baker, & Marshall, 1995), with a pod and seed resistance mechanism being implicated (Clement, Hardie, & Elberson, 2002), it was attempted to introduce it into cultivated pea. Crosses were used to transfer the powdery mildew (Fondevilla, Torres, Moreno, & Rubiales, 2007) and bruchid (Byrne, Hardie, Khan, & Yan, 2008) resistances from *Pisum fulvum* into cultivated pea as well as incorporation of PSbMV and *Fusarium* resistances from primitive landraces (McPhee, Tullu, Kraft, & Muehlbauer, 1999; Provvidenti, 1990; Provvidenti & Alconero, 1988). The value of wild crop relatives has been illustrated by novel *Er3* gene, conferring dominant resistance to *E. pisi*, identified in *Pisum fulvum* (Fondevilla et al., 2008). Similarly, *Pisum fulvum* has been found to provide resistance to bruchids (Byrne et al., 2008; Clement, McPhee, Elberson, and Evans 2009) and both traits could be introgressed in cultivated pea germplasm. Resistances to *Mycosphaerella pinodes* and *Orobanche crenata* have been identified in some *P. fulvum* accessions and crossed into cultivated pea (Fondevilla, ÅVila, Cubero, & Rubiales 2005; Rubiales, Moreno, & Sillero 2005). Valuable resistance can be found in *Lathyrus* species of the tertiary pea gene pool (Vaz Patto, Fernåndez-Aparicio, Moral, & Rubiales 2007; Vaz Patto & Rubiales, 2009). However it is difficult to introduce this by conventional method including *in vitro* culture, embryo rescue or protoplast fusion (Ochatt et al., 2004); moreover, it is not known if such resistance is due to pathogen–host specialization. The use of wide crosses to source key traits results in breeding difficulties as wild-type traits are introduced and crop productivity requires many years to be restored by backcrosses. As shown by Byrne (2005), two backcrosses are sufficient to restore much of the seed and plant architecture (pod, branching, flowering time), while maintaining a desired introgressed trait.

3.5 Germplasm Maintenance

3.5.1 Pea Core Collections

Core collections can also be focused on particular traits, according to the breeding objectives. One of the greatest variables in this process is the choice of method to assess genetic differences among individuals within the wider materials. Once established, cores may be screened for traits such as disease reactions and adaptation to new environments and thus to direct germplasm users toward sections of the entire collection for further in-depth assessment (Upadhyaya et al., 2011). Also, cores may be used to highlight specific geographic areas for deeper trait mining. Recently developed Core Hunter software specifically dedicated to addressing the issue of selection of accessions into representative core collections of various sizes and based on different selection criteria has been applied to establish a European pea core collection based on 3020 JIC pea accessions (containing 1200 *P. sativum* cultivars, 600 traditional landraces and 750 wild *Pisum* samples, together with genetic stocks and reference lines from other collections) previously analysed by Jing et al. (2010), with an additional 1518 *Pisum* accessions selected from other major European collections (Jing et al., 2010). This analysis led to the identification of novel genetic materials from northern Pakistan originating from Centre for Genetic Resources (CGN) germplasm (the Netherlands) as well as diverse *P. abyssinicum* accessions from a Polish germplasm collection. With the addition of a mini-core collection of pea landraces from China, Smýkal et al. (2011) applied BAPS analysis to demonstrate that this added new diversity to *Pisum*; they also applied two approaches to identify subsets of accessions that represent the genetic diversity present in the germplasm. The first combined structural and multifactorial analysis. Six accessions strongly assigned to each of these 23 clusters were selected for their high corresponding Q values (corresponding to 138 accessions). These were augmented with the 7 outliers in the multifactorial plot discussed to maximize the represented diversity, giving 141 accessions. The second approach used the Core Hunter programme (Thachuk et al., 2009), which identified subsets of representative accessions on the basis of maximizing average genetic distance. This resulted in core collections of size 5%, 10%, 20% and 30% of the original. These selections generally overrepresent rare alleles, and a tendency to equalize allele frequencies would be expected for methods sampling distinct haplotypes equally. Further improvement to the Core Hunter algorithm has led to the development of an advanced Mixed Replica Search algorithm, using minimum (instead of the default mean) distance measures and simpler heuristics (De Beukelaer, Smykal, Davenport, & Fack, 2012). Further work is needed to test and adapt these methods also for phenotypic data.

3.5.2 Genetic Resource Databases

To be able effectively to exploit conserved diversity, it is crucial to know what diversity exists for traits and where it is conserved. There is currently no single universal database resource providing worldwide representation for a given crop, including pea (Smýkal, Coyne, et al., 2008; Smýkal et al., 2011). However there are several well-maintained international collection databases which possess information also for pea, such as European Cooperative Programme on Plant Genetic Resources, Genetic Resources Information Network and Systemwide Information Network for Genetic Resources databases. All together, these databases provide information on around two million accessions. The deposition and availability of molecular, agronomic and morphological trait data is a very critical issue. So far, data held at the national level has not been broadly accessible. Searchable databases are indispensable tools for the principal clients of gene banks, plant breeders and germplasm enhancement scientists, to search for accessions that meet multitrait criteria such as disease resistance, seed weight and grain yield expressions, or even the accessions originating from various environments (Lee et al., 2005). Combining passport, morphological and genotypic data of many gene banks will both improve germplasm management and enable search/query data exploration for germplasm with multiple traits from a virtual world pea collection online (Furman, Ambrose, Coyne, & Redden, 2006; Smýkal, Coyne, et al., 2008; Smýkal et al., 2011). The value of a gene bank depends on the representation of diversity in the species, its characterization for agricultural phenotypes and on identification of interesting genes and alleles. Initially only core collections are expected to be fully characterized phenotypically and genetically, but a long-term goal will be the detailed characterization of germplasm diversity. Inadvertent duplication of effort can be avoided with full documentation of synonyms of accessions and the pathways for sharing germplasm among gene banks. Sharing characterization data worldwide maximizes the benefit for all and spreads the cost, provided there is agreement on the technology for genotypic characterization and on comparable protocols for phenotyping. A coordinated effort to characterize germplasm collections could be achieved through an international consortium for pea genetic resources, and advanced analytical methods allowing three-way testing of diversity of genotypes, locations and quantitative traits to provide dynamic characterization of genotypic and phenotypic diversity in a molecular/ecogeographic diversity core collection for pea, as has been achieved for an azuki bean (*Vigna angularis*) core collection from China (R. J. Redden, Kroonenberg, & Basford, 2012). This approach could be used to study adaptation in pea across a range of different ecological locations of countries from the Middle East across Central Asia, where pea is a significant crop.

3.5.3 Bioinformatics of Germplasm Evaluation Data Sets

Improvements in marker methods have been accompanied by refinements in computational methods to convert original data into useful representations of diversity and genetic structure. Initial distance-based methods have been challenged by model-based Bayesian approaches (Beaumont & Rannala, 2004). Incorporation of probability, measures of support, accommodation of complex models, and various data types make them more attractive and powerful. The utility and complementarity of these approaches has been shown (Corander, Waldmann, Marttinen, & Sillanpää 2004; Rosenberg, 2002; Smýkal, Hýbl, et al., 2008). While additional computing is needed to provide support for distance-based clustering, all these parameters are directly provided by model-based approaches (Corander et al., 2004; Rosenberg 2002). Another very important issue favoring Bayesian approaches is the incorporation and combination of different data types (Corander, Gyllenberg, & Koski, 2007; Corander & Martiinen, 2006; Smýkal, Hýbl, et al., 2008; Smýkal et al., 2011). An agreed international core set for genetic diversity would provide a useful and powerful resource for next-generation markers such as SNPs or whole genome sequencing (WGS) and, more importantly, for phenotypic analysis of agronomic traits. The molecularly analysed major world pea collections and formulated core collections (Table 3.1) might act as toolkits for association mapping, a strategy to gain insight into genes and genomic regions underlying desired traits. Recent advances in genomic technology, the impetus to exploit natural diversity, and development of robust statistical analysis methods make association mapping affordable to pea research programmes (reviewed in Smýkal et al., 2012). The ability to map QTLs in collections of breeding lines, landraces or samples from natural populations has potential for future trait improvement and germplasm security. The choice of germplasm, extent of genome-wide linkage disequilibrium (LD) and relatedness within the population determine the mapping resolution, which together with marker density and statistical methods are critical to the success of association analysis. Estimates of the rate of LD decay in pea within progressively more distantly related accessions tentatively suggest high LD among cultivars (Jing et al., 2007), comparable to rice and maize. This estimate should be considered preliminary, but would imply that a greater number of SNPs than currently available might be required for effective genome-wide association mapping and marker-assisted breeding.

With a wide range of approaches now available for genotyping and declining cost for WGS, the greatest limitation for gene banks is precise phenotyping, not only for descriptive traits, but agriculturally relevant quantitative traits relating to expression of yield, crop growth and disease resistance. To increase precision, a single seed should be used for self-pollination to provide genetically uniform progeny for genotypic and phenotypic analysis. The genetic diversity within landrace accessions is purposely neglected, but hopefully compensated for by a wide survey across germplasm diversity. This level of precision is desirable if the key alleles of genes for important agronomic traits are to be identified, but broad characterization of diversity in pea germplasm can be based on a pooled DNA sample and phenotyping done on the bulked landrace mixture. Quantitative trait and disease resistance characterization has generally been done in field nurseries and for only one year. Multi-environment analysis of quantitative variation involving multitrait evaluation is far more informative than a single environment trial and potentially provides some prediction for performance in other environments (Redden et al., 2012). The challenge for gene bank curators is to strategically sample collections and maximize information from costly evaluation trials. One approach is to use core collections, geographically sub-sampled or sampled using molecular marker diversity to characterize species diversity, or to sample based on priority traits. This has led to using climatic site descriptors for characterization of natural selection and hence abiotic stress response and to provide lists of prospective germplasm with potential tolerances to heat, frost and drought stresses (Li et al., 2012). Differential sets of germplasm with specific responses to races of pathogen also can be tested with germplasm collections either in controlled inoculations or in different field locations, to evaluate genetic diversity for disease resistance.

3.6 Limitations in Germplasm Use

Vast Pisum germplasm collections are accessible (Table 3.1), but their use for crop improvement is limited, since accessing genetic diversity is still a challenge. Unfortunately, efficient extraction and exploitation of the adaptive variation and valuable traits maintained in gene banks has yet to be fully achieved, though it remains a high priority of gene bank managers (Glaszmann, Kilian, Upadhyaya, & Varshney, 2010). Traditional methods, which screen large, heterogeneous collections for phenotypic variation in agricultural traits, are not only logistically challenging, but they may overlook valuable genotypic variation concealed by epistasis in non-elite genetic backgrounds (Tanksley & McCouch, 1997). The core collection, a representative subset of the complete collection that has been optimized to contain maximal diversity in a minimal number of accessions, has been the primary solution proposed for facilitating the utilization of diverse germplasm collections (Frankel & Brown, 1984; Brown & Spillane, 1999) as well as trait core collections (Li Ling et al., 2013). As suggested also for pea (Smýkal et al., 2008b), implementing the core collection concept through 'core reference sets' would allow orchestrated and cost-efficient genotyping as well as integration of extensive phenotypic assess*m*ent (Glaszmann et al., 2010; Upadhyaya et al., 2011). This approach was applied to several grain legumes, namely chickpea, pigeon pea, and lentil, by the Generation Challenge Programme (GCP) (Upadhyaya et al., 2011). The potential improvement in screening efficiency offered by the core collection concept to conventional breeding is equally applicable to modern allele mining efforts (Reeves, Panella, & Richards, 2012) to recover useful adaptations from gene banks. Agronomic loci have been identified using a variety of approaches including mutant screens, QTL analysis, association mapping, and genome-wide surveys for the signature of artificial selection. Novel alleles recovered at loci of agronomic importance can be integrated into crop breeding programmes using conventional or molecular approaches and might be utilized to combat disease, to promote yield increases, to produce better storage and nutritional properties, or to improve stress tolerance (reviewed in Reeves et al., 2012).

The success of allele mining is dependent on the availability of diverse germplasm collections. The majority of allelic variation at any given locus is predicted to occur in the wild relatives of a crop and not the crop itself, due to the inevitable loss of variation at the domestication bottleneck, as shown in numerous recent studies. However, utilization of diversity and of trait-specific core collections should accelerate the extraction of beneficial adaptations from gene banks by making the exploration of large germplasm collections for novel alleles more efficient. Inexpensive genotyping has made marker-based core collection optimization popular. New DNA sequencing and genotyping technologies provide the power to interrogate thousands to millions of diagnostic polymorphisms, across hundreds to thousands of genotypes, thus facilitating the analysis of genetic structure and providing a rational basis to identify and select among genotypes. Another form of molecular characterization is allele resequencing in diverse materials, as documented recently in pea flower colour A gene (Hellens et al., 2010).

Further, ecogeographical information concerning the materials (ideally included in the passport information in germplasm banks, but largely missing) is essential for locating and identifying unique variants for specific adaptation. Such information might be effectively used to uncover alleles of gene of interest through the Focused Identification of Germplasm Strategy (FIGS) (Bari et al., 2011). It is likely that renewed sampling outside of existing pea collections will still be necessary. The adaptive potential of these materials can also be grasped through accurate description of their environments of origin. The availability and quality of ecogeographical/passport information will be the key to a more ecological approach to germplasm management (Li Ling et al., 2013).

3.7 Germplasm Enhancement Through Wide Crosses

Plant breeders have tried to use interspecific crosses in the Leguminosae to increase the size and diversity of the gene pool. Wide intergeneric legume hybrids have been critically reviewed by McComb (1975), who concluded that there is insufficient evidence for all reported crosses, due to misleading paper titles, confusion of vegetative with generic hybrids, the occurrence of patrocliny, and very often misplaced generic boundaries. Sobolev and Bugrii (1970), Sobolev, Agarkova, and Adamchuk (1971a, 197lb) reported hybrids between *Vicia faba* and pea with chromosome numbers between $2n=12$ and 16. The most common type, $2n=14$, had four satellite chromosomes as in the pea karyotype. The non-homologous chromosomes of peas and faba beans formed bivalents, which separated to give two groups; F_1 hybrids had low fertility and segregated sterile forms. This result is doubtful today in light of unsuccessful hybridization attempts between *V. faba* and any of its closest relatives such as *V. narbonesis*, *V. johannies* and *V. paucijuga*. In contrast, a well-documented example of successful intergeneric cross has been reported by Golubev (1990) between *Vavilovia formosa* and *P. sativum*. The hybridization of maternal *V. formosa* × paternal *P. sativum* was successful, resulting in several normally developed F_1 seeds. However, only one produced a hybrid plant. This plant had several stems, or basal branches, with long internodes and none of the lateral branches typical for *Vavilovia*. Its leaves were compound, with one pair of leaflets and, instead of the rachis present in *Vavilovia*, a third and smaller leaflet, resembling the trifoliate leaves of *Medicago* or *Trifolium* species. This, the one and only ever received F_1 plant, eventually withered due to chlorosis. However a reciprocal combination of maternal *P. sativum* × paternal *V. formosa* also resulted in one F_1 hybrid plant which had much greater height in comparison to both pea and *Vavilovia* and numerous basal and lateral branches. Flowers and pods were produced, but the F_2 seeds either aborted or remained immature (Golubev, 1990). According to unpublished data based upon personal communication from Golubev, the hybridization between *Vavilovia* and its closest relatives, such as *P. fulvum*, is possible if *Vavilovia* is used as the male parent (Mikič et al., 2009; Akopian et al., 2010). Considering the perenniality and winter hardiness of *Vavilovia*, such an interspecific hybrid could be of practical importance. Ochatt et al. (2004) confirmed the strong cross-incompatibility existing between *P. sativum* and *L. sativus* as first described by Campbell (1997), while successful although low fertility hybrids were obtained between *P. sativum* and *P. fulvum*, similarly to Errico, Conicella, De Martino, Ercolano, and Monti (1996). Durieu and Ochatt (2000) have also tested protoplast fusion and regeneration of calli between pea and *Lathyrus*. Although the heterokaryons were detected and up to six cell divisions were observed, no further growth or plant regeneration could be achieved. Although not aimed specifically to produce hybrids for further study, pioneering work of Ben-Ze'ev and Zohary (1973) on crosses among different *Pisum* species and subspecies has not only contributed to taxonomy but also can be considered as a first attempt at wider hybridization. Some hybridization barriers were indicated, as between some genotypes of '*P. humile*' there were five bivalents and one quadrivalent in meiosis (instead of the normal seven bivalents), indicating translocation difference. Similarly hybrids between genotypes of '*P. humile*' and *P. sativum* subsp. *elatius*, and between '*P. humile*' and *P. sativum*, either showed seven bivalents or indicated a translocation. Importantly, the F_1 hybrids of these crosses were highly fertile and produced seeds. Contrary to this, crosses of *P. fulvum* with *P. sativum* subsp. *elatius*, '*P. humile*' and *P. sativum* produced seeds only when *P. fulvum* was a male parent. The F_1 hybrids showed a reduction in chiasmata formation, with common univalent and multivalents. The hybrids were semi-sterile and produced few seeds. The karyotype of *P. fulvum* differed considerably from the other three taxa (Ben-Ze'ev & Zohary, 1973). The synthesis of exotic libraries, such as introgression lines (ILs) and near isogenic lines, containing chromosome segments defined by molecular markers from wild species in a constant genetic background of the related cultivated species has made the use of alien genomes more precise and efficient. Such an approach was pioneered on tomato and rice (Gur and Zamir, 2004; McCouch, 2004; Zamir, 2001), and it clearly has the potential for genetic improvement of most crop plants from incorporation of traits from related wild species and other exotic germplasm sources. Development of backcross recombinant inbred lines containing chromosome segments of wild pea (*P. fulvum* WL2140) genome in cultivated pea (*P. sativum* WL1238 or cv. Terno) genetic background defined by molecular markers is currently performed by Smýkal and Kosterin (2010). An identical approach has been started with two selected *P. sativum* subsp. *elatius* accessions (Smýkal, unpublished results). As of autumn 2012 the project of *P. fulvum* × *P. sativum* cross is in $BC_{2-3}F_2$ generations of around 200 lines and aims to establish a permanent introgression library with characterized genomic fragments of wild pea in a defined genetic background. This would allow phenotypic characterization of an unlimited number of target traits; coupled with molecular tools this will provide the means for final gene identification and its subsequent incorporation, pyramiding in desired genotypes ultimately leading to better performing commercial varieties (Upadhyaya et al., 2011).

3.8 Pea Genomic Resources

The standard pea karyotype comprises seven chromosomes: five acrocentric chromosomes and two (4 and 7) with a secondary constriction corresponding to the 45S rRNA gene cluster. The numbering of pea chromosomes is unconventional in that the largest chromosome, traditionally named Chromosome 1, is actually Chromosome 5 in pea and aligns with linkage group (LG) III. The current chromosome naming scheme arises from an early attempt to coordinate the names of linkage groups and chromosomes (Folkeson, 1990a, 1990b). There is no simple solution to this inconsistency in pea, because the two small, submetacentric chromosomes (1 and 2) are statistically impossible to distinguish in terms of relative size and arm length ratios, except of *in situ* hybridization. A set of translocation stocks was generated, but there was considerable disagreement about which linkage groups and chromosomes were involved (Hall et al., 1997; Lamm & Miravalle, 1995). Pea chromosome names should be redefined, but no systematic renaming has been agreed upon. For this reason the chromosome numbers and linkage group numbers are referred to using Arabic and Roman numerals respectively (1=VI, 2=I, 3=V, 4=IV, 5=III, 6=II and VII=7). Nuclear genome size was estimated to be 9.09pg DNA/2C, which corresponds to the haploid genome size (1C) of 4.45Gbp (Dolezel & Greilhuber, 2010). Recent investigations using next-generation sequencing data confirmed the occurrence of highly diverse families of repeats and revealed that about 50–60% of pea nuclear DNA is made up of highly to moderately repeated sequences. *Ty3/gypsy* LTR-retrotransposon has been identified as the main component of the pea repeats. Ogre elements alone were estimated to represent 20–33% of the pea genome (Macas, Neumann, & Navrátilová, 2007). Pea repeats have been the subject of a number of studies focusing on individual elements; some of the satellites provide useful cytogenetic markers, allowing discrimination of individual chromosomes (reviewed in Smýkal et al., 2012). Different types of polymorphisms were successively used for genetic mapping studies in pea: morphological markers, isozymes, molecular markers like RFLP, RAPD, SSR, EST-based and PCR-based techniques and, more recently, high-throughput parallel genotyping, resulting in a genetic map (reviewed in Smýkal et al., 2012). Later, a pea consensus linkage map was built up comprising 239 microsatellite markers (Loridon et al., 2005). These markers are quite evenly distributed throughout the seven linkage groups of the map, with 85% of intervals between the adjacent SSR markers being smaller than 10cM. This map was used to localize QTLs for disease resistance as well as quality and morphological traits. More recently, functional maps, that is composed of genes of known function, have been developed (Aubert et al., 2006; Bordat et al., 2011; Deulvot et al., 2010; Gilpin et al., 1997; Timmerman-Vaughan, Frew, & Weeden, 2000). The latest consensus map provides a comprehensive view built from data obtained for 1022 Recombinant Inbred Lines (RILs) belonging to six RIL populations (Bordat et al., 2011), providing a framework for translational genomic approaches among legumes. The map includes 214 functional markers, representing genes from diverse functional classes such as development, carbohydrate metabolism, amino acid metabolism, transport and transcriptional regulation. It also includes 180 SSR, 133 RAPD and three morphological markers and is thus intrinsically related to previous maps. However, as compared to other economically important food crops, fewer QTL mapping studies for agronomical traits have been reported in pea (reviewed in Smýkal et al., 2012). In order to support comparative legume biology, several databases were developed, integrating genetic and physical map data and enabling *in silico* analysis (reviewed in Smýkal et al., 2012). Colinearity of the genome sequences among legumes allows faster identification and isolation of genes involved in symbiosis with rhizobia and arbuscular mycorrhiza, as well as flowering time control and flower organization (reviewed in Smýkal et al., 2012).

Further, inheritance studies of the dehiscent pod character led to the identification of three regions. The region on LG III corresponded to the expected position of *Dpo*, a gene known to influence pod dehiscence. A locus on LG V appeared to have a slightly smaller effect on expression of the phenotype. The third region, observed only in one cross, had a greater effect than *Dpo* and was postulated to be yellow pod allele at the *Gp* locus (Swiecicki & Timmerman-Vaughan, 2005; Weeden, 2007). Lateral branching was probably suppressed in the pea domestication process, leading to currently grown determinate varieties essentially not branching. On the other hand, most of the wild *Pisum* accessions display proliferation of lateral meristems. Several genes regulating this process were isolated, with one identified as a novel carotenoid-derived phytohormone, strigolactone (Gomez-Roldan et al., 2008). Pea plants were original tall climbing vines. In order to minimize lodging, gradually all field pea types were selected for shorter vines, owing to a mutation at the *Le* gene (GA3-oxidase) active in gibberellin biosynthesis. Agronomically, the recessive *le* allele is required in the modern dry pea cultivars in combination with the semileafless trait to minimize crop lodging. Possibly there was a single introduction of this dwarf *le* trait for breeding of cultivars (Lester, Ross, Davies, & Reid, 1997). The *afila* trait, converting all leaflets to tendrils, was found in germplasm in the 1950s (Kujala, 1953; Solovieva, 1955), but its value was not recognized by breeders until the 1970s (Kielpinski & Blixt, 1982). Its first application was the development of the fully 'leafless' pea ideotype within a pea breeding programme at the JIC and the release of the first UK 'leafless' cultivar Filby (JI 1768) in 1978. However, the 'leafless' trait limited the total biomass of plants and the crop itself at low planting densities (Goldman & Gritton, 1992). The different loci affecting the expression of semileafless and stipule traits were described by Berry (1981). Introduction of the *afila* mutation with retained wild-type stipules led to the development of semileafless pea cultivars that proved superior to leafless in photosynthetic capacity, similar to that of the wild type (Snoad & Gent, 1976). This is considered perhaps the greatest achievement in pea breeding (Duparque, 1996). The significantly increased standing ability of semileafless dwarf pea cultivars reduced grain yield losses and the associated reduction in canopy disease severity increased the interest in cultivating pea as a quality food and feed. Its genetic background is well studied and provides breeding and other applied research with diverse beneficial possibilities (Mikić et al., 2011).

3.9 Conclusions

We have shown that in spite of being a small genus with two to three recognized species, pea is remarkably diverse and existing germplasm collections with approximately 90,000 accessions capture relatively well genetic diversity of cultivated type, yet substantially less in the case of wild materials. Unfortunately pea suffers largely from lack of international support, as compare to other grain legumes. There is an urgent need to capture and conserve wild pea diversity both *in situ* and *ex situ*. The genetic diversity of major collections has been revealed by molecular markers and led to formulation of several core collections, which facilitate the further phenotypic screening and agronomic evaluation. Furthermore, current genomic resources allow initiation of association mapping also for pea, linking genetic diversity preserved in germplasm with trait manifestation. Only a small part of the enormous potential has been exploited in breeding of biotic and abiotic stresses or novel agronomical traits. Current genomic knowledge and technologies can substantially facilitate allele mining and its incorporation in desired genetic background. Once agricultural policies recognize again the value of legumes as protein crops as well as nitrogen fixers, as well as investing in related research, there should be a bright future also for pea, particularly for temperate regions to fill the gap between soybean, chickpea and common beans.

References

Abbo, S., Lev-Yadun, S., & Gopher, A. (2010). Agricultural origins: Centers and noncenters; a near eastern reappraisal. *Critical Reviews in Plant Sciences*, *29*(5), 317–328.

Akopian, J., Sarukhanyan, N., Gabrielyan, I., Vanyan, A., Mikić, A., Smýkal, P., et al. (2010). Reports on establishing an *ex situ* site for 'beautiful' vavilovia (*Vavilovia formosa*) in Armenia. *Genetic Resources and Crop Evolution*, *57*(8), 1127–1134.

Ali, Z., Qureshi, A. S., Ali, W., Gulzar, H., & Nisar, M. (2007). Evaluation of Genetic diversity present in pea (*Pisum sativum* L.). Germplasm based on morphological traits, resistance to powdery mildew and molecular characteristics. *Pakistanian Journal of Botany*, *39*(7), 2739–2747.

Ambrose, M. J. (1995). From Near East centre of origin the prized pea migrates throughout world. *Diversity*, *11*(1–2), 118–119.

Atlagić, J., Mikić, A., Terzić, S., Zorić, L., Zeremski-Škorić, T., Mihailović, V., et al. (2010). Contributions to the characterization of *Vavilovia formosa* (syn. *Pisum formosum*). II. Morphology of androecium and gynaecium and mitosis. *Pisum Genetics*, *42*, *25–27*.

Aubert, G., Morin, J., Jacquin, F., Loridon, K., Quillet, M. C., Petit, A., et al. (2006). Functional mapping in pea, as an aid to the candidate gene selection and for investigating synteny with the model legume *Medicago truncatula*. *Theoretical and Applied Genetics*, *112*(6), 1024–1041.

Azmat, M. A., Ali Khan, A., Asif, S., Muhammad, A., & Shahid, N. (2012). Screening pea germplasm against *Erysiphe polygoni* for disease severity and latent period. *International Journal of Vegetable Science*, *18*(*2*), *153–160*.

Baldev, B. (1988). Origin, distribution, taxonomy, and morphology. In B. Baldev, S. Ramanujam, & H. K. Jain (Eds.), *Pulse crops* (pp. 3–51). New Delhi, India: Oxford & IBH Publishing Co.

Baranger, A. G., Aubert, G., Arnau, G., Lainé, A. L., Deniot, G., Potier, J., et al. (2004). Genetic diversity within *Pisum sativum* using protein- and PCR-based markers. *Theoretical and Applied Genetics*, *108*(7), 1309–1321.

Bari, A., Street, K., Mackay, M., Endresen, D., DePauw, E., & Amri, A. (2011). Focused identification of germplasm strategy (FIGS) detects wheat stem rust resistance linked to environmental variables. *Genetic Resources and Crop Evolution*, *59*, 1–17.

Bastianelli, D., Grosjean, F., Peyronnet, C., Duparque, M., & Régnier, J. M. (1998). Feeding value of pea (*Pisum sativum* L.) 1. Chemical composition of different categories of pea. *Animal Science*, *67*(3), 609–619.

Beaumont, M. A., & Rannala, B. (2004). The Bayesian revolution in genetics. *Nature Reviews in Genetics*, *5*(4), 251–261.

Ben-Ze'ev, N., & Zohary, D. (1973). Species relationships in the genus *Pisum* L. *Israeli Journal of Botany*, *22*(2), 73–91.

Berger, J. D., Kumar, S., Nayyar, H., Street, K., Sandhu, J. S., Henzell, J. M., et al. (2012). Temperature-stratified screening of chickpea (*Cicer arietinum* L.) genetic resource collections reveals very limited reproductive chilling tolerance compared to its annual wild relatives. *Field Crops Research, 126*, 119–129.

Berry, G. J. (1981). Genotype environment studies in *Pisum sativum* L. in relation to breeding objectives. PhD thesis, University of Melbourne.

Bisby, F. A., Buckingham, J., & Harborne, J. B. (Eds.), (1994). *Phytochemical dictionary of the Leguminosae Vol. 1: Plants and their constituents*. London: Chapman & Hall.

Blixt, S. (1972). Mutation genetics in *Pisum*. *Agri Hortique Genetica, 30*, 1–293.

Blixt, S., & Williams, J. T. (1982). *The pea model*. Rome: IBPGR.

Bogdanova, V. S., Galieva, E. R., & Kosterin, O. E. (2009). Genetic analysis of nuclear-cytoplasmic incompatibility in pea associated with cytoplasm of an accession of wild subspecies *Pisum sativum* subsp. *elatius* (Bieb.) Schmahl. *Theoretical and Applied Genetics, 118*(4), 801–809.

Bordat, A., Savois, V., Nicolas, M., Salse, J., Chauveau, A., Bourgeois, M., et al. (2011). Translational genomics in legumes allowed placing *in silico* 5460 unigenes on the pea functional map and identified candidate genes in *Pisum sativum* L. *G3. Genes-Genomes-Genetics, 1*(2), 93–103.

Brown, A. H. D., & Spillane, C. (1999). Implementing core collections – principles, procedures, progress, problems and promise. In R. C. Johnson & T. Hodgkin (Eds.), *Core collections for today and tomorrow* (pp. 1–9). Rome: International Plant Genetic Resources Institute.

Byrne, O.M.T. (2005). Incorporation of pea weevil resistance from wild pea (*Pisum fulvum*) into cultivated field pea (*Pisum sativum*). PhD thesis, The University of Western Australia, p. 150.

Byrne, O. M., Hardie, D. C., Khan, T., & Yan, G. (2008). Genetic analysis of pod and seed resistance to pea weevil in a *Pisum sativum* × *P. fulvum* interspecific cross. *Australian Journal of Agricultural Research, 59*(9), 854–862.

Campbell, C. G. (1997). *Grass pea, Lathyrus sativus L.* Rome: Gatersleben/IPGRI.

Chimwamurombe, P. M., & Khulbe, R. K. (2011). Domestication. In A. Pratap & J. Kumar (Eds.), *Biology and Breeding of Food Legumes* (pp. 19–34). Cambridge, MA: CABI.

Cieslarová, J., Hanáček, P., Fialová, E., Hýbl, M., & Smýkal, P. (2011). Estimation of pea (*Pisum sativum* L.) microsatellite mutation rate based on pedigree and single-seed descent analyses. *Journal of Applied Genetics, 52*, 391–401.

Cieslarová, J., Smýkal, P., Dočkalová, Z., Hanáček, P., Procházka, S., Hýbl, M., et al. (2010). Molecular evidence of genetic diversity changes in pea (*Pisum sativum* L.) germplasm after long-term maintenance. *Genetic Resources and Crop Evolution, 58*(3), 439–451.

Clement, S. L., Hardie, D. C., & Elberson, L. R. (2002). Variation among accessions of *Pisum fulvum* for resistance to Pea Weevil. *Crop Science, 42*(6), 2167–2173.

Clement, S. L., McPhee, K. E., Elberson, L. R., & Evans, M. A. (2009). Pea weevil, *Bruchus pisorum* L. (Coleoptera: Bruchidae), resistance in *Pisum sativum* × *Pisum fulvum* interspecific crosses. *Plant Breeding, 12*(5), 478–486.

Cooper, S. R., & Cadger, C. A. (1990). Germination of *Vavilovia formosa* (Stev.) Davis in the laboratory. *Pisum Newsletter, 22*, 5.

Corander, J., Gyllenberg, M., & Koski, T. (2007). Random partition models and exchangeability for Bayesian identification of population structure. *The Bulletin of Mathematical Biology, 69*(3), 797–815.

Corander, J., & Martiinen, P. (2006). Bayesian identification of admixture events using multilocus molecular markers. *Molecular Ecology, 15*(10), 2833–2843.

Corander, J., Waldmann, P., Marttinen, P., & Sillanpää, M. J. (2004). BAPS 2: Enhanced possibilities for the analysis of genetic population structure. *Bioinformatics, 20*(15), 2363–2369.

Coyne, C. J., McGee, R. J., Redden, R. J., Ambrose, M. J., Furman, B. J., & Miles, C. A. (2011). Genetic adjustment to changing climates: Pea. In S. S. Yadav, B. Redden, J. L. Hatfield, & H. Lotze-Campen (Eds.), *Crop adaptation to climate change* (pp. 238–249). Ames, IA: Wiley-Blackwell.

Cubero, J. L., Perez de la Varga, M., & Fratini, R. (2009). Origin, phylogeny and spread. In W. Erskine, F. J. Muehlbauer, A. Sarker, & B. Sharme (Eds.), *The lentil, botany, production and uses* (pp. 13–33). Wallingford: CABI – ICARDA. Chapter 3.

Cupic, T., Tucak, M., Popovic, S., Bolaric, S., Grljusic, S., & Kozumplik, V. (2009). Genetic diversity of pea (*Pisum sativum* L.) genotypes assessed by pedigree, morphological and molecular data. *Journal of Food, Agriculture & Environment*, *7*(3–4), 343–348.

Davidson, J. A., Krysinska-Kaczmarek, M., Kimber, R. B. E., & Ramsey, M. D. (2004). Screening field pea germplasm for resistance to downy mildew (*Peronospora viciae*) and powdery mildew (*Erysiphe pisi*). *Australas Plant Pathology*, *33*(3), 413–417.

Davis, P. H. (1970). *Pisum* L. In P. H. (1970). Davis (Ed.), *Flora of turkey and east Aegean Islands* (Vol. 3, pp. 370–373). Edinburg: Edinburg University Press.

De Beukelaer, H., Smykal, P., Davenport, G., & Fack, V. (2012). Fast core subset selection based on multiple genetic diversity measures using mixed replica search. *BMC Bioinformatics*, *13*, 312.

DeCaire, J., Coyne, C. J., Brumett, S., & Schultz, J. L. (2012). Additional pea EST-SSR markers for comparative mapping in pea (*Pisum sativum* L.). *Plant Breeding*, *131*(1), 222–226.

De Candolle, A. (1882). *Origin of cultivated plants*. Whitefish, MT: Kesinger Publishing LCC. 2006.

Deulvot, C., Charrel, H., Marty, A., Jacquin, F., Donnadieu, C., Lejeune-Henaut, I., et al. (2010). Highly-multiplexed SNP genotyping for genetic mapping and germplasm diversity studies in pea. *BMC Genomics*, *11*, 468.

Dixon, R. A., & Sumner, L. W. (2003). Legume natural products: Understanding and manipulating complex pathways for human and animal health. *Plant Physiology*, *131*(3), 878–885.

Dolezel, J., & Greilhuber, J. (2010). Nuclear genome size: Are we getting closer? *Cytometry Part A*, *77A*(7), 635–642.

Doyle, J. J., Doyle, J. L., Ballenger, J. A., Dickson, E. E., Kajita, T., & Ohashi, H. (1997). A phylogeny of the chloroplast gene *rbcL* in the Leguminosae: taxonomic correlations and insights into the evolution of nodulation. *American Journal of Botany*, *84*(4), 541–554.

Duparque, M. (1996). Main history steps of the pea improvement. *Grain Legumes*, *12*, 18.

Durieu, P., & Ochatt, S. J. (2000). Efficient intergeneric fusion of pea (*Pisum sativum* L.) and grass pea (*Lathyrus sativus* L.) protoplasts. *Journal of Experimental Botany*, *51*(348), 1237–1242.

Ellis, T. H. N. (2011). Pisum. In C. Kole (Ed.), *Wild crop relatives: genomic and breeding resources* (pp. 237–248). Berlin: Springer-Verlag.

Ellis, T. H. N., Poyser, S. J., Knox, M. R., Vershinin, A. V., & Ambrose, M. J. (1998). Polymorphism of insertion sites of *Ty1-copia* class retrotransposons and its use for linkage and diversity analysis in pea. *Molecular Genomics & Genetics*, *260*(1), 9–19.

Errico, A., Conicella, C., De Martino, T., Ercolano, R., & Monti, L. M. (1996). Chromosome reconstructions in *P. sativum* through interspecific hybridisation with *P. fulvum*. *Journal of Genetics & Breeding*, *50*(4), 309–313.

Food and Agriculture Organization (1996). FAOSTAT statistical database of the United Nations. Food and Agriculture Organization (FAO). Rome. Available at http://faostat.fao.org/site.

Food and Agriculture Organization (2011). FAOSTAT Statistical database of the United Nations Food and Agriculture Organization (FAO), Rome. Available at http://faostat.fao.org/site.

Folkeson, D. (1990a). Assignment of linkage segments to chromosomes 3 and 5 in *Pisum sativum*. *Hereditas*, *112*(3), 249–255.

Folkeson, D. (1990b). Assignment of linkage segments to the satellite chromosomes 4 and 7 in *Pisum sativum*. *Hereditas*, *112*(3), 257–263.

Fondevilla, S., ÅVila, C. M., Cubero, J. I., & Rubiales, D. (2005). Response to *Mycosphaerella pinodes* in a germplasm collection of *Pisum* spp. *Plant Breeding*, *124*(3), 313–315.

Fondevilla, S., Rubiales, D., Moreno, M. T., & Torres, A. M. (2008). Identification and validation of RAPD and SCAR markers linked to the gene *Er3* conferring resistance to *Erysiphe pisi* DC in pea. *Molecular Breeding*, *22*(2), 193–200.

Fondevilla, S., Torres, A. M., Moreno, M. T., & Rubiales, D. (2007). Identification of a new gene for resistance to powdery mildew in *Pisum fulvum*, a wild relative of pea. *Breeding Science*, *57*(2), 181–184.

Foucher, F., Morin, J., Courtiade, J., Cadioux, S., Ellis, N., Banfield, M. J., et al. (2003). *DETERMINATE* and *LATE FLOWERING* are two *TERMINA FLOWER1/ CENTRORADIALIS* homologs that control two distinct phases of flowering initiation and development in pea. *Plant Cell*, *15*(11), 2742–2754.

Frankel, O. H., & Brown, A. H. D. (1984). Plant genetic resources today: A critical appraisal. In J. H. W. Holden & J. T. Williams (Eds.), *Crop genetic resources conservation and evaluation* (pp. 249–257). Winchester, MA: Allen & Unwin.

Franssen, S. U., Shrestha, R. P., Bräutigam, A., Bornberg-Bauer, E., & Weber, A. P. M. (2011). Comprehensive transcriptome analysis of the highly complex *Pisum sativum* genome using next generation sequencing. *BMC Genomics*, *12*, 227.

Furman, B. J., Ambrose, M., Coyne, C. J., & Redden, B. (2006). Formation of PeaGRIC: An international consortium to co-ordinate and utilize the genetic diversity and agro ecological distribution of major collections of *Pisum*. *Pisum Genetics*, *38*, 32–34.

Gilpin, B. J., McCallum, J. A., Frew, T. J., & Timmerman-Vaughan, G. M. (1997). A linkage map of the pea (*Pisum sativum* L.) genome containing cloned sequences of known function and expressed sequence tags (ESTs). *Theoretical and Applied Genetics*, *95*, 1289–1299.

Glaszmann, J. C., Kilian, B., Upadhyaya, H. D., & Varshney, R. K. (2010). Accessing genetic diversity for crop improvement. *Current Opinion in Plant Biology*, *13*(2), 167–173.

Goldman, I. L., & Gritton, E. T. (1992). Seasonal variation in leaf component allocation in normal, afila, and afila-tendrilled acacia pea foliage near-isolines. *American Society of Horticultural Sciences*, *117, 1017–1020*.

Golubev, A. A. (1990). Habitats, collection, cultivation and hybridization vavilovia (*Vavilovia formosa* Fed.). *Bulletin Applied Botany Genetics & Plant Breeding*, *135*, 67–75. in Russian.

Gomez-Roldan, V., Fermas, S., Brewer, P. B., Puech-Pages, V., Dun, E. A., Pillot, J. P., et al. (2008). Strigolactone inhibition of shoot branching. *Nature*, *455*, 189–194.

Goss, J. (1824). On the variation in the colour of peas, occasioned by cross-impregnation. *Transactions of the Horticultural Society of London*, *5*, 234.

Gottlieb, O. R. (1982). *Micromolecular evolution, systematics, and ecology: An essay into a novel botanical discipline*. Berlin: Springer.

Govorov, L. I. (1937). Pisum. In N. I. Vavilov & E. V. Wulff (Eds.), *Flora of cultivated plants IV: Grain leguminosae* (pp. 231–336). Moscow: State Agricultural Publishing Company.

Green, F. N. (2008). Classification of *Pisum sativum* subsp. *sativum* cultivars into groups and subgroups using simply inherited characters. *Acta Horticulturae*, *799*, 155–162.

Gur, A., & Zamir, D. (2004). Unused natural variation can lift yield barriers in plant breeding. *PLoS Biology*, *2*(10), e245.

Hagedorn, J., & Gritton, E. T. (1973). Inheritance of resistance to pea seed-borne mosaic virus. *Phytopathology*, *63*, 1130–1133.

Hall, K. J., Parker, J. S., Ellis, T., Turner, L., Knox, M. R., Hofer, J. M. I., et al. (1997). The relationship between genetic and cytogenetic maps of pea. II. Physical maps of linkage mapping populations. *Genome*, *40*(5), 755–769.

Hance, S. T., Grey, W., & Weeden, N. F. (2004). Identification of tolerance to *Fusarium solani* in *Pisum sativum* ssp. *elatius*. *Pisum Genetics*, *36*, 9–13.

Harborne, J. B. (1971). Distribution of flavonoids in the Leguminosae. In J. B. Harborne, D. Boulter, & B. L. Turner (Eds.), *Chemotaxonomy of the leguminosae* (pp. 31–71). London: Academic Press.

Hardie, D. C., Baker, G. J., & Marshall, D. R. (1995). Field screening of *Pisum* accessions to evaluate their susceptibility to the pea weevil (Coleoptera: Bruchidae). *Euphytica*, *84*(2), 155–161.

Hecht, V., Knowles, C. L., Vander Schoor, J. K., Liew, L. C., Jones, S. E., Lambert, M. J. M., et al. (2007). Pea LATE BLOOMER1 is a GIGANTEA ortholog with roles in photoperiodic flowering, deetiolation, and transcriptional regulation of circadian clock gene homologs. *Plant Physiology*, *144*(2), 648–661.

Hellens, R. P., Moreau, C., Lin-Wang, K., Schwinn, K. E., Thomson, S. J., Fiers, M. W. E. J., et al. (2010). Identification of Mendel's white flower character. *PLoS One*, *5*(10), e13230.

Heywood, V. H., Kell, S. P., & Maxted, N. (2008). Towards a global strategy for the conservation and use of crop wild relatives. In N. Maxted, B. V. Ford-Lloyd, S. P. Kell, J. Iriondo, E. Dulloo, & J. Turok (Eds.), *Crop wild relative conservation and use* (pp. 653–662). Wallingford: ABI Publishing.

Hoey, B. K., Crowe, K. R., Jones, V. M., & Polans, M. O. (1996). A phylogenetic analysis of *Pisum* based on morphological characters, allozyme and RAPD markers. *Theoretical and Applied Genetics*, *92*(1), 92–100.

Hofer, J., Turner, L., Moreau, C., Ambrose, M., Isaac, P., Butcher, S., et al. (2009). Tendril-less regulates tendril formation in pea leaves. *Plant Cell*, *21*(2), 420–428.

Jha, A. B., Arganosa, G., Tar'an, B., Diederichsen, A., & Warkentin, T. D. (2012). Characterization of 169 diverse pea germplasm accessions for agronomic performance, *Mycosphaerella* blight resistance and nutritional profile. *Genetic Resources and Crop Evolution*. doi: 10.1007/s10722-012-9871-1.

Jing, R., Johnson, R., Seres, A., Kiss, G., Ambrose, M. J., Knox, M. R., et al. (2007). Gene-based sequence diversity analysis of field pea (*Pisum*). *Genetics*, *177*(4), 2263–2275.

Jing, R., Knox, M. R., Lee, J. M., Vershinin, A. V., Ambrose, M., Ellis, T. H. N., et al. (2005). Insertional polymorphism and antiquity of PDR1 retrotransposon insertions in *Pisum* species. *Genetics*, *171*(2), 741–752.

Jing, R., Vershinin, A., Grzebyta, J., Shaw, P., Smýkal, P., Marshall, D., et al. (2010). The genetic diversity and evolution of field pea (*Pisum*) studied by high throughput retrotransposon based insertion polymorphism (RBIP) marker analysis. *BMC Evolutionary Biology*, *10*, 44.

Kaya, Z., Kün, E., & Güner, A. (1998). National plan for *in situ* conservation of plant genetic diversity in Turkey. In N. Zencirci, Z. Kaya, Y. Anikster, & W. T. Adams (Eds.), *The Proceedings of International Symposium on* In Situ *Conservation of Plant Genetic Diversity*. Turkey: Central Research Institute for Field Crops.

Keiša, A., Maxted, N., & Ford-Lloyd, B. V. (2007). The assessment of biodiversity loss over time: Wild legumes in Syria. *Genetic Resources and Crop Evolution*, *55*(4), 603–612.

Kielpinski, M., & Blixt, S. (1982). The evaluation of afila character with regard to its utility in new cultivars of dry pea. *Agri Hortique Genetica XL, 51–74.*

Kislev, M. E., & Bar-Yosef, O. (1988). The legumes: The earliest domesticated plants in the Near East? *Current Anthropology*, *29*(1), 175–179.

Kloz, J. (1971). Serology of the leguminosae. In J. B. Harborne, D. Boulter, & B. L. Turner (Eds.), *Chemotaxonomy of the Leguminosae* (pp. 309–365). London: Academic Press.

Knight, T. (1799). An account of some experiments on the fecundation of vegetables. *Philosophical Transactions of Royal Society*, *89*, 195–204.

Kosterin, O. E., & Bogdanova, V. S. (2008). Relationship of wild and cultivated forms of *Pisum* L. as inferred from an analysis of three markers, of the plastid, mitochondrial and nuclear genomes. *Genetic Resources and Crop Evolution*, *55*(5), 735–755.

Kosterin, O. E., Zaytseva, O. O., Bogdanova, V. S., & Ambrose, M. J. (2010). New data on three molecular markers from different cellular genomes in Mediterranean accessions reveal new insights into phylogeography of *Pisum sativum* L. subsp. *elatius* (Bieb.) Schmalh. *Genetic Resources and Crop Evolution*, *57*(5), 733–739.

Kujala, V. (1953). Felderbse, bei welcher die ganz Blattspreite in Ranken umgewandelt ist. *Arch. Soc. Zool. Bot. Fenn. Vanamo*, *8*, *44–45*.

Künne, C., Lange, M., Funke, T., Miehe, H., Thiel, T., Grosse, I., et al. (2005). CR-EST: A resource for crop ESTs. *Nucleic Acids Research*, *33*, D619–D621.

Kupicha, F. K. (1981). Vicieae (Adans.) DC. (1825) nom conserv prop. In R. M. Polhill & P. H. Raven (Eds.), *Advances in legume systematics* (Vol. 1, pp. 377–381). Kew: The Royal Botanical Gardens.

Lamm, R. (1951). Cytogenetical studies on translocations in *Pisum*. *Hereditas*, *37*, 356–372.

Lamm, R., & Miravalle, J. R. (1995). A translocation tester set in *Pisum*. *Hereditas*, *45*, 417–440.

Lamprecht, H. (1963). Zur Kenntnis von *Pisum arvense* L. *oect. abyssinicum* Braun, mit genetischen und zytologischen Ergebnissen. *Agri Hortique Genetica*, *21*, 35–55.

Lamprecht, H. (1964). Partial sterility and chromosome structure in *Pisum*. *Agri Hortique Genetica*, *22*, 56–148.

Lamprecht, H. (1966). *Die Enstehung der Arten und hoheren Kaategorien*. Wien: Springer-Verlag.

Lavin, M., Herendeen, P. S., & Wojciechowski, M. (2005). Evolutionary rates analysis of Leguminosae implicates a rapid diversification of lineages during the Tertiary. *Systematic Biology*, *54*(4), 575–594.

Lee, M. J., Davenport, G. F., Marshall, D., Ellis, T. H. N., Ambrose, M. J., Dicks, J., et al. (2005). GERMINATE. A generic database for integrating genotypic and phenotypic information for plant genetic resource collections. *Plant Physiology*, *139*(2), 619–631.

Lehman, C. (1954). Das morphologische System der Saaterbsen (*Pisum sativum* L. sensu lato Gov. subsp. *sativum*). *Der Züchter*, *24*(11–12), 316–337.

Lejeune-Henaut, I., Hanocq, E., Bethencourt, L., Fontaine, V., Delbreil, B., Morin, J., et al. (2008). The flowering locus Hr colocalizes with a major QTL affecting winter frost tolerance in *Pisum sativum* L. *Theoretical and Applied Genetics*, *116*(8), 1105–1116.

Leonforte, A., Forster, J., Redden, R., Nicolas, M., & Salisbury, P. A. (2012). Sources of high tolerance to salinity in pea (*Pisum sativum* L.). *Euphytica, 189*, 203–216.

Lester, D. R., Ross, J. J., Davies, P. J., & Reid, J. B. (1997). Mendel's stem length gene (*Le*) encodes a gibberellin 3[beta]-hydroxylase. *Plant Cell*, *9*(8), 1435–1443.

Lewis, G., Schrirer, B., Mackinder, B., & Lock, M. (Eds.), (2005). *Legumes of the world*. Kew: The Royal Botanical Gardens.

Li, L., Redden, R. J., Zong, X., Berger, J. D., & Bennett, S. J. (2013). Ecogeographic analysis of pea collection sites from China to determine potential sites with abiotic stresses. *Genetic Resources and Crops Evolution*. http://dx.doi.org/10.1007/s10722-013-9955-6.

Liang, D., Wong, C. E., Singh, M. B., Beveridge, C. A., Phipson, B., Smyth, G. K., et al. (2009). Molecular dissection of the pea shoot apical meristem. *Experimental Botany*, *60*, 4201–4213.

Loridon, K., McPhee, K. E., Morin, J., Dubreuil, P., Pilet-Nayel, M. L., Aubert, G., et al. (2005). Microsatellite marker polymorphism and mapping in pea (*Pisum sativum* L.). *Theoretical and Applied Genetics*, *111*(6), 1022–1031.

Macas, J., Neumann, P., & Navrátilová, A. (2007). Repetitive DNA in the pea (*Pisum sativum* L.) genome: Comprehensive characterization using 454 sequencing and comparison to soybean and *Medicago truncatula*. *BMC Genomics*, *8*, 427.

Majeed, M., Safdar, W., Ali, B., Mohammad, A., Ahmad, I., & Mumtaz, A. S. (2012). Genetic assessment of the genus *Pisum* L. based on sequence specific amplification polymorphism data. *Journal of Medicinal Plants Research*, *6*(6), 959–967.

Makasheva, R. K. (1973). Perennial pea. *Bulletin of Applied Botany of Genetics and Plant Breeding*, *51*, 44–56.

Makasheva, R. K. (1979). Gorokh (pea). In O. N. Korovina (Ed.), *Kulturnaya flora SSR* (pp. 1–324). Leningrad: Kolos Publishers. in Russian.

Martin-Sanz, A., Caminero, C., Jing, R., Flavell, A. J., & Perez de la Vega, M. (2011). Genetic diversity among Spanish pea (*Pisum sativum* L.) landraces, pea cultivars and the world *Pisum* sp. core collection assessed by retrotransposon-based insertion polymorphisms (RBIPs). *Spanish Journal of Agricultural Research*, *9*(1), 166–178.

Martín-Sanz, A., Pérez de la Vega, M., & Caminero, C. (2012). Resistance to *Pseudomonas syringae* in a collection of pea germplasm under field and controlled conditions. *Plant Pathology*, *61*(2), 375–387.

Marx, G. A. (1977). Classification, genetics and breeding. In J. F. Sutcliffe & J. S. Pate (Eds.), *Physiology of the Garden Pea* (pp. 21–43). New York, NY: Academic Press.

Maxted, N., & Ambrose, M. (2001). Peas (*Pisum* L.. In N. Maxted & S. J. Bennett (Eds.), *Plant Genetic Resources of Legumes in the Mediterranean* (pp. 181–190). Dordrecht: Kluwer Academic Publishers.

Maxted, N., Dulloo, E., Ford-Lloyd, B. V., Iriondo, J., & Jarvis, A. (2008). Genetic gap analysis: A tool for more effective genetic conservation assessment. *Diversity and Distributions*, *14*, 1018–1030.

Maxted, N., & Kell, S. (2009). Establishment of a global network for the *in situ* conservation of crop wild relatives: Status and needs: *FAO Consultancy Report*. Rome: FAO. pp. 1–265.

Maxted, N., & Kell, S. (2012). PGR Secure: Enhanced use of traits from crop wild relatives and landraces to help adapt crops to climate change. *Crop Wild Relative*, *8*(4–8) ISSN 1742-3694 (online).

Maxted, N., Kell, S., Toledo, A., Dulloo, E., Heywood, V., Hodgkin, T., et al. (2010). A global approach to crop wild relative conservation: Securing the gene pool for food and agriculture. *Kew Bulletin*, *65*, 561–576.

Maxted, N., Shelagh, K., Ford-Lloyd, B., Dulloo, E., & Toledo, A. (2012). Toward the systematic conservation of global crop wild relative diversity. *Crop Science*, *52*(2), 774–785.

Maxted, N., van Slageren, M. W., & Rihan, J. (1995). Ecogeographic surveys. In L. Guarino, V. R. Rao, & R. Reid (Eds.), *Collecting plant genetic diversity: Technical guidelines* (pp. 255–286). Wallingford: CAB International.

McComb, J. A. (1975). Is intergeneric hybridization in the leguminosae possible? *Euphytica*, *24*(2), 497–502.

McCouch, S. (2004). Diversifying selection in plant breeding. *PLoS Biology*, *2*(10), e347.

McPhee, K. (2005). Variation for seedling root architecture in the core collection of pea germplasm. *Crop Science*, *45*, 1758–1763.

McPhee, K. E., Tullu, A., Kraft, J. M., & Muehlbauer, F. J. (1999). Resistance to fusarium wilt race 2 in the *Pisum* core collection. *Journal of American Society of Horticulture Science, 124*(1), 28–31.

Mendel, G. (1866). Versuche uber pflanzen-hybriden. [Experiments on Plant Hybrids]. *Verhandlungen der naturfoschung Vereins, 4*, 3–47. (see also http://www.mendelweb.org/).

Mikič, A. (2012). Origin of the words denoting some of the most ancient old world pulse crops and their diversity in modern European languages. *PLoS One, 7*(9), e44512.

Mikic, A., Smýkal, P., Kenicer, G. J., et al. (2013). The bicentenary of the research on 'beautiful' vavilovia (Vavilovia formosa), a legume crop wild relative with taxonomic and agronomic potential. *Botanical Journal of the Linnean Society* (in press).

Mikić, A., Mihailović, V., Ćupina, B., Kosev, V., Warkentin, T., McPhee, K., et al. (2011). Genetic background and agronomic value of leaf types in pea (*Pisum sativum*). *Field and Vegetable Crops Research,, 48*, 275–284.

Mikič, A., Smýkal, P., Kenicer, G., Vishnyakova, M., Sarukhanyan, N., Akopian, J., et al. (2009). A revival of research on beautiful vavilovia (*Vavilovia formosa* syn. *Pisum formosanum*). *Pisum Genetics, 41*, 11–14.

Mishra, R. K., Gangadhar, B. H., Nookaraju, A., Kumar, S., & Park, S. W. (2012). Development of EST-derived SSR markers in pea (*Pisum sativum*) and their potential utility for genetic mapping and transferability. *Plant Breeding, 131*(1), 118–124.

Mittal, V., & Ujagir, U. (2005). Field screening of pea, *Pisum sativum* L. germplasm for resistance against major insect pests. *Journal of Plant Protection and Environment, 2*(1), 50–58.

Mourão, I., Freitas, M. C., Brito, L. M., Queiroz, A., & Ferreira, R. B. (2010). Pea landraces and cultivars response to acclimation at high temperatures and its correlation to morphological and agronomic characteristics. *Acta Horticulturae, 918*(2), 565–573.

Mumtaz, A. S., Shehadeh, A., Ellis, T. H. N., Ambrose, M. J., & Maxted, N. (2002). The collection and ecogeography of non-cultivated peas (*Pisum* L.) from Syria. *Plant Genetic Resources Newsletter, 146*, 3–8.

Murfet, I. C. (1973). Flowering in *Pisum*. Hr, a gene for high response to photoperiod. *Heredity, 31*(2), 157–164.

Murfet, I. C., & Reid, J. B. (1993). Developmental mutants. In R. Casey & D. R. Davies (Eds.). *Peas: Genetics, molecular biology & biotechnology* (pp. 165–216). CAB. Wallingford UK.

Nasiri, J., Haghnazari, A., & Saba, J. (2010). Genetic diversity among varieties and wild species accessions of pea (*Pisum sativum* L.) based on SSR markers. *African Journal of Biotechnology, 15*(8), 3405–3417.

Ochatt, S. J., Benabdelmouna, A., Marget, P., Aubert, G., Moussy, F., Pontécaille, C., et al. (2004). Overcoming hybridization barriers between pea and some of its wild relatives. *Euphytica, 137*(3), 353–359.

Oskoueiyan, R., Osaloo, S. K., Maassoumi, A. A., Nejadsattari, T., & Mozaffarian, V. (2010). Phylogenetic status of *Vavilovia formosa* (Fabaceae-Fabeae) based on nrDNA ITS and cpDNA sequences. *Biochemical Systematics and Ecology, 38*(3), 313–319.

Palmer, J. D., Jorgensen, R. A., & Thompson, W. F. (1985). Chloroplast DNA variation and evolution in *Pisum*: Patterns of change and phylogenetic analysis. *Genetics, 109*(1), 195–213.

Pate, J. S. (1975). Pea. In L. T. Evans (Ed.), *Crop physiology: Some case histories* (pp. 191–224). London: Cambridge University Press.

Pavan, S., Schiavulli, A., Appiano, M., Marcotrigiano, A. R., Cillo, F., Visser, R. G. F., et al. (2011). Pea powdery mildew er1 resistance is associated to loss-of-function mutations at a MLO homologous locus. *Theoretical and Applied Genetics, 123*(8), 1425–1431.

Pearce, S. R., Knox, M., Ellis, T. H., Flavell, A. J., & Kumar, A. (2000). Pea Ty1-copia group retrotransposons: Transpositional activity and use as markers to study genetic diversity in *Pisum*. *Molecular and General Genetics, 263*(6), 898–907.

Peas: Genetics, molecular biology and biotechnology. (pp. 165–216). Wallingford: CAB International.

Phillips, D. A. (1980). Efficiency of symbiotic nitrogen fixation in legumes. *Annual Review in Plant Physiology*, *31*, 29–49.

Planta Europa (2001). *European Plant Conservation Strategy. Plantlife International (Salisbury, UK) and the Council of Europe*. Strasbourg, France.

Planta Europa (2008). *A Sustainable Future for Europe; the European Strategy for Plant Conservation 2008–2014. Plantlife International (Salisbury, UK) & the Council of Europe*. Strasbourg, France.

Plitman, U., & Kislev, M. E. (1989). Reproductive changes induced by domestication. In C. H. Stirton & J. L. Zarucchi (Eds.), *Advances in legume biology* (pp. 487–503). St. Louis, MO: Missouri Botanical Garden Press.

Polans, N. O., & Saar, D. E. (2002). ITS sequence variation in wild species and cultivars of pea. *Pisum Genetics*, *34*, 9–13.

Porter, L. D. (2010). Identification of tolerance to *Fusarium* root rot in wild pea germplasm with high levels of partial resistance. *Pisum Genetics*, *42*, 1–6.

Priliouk, L., Vavilov, N. I., Robertson, L., Francis, C., Khan, T., Gorfu, D., et al. (1999). Genetic diversity in pea germplasm from Vavilov Institute and ICARDA collections for black spot resistance and agronomic merit. *Pisum Genetics*, *31*, 53.

Provvidenti, R. (1990). Inheritance of resistance to pea seedborne mosaic virus in *Pisum sativum*. *Journal of Heredity*, *81*, 143–145.

Provvidenti, R., & Alconero, R. (1988). Inheritance of resistance to a lentil strain of pea seedborne mosaic virus in *Pisum sativum*. *Journal of Heredity*, *79*, 45–47.

Redden, B., Leonforte, T., Ford, R., Croser, J., & Slattery, J. (2005). Pea (*Pisum sativum* L.). In R. J. Singh & P. P. Jauhar (Eds.), *Genetic resources, chromosome engineering, and crop improvement Vol. 1: Grain legumes* (pp. 49–83). Boca Raton, FL: CRC Press.

Redden, R. J., Kroonenberg, P. M., & Basford, K. E. (2012). Adaptation analysis of diversity in Adzuki germplasm introduced into Australia. *Crop & Pasture Science*, *63*, 142–154.

Reeves, P. A., Panella, L. W., & Richards, C. M. (2012). Retention of agronomically important variation in germplasm core collections: implications for allele mining. *Theoretical and Applied Genetics*, *124*(6), 1155–1171.

Rosenberg, N. A. (2002). Genetic structure of human populations. *Science*, *298*(5602), 2381–2385.

Rubiales, D., Moreno, M. T., & Sillero, J. C. (2005). Search for resistance to crenata broomrape (*Orobanche crenata*) in pea germplasm. *Genetic Resources and Crop Evolution*, *52*(7), 853–861.

Saar, D. E., & Polans, N. O. (2000). ITS sequence variation in selected taxa of *Pisum*. *Pisum Genetics*, *32*, 42–45.

Sardana, S., Mahajan, R. K., Gautam, N. K., & Ram, B. (2007). Genetic variability in pea (*P. sativum* L.) germplasm for utilization. *SABRAO Journal of Breeding and Genetics*, *39*(1), 31–41.

Sarikamiş, G., Yanmaz, R., Ermiş, S., Bakır, M., & Yüksel, C. (2010). Genetic characterization of pea (*Pisum sativum*) germplasm from Turkey using morphological and SSR markers. *Genetics and Molecular Research*, *9*(1), 591–600.

Schaefer, H., Hechenleitner, P., Santos-Guerra, A., Menezes de Sequeira, M., Pennington, R. T., Kenicer, G., et al. (2012). Systematics, biogeography, and character evolution of the legume tribe Fabeae with special focus on the middle-Atlantic island lineages. *BMC Evolutionary Biology*, *12*, 250.

Smartt, J. (1990). *Grain legumes: Evolution and genetic resources*. Cambridge: Cambridge University Press.

Smýkal, P. (2006). Development of an efficient retrotransposon-based fingerprinting method for rapid pea variety identification. *Journal of Applied Genetics*, *47*(3), 221–230.

Smýkal, P., Aubert, G., Burstin, J., Coyne, C., Ellis, N., Flavell, A., et al. (2012). Pea (*Pisum sativum* L.) in the genomic era. *Agronomy*, *2*(2), 74–115.

Smýkal, P., Coyne, C. J., Ford, R., Redden, R., Flavell, A. J., Hybl, M., et al. (2008). Effort towards a world pea (*Pisum sativum* L.) germplasm core collection: The case for common markers and data compatibility. *Pisum Genetics*, *40*, 11–14.

Smýkal, P., Hýbl, M., Corander, J., Jarkovský, J., Flavell, A., & Griga, M. (2008). Genetic diversity and population structure of pea (*Pisum sativum* L.) varieties derived from combined retrotransposon, microsatellite and morphological marker analysis. *Theoretical and Applied Genetics*, *117*(3), 413–424.

Smýkal, P., Kalendar, R., Ford, R., Macas, J., & Griga, M. (2009). Evolutionary conserved lineage of *Angela*-family retrotransposons as a genome-wide microsatellite repeat dispersal agent. *Heredity*, *103*, 157–167.

Smýkal, P., Kenicer, G., Flavell, A. J., Kosterin, O., Redden, R. J., Ford, R., et al. (2011). Phylogeny, phylogeography and genetic diversity of the *Pisum* genus. *Plant Genetic Resources*, *9*(1), 4–18.

Smýkal, P., & Kosterin, O. (2010). Towards introgression library carrying wild pea (*Pisum fulvum*) segments in cultivated pea (*Pisum sativum*) genome background. In *Book of Abstracts of Vth International Congress on Legume Genetics and Genomics* (p. 123). Asilomar, USA, 2–8 July 2010.

Smýkal, P., Šafářová, D., Navrátil, M., & Dostálová, R. (2010). Marker assisted pea breeding: eIF4E allele specific markers to pea seed-borne mosaic virus (PSbMV) resistance. *Molecular Breeding*, *26*(3), 425–438.

Snoad, B., & Gent, G. P. (1976). The evaluation of leafless and semileafless peas. *J.I. Annual Report Norwich*, 35–37.

Sobolev, N. A., Agarkova, S. N., & Adamchuk, G. K. (1971a). Overcoming cross incompatibility between pea and broad beans. 1. The effect of X irradiation on incompatibility. *Nauchnye trudy zernobobovych kultur*, *3*, 136–146. in Russian.

Sobolev, N. A., Agarkova, S. N., & Adamchuk, G. K. (1971b). Overcoming incompatibility between pea and broad bean. II. Characteristics of progeny raised from seeds in pea–bean crosses. *Nauchnye trudy zernobobovych kultur*, *3*, 147–156. in Russian.

Sobolev, N. A., & Bugrii, V. P. (1970). A spontaneous intergeneric hybrid in the vetch tribe and its characteristics. (Ru.). Moscow, USSR, Kolos. *Otdalenaja gibridizatsiya rastenij*, 415–421 (in Russian).

Solovieva, W. K. (1955). *Vining pea*. Moscow (in Russian).

Swiecicki, W.K., & Timmerman-Vaughan, G. (2005). Localization of important traits: The example of pea. In T. Nagata, H. Lorz, & J. M. Widholm (series Eds.) H. Lörz, & G. Wenzel (Vol. Eds.), *Molecular marker systems in plant breeding and crop improvement: Biotechnology in Agriculture and Forestry Vol. 55* (pp. 155–169). Berlin: Springer-Verlag.

Tanksley, S. D., & McCouch, S. R. (1997). Seed banks and molecular maps: Unlocking genetic potential from the wild. *Science*, *277*(5329), 1063–1066.

Tar'an, B., Zhang, C., Warkentin, T., Tullu, A., & Vandenberg, A. (2005). Genetic diversity among varieties and wild species accessions of pea (*Pisum sativum* L.) based on molecular markers, and morphological and physiological characters. *Genome*, *48*(2), 257–272.

Thachuk, C., Crossa, J., Franco, J., Dreisigacker, S., Warburton, M., & Davenport, G. F. (2009). Core hunter: An algorithm for sampling genetic resources based on multiple genetic measures. *BMC Bioinformatics*, *10*, 243.

Timmerman-Vaughan, G. M., Frew, T. J., & Weeden, N. F. (2000). Characterization and linkage mapping of R-gene analogous DNA sequences in pea (*Pisum sativum* L.). *Theoretical and Applied Genetics*, *101*, 241–247.

Upadhyaya, H. D., Dwivedi, S. L., Ambrose, M., Ellis, N., Berger, J., Smýkal, P., et al. (2011). Legume genetic resources: Management, diversity assessment, and utilization in crop improvement. *Euphytica*, *180*, 27–47.

Vavilov, N. I. (1949). The phytogeographic basis of plant breeding. In K. S. Chester (Ed.), *The origin, variation, immunity, and breeding of cultivated plants* (pp. 13–54). Waltham, MA: Chronica Botanica.

Vaz Patto, M. C., Fernández-Aparicio, M., Moral, A., & Rubiales, D. (2007). Resistance reaction to powdery mildew (*Erysiphe pisi*) in a germplasm collection of *Lathyrus cicera* from Iberian origin. *Genetic Resources and Crop Evolution*, *54*(7), 1517–1521.

Vaz Patto, M. C., & Rubiales, D. (2009). Identification and characterization of partial resistance to rust in a germplasm collection of *Lathyrus sativus*. *Plant Breeding*, *128*(5), 495–500.

Vershinin, A. V., Allnutt, T. R., Knox, M. R., Ambrose, M. J., & Ellis, N. T. H. (2003). Transposable elements reveal the impact of introgression, rather than transposition, in *Pisum* diversity, evolution, and domestication. *Molecular Biology and Evolution*, *20*(12), 2067–2075.

Vilmorin, P.. (1913). Étude sur le Caractère "Adhérence des Grains entre eux" chez le Pois Chenille. *IV. Conf. intern. Génétique* (pp. 368–372). Paris.

Waines, J. G. (1975). The biosystematics and domestication of peas (*Pisum* L.). *Bulletin of the Torrey Botanical Club*, *102*(6), 385–395.

Wang, Z., Luo, Y., Li, X., Wang, L., Xu, S., Yang, J., et al. (2008). Genetic control of floral zygomorphy in pea (*Pisum sativum* L.). *Proceedings of National Academy of Sciences USA*, *105*(30), 10414–10419.

Weeden, N. F. (2007). Genetic changes accompanying the domestication of *Pisum sativum*: Is there a common genetic basis to the 'Domestication Syndrome' for legumes? *Annals of Botany*, *100*(5), 1017–1025.

Weeden, N. F., & Wolko, B. (2001). Allozyme analysis of *Pisum sativum* spp. *abyssinicum* and the development of genotypic definition for this subspecies. *Pisum Genetics*, *33*, 21–25.

Weller, J. L., Hecht, V., Liew, L. C., Sussmilch, F. C., Wenden, B., Knowles, C. L., et al. (2009). Update on the genetic control of flowering in garden pea. *Journal of Experimental Botany*, *60*(9), 2493–2499.

Yarnell, S. H. (1962). Cytogenetics of the vegetable crops III. Legumes. A. Garden peas, *Pisum sativum* L. *Botanical Reviews*, *28*(4), 465–573.

Zamir, D. (2001). Improving plant breeding with exotic genetic libraries. *Nature Review in Genetics*, *2*(12), 983–989.

Zaytseva, O. O., Bogdanova, V. S., & Kosterin, O. E. (2012). Phylogenetic reconstruction at the species and intraspecies levels in the genus *Pisum* (L.) (peas) using a histone H1 gene. *Gene*, *504*(2), 192–202.

Zhukovskyi, P. M. (1971). *Cultivated plants and their congeners*. Leningrad.

Zohary, D., & Hopf, M. (2000). *Domestication of plants in the old world*. Oxford: Oxford University Press.

Zong, X., Guan, J. P., Wang, S. M., Liu, Q., Redden, R., & Ford, R. (2008). Genetic diversity and core collection of alien *Pisum sativum* L. germplasm. *Acta Agronomica Sinica*, *34*, 1518–1528.

Zong, X., Redden, R., Liu, Q., Wang, S., Guan, J., Liu, J., et al. (2009). Analysis of a diverse *Pisum* sp. collection and development of a Chinese *P. sativum* core collection based on microsatellite markers. *Theoretical and Applied Genetics*, *118*(2), 193–204.

Zorić, L., Mikić, A., Akopian, J., Sarukhanyan, N., Gabrielyan, I., Vanyan, A., et al. (2010). Contributions to the characterization of *Vavilovia formosa* (syn. *Pisum formosum*). I. Anatomy of stem, leaf and calyx. *Pisum Genetics*, *42*, 21–24.

4 Chickpea

Shivali Sharma, Hari D. Upadhyaya, Manish Roorkiwal, Rajeev K. Varshney and C.L. Laxmipathi Gowda

International Crops Research Institute for the Semi-Arid Tropics (ICRISAT), Patancheru, Hyderabad, India

4.1 Introduction

Chickpea (*Cicer arietinum* L.) is a self-pollinated true diploid ($2n=2x=16$) cool season leguminous crop that ranks second among food grain legumes in the world after common bean (FAOSTAT, 2011). It is grown in a wide range of environments in over 50 countries in subtropical and temperate regions of the world, mainly in the Indian subcontinent, West Asia, North Africa, the Americas and Australia (FAOSTAT, 2011). Based on seed shape, size and colour, two distinct forms of cultivated chickpea are known (Cubero, 1975); namely, the *desi* type, characterized mostly by pink flowers, angular. brown, small seeds with a high percentage of fibre, primarily grown in South Asia and Africa; and *kabuli* type, having white flowers and owl-head-shaped, beige, large seeds with a low percentage of fibre, grown in Mediterranean countries. A third type, designated as intermediate or pea-shaped, is characterized by medium to small size and round, pea-shaped seeds. Kabuli types are grown in about two-thirds of chickpea-growing countries, but desi type predominates in chickpea production and accounts for about 85%, while kabuli accounts for about 15% of the world chickpea production.

It is grown primarily for its protein-rich seeds. In addition, chickpea seeds are also rich in minerals (calcium, potassium, phosphorus, magnesium, iron and zinc), fibre, unsaturated fatty acids, and β-carotene (Jukanti, Gaur, Gowda, & Chibbar, 2012). Owing to its high nutritional qualities, chickpea is considered one of the most nutritious food grain legumes for human consumption, with potential health benefits. For example, high fibre content in chickpea has the ability to lower the cholesterol level as well as prevent blood sugar levels from rising too rapidly after a meal, thus making it a healthy food for diabetic patients (McIntosh & Miller, 2001; Pittaway et al., 2006). Further, chickpea does not contain any antinutritional factors except the raffinose-type oligosaccharides, which cause flatulence (Williams & Singh, 1987) and can be neutralized by boiling or mere soaking in water (Queiroz, de Oliveira, & Helbig, 2002). Chickpea plant is an efficient symbiotic nitrogen fixer, improving soil fertility by fixing atmospheric nitrogen, meeting up to 80% of its nitrogen requirement and playing an important role in crop diversification and sustainability of farming systems. However, chickpea is cultivated mostly in marginal lands under rain-fed conditions, with low and unstable productivity (Kumar & van

Genetic and Genomic Resources of Grain Legume Improvement. DOI: http://dx.doi.org/10.1016/B978-0-12-397935-3.00004-9

Rheenen, 2000). Development of high-yielding, early-maturing cultivars that fit well into the short cropping season is one of the major objectives of chickpea improvement programmes. But the narrow genetic base of cultivated chickpea is one of the major obstacles to sustaining and improving its productivity and renders the crop vulnerable to new biotic and abiotic stresses. The narrow genetic base of chickpea is particularly due to the restricted distribution of its wild progenitor, *Cicer reticulatum*, the founder effect associated with domestication, the shift from winter to summer cropping and the replacement of locally adapted landraces by the genetically uniform modern varieties (Abbo, Berger, & Turner, 2003). Plant genetic resources comprising landraces, obsolete varieties and crop wild relatives are the reservoirs of natural genetic variations, but general reluctance of the breeders to use exotic germplasm has severely restricted the introgression of useful variation present in the exotic germplasm. This chapter will provide information about the nature and extent of chickpea genetic resources conserved across gene banks globally, the pattern of diversity in cultivated and wild *Cicer* species, and various approaches including genomic tools to promote utilization of genetic resources to broaden the genetic base for sustainable chickpea crop production.

4.2 Origin, Distribution, Diversity and Taxonomy

Chickpea is one of the earliest grain crops domesticated in the Old World at Tell el-Kerkh (tenth millennium BC) in Syria, Cayönü (7250–6750 BC), and Hacilar (ca 6700 BC) in Turkey, and Jericho (8350–7370 BC) in the West Bank. The earliest to date is Tell el-Kerkh, where both *Cicer arietinum* and its immediate progenitor *Cicer reticulatum* were clearly identified. Since Tell el-Kerkh is at a considerable distance from the native lands of the wild chickpea, *C. reticulatum* in southeast Turkey, it is suggested that the domestication took place somewhat earlier than that (Tanno & Willcox, 2006). However, the cultivation of chickpea is well documented from 3300 BC onwards in Egypt and the Middle East (van der Maesen, 1972). Most probably, it originated in an area of present-day southeastern Turkey and Syria, where three wild annual *Cicer* species are found, namely, *C. bijugum*, *C. echinospermum* and *C. reticulatum*, closely related to chickpea. From here, chickpea spread with human migration toward the West and South via the Silk Route (Singh et al., 1997). Four centres of diversity have been identified in the Mediterranean, Central Asia, the Near East and India, as well as a secondary centre of origin in Ethiopia (Vavilov, 1951).

Presently, *Cicer* species occur from sea level to over 5000 m near glaciers in the Himalayas. The cultivated species *C. arietinum* is found only in cultivation and cannot colonize successfully without human intervention. The wild *Cicer* species occur in weedy habitats (fallow or disturbed habitats, roadsides, cultivated fields of wheat, places not touched by man or cattle), mountain slopes among rubble and also naturally in inhospitable areas of the Himalayas in India (Chandel, 1984).

Globally, chickpea is grown on about 13.2 million hectare area with a production of 11.62 million metric tons and an average productivity of 880.4 kg/ha (FAOSTAT, 2011). The developing countries account for 90% of the global chickpea cultivation and South and Southeast Asia (SSEA) contribute about 79% of the global chickpea production. India is the principal chickpea-producing country, with a 68% share in the global chickpea area and production. Other countries producing substantial amounts of chickpea include Australia, Pakistan, Turkey, Myanmar, Ethiopia, Iran, Mexico, Canada, USA, Morocco and Yemen (FAOSTAT, 2011). Chickpea is the only domesticated species under the genus *Cicer*, family Fabaceae and subfamily Papilionoideae. Earlier, the genus *Cicer* was classified in the tribe Vicieae Alef., which was later reported to belong to its own monogeneric tribe, Cicereae Alef. (Kupicha, 1981). The tribe Cicereae is closer to the tribe Trifolieae, which differs from the former in having hypogeal germination, tendrils, stipules free from the petiole, and nonpapillate unicellular hairs. The genus *Cicer* currently comprises 44 species, including 35 wild perennials, 8 wild annuals and the cultivated annual (Muehlbauer, 1993; van der Maesen, 1972) (Table 4.1). The infragenic classification of genus *Cicer* includes two subgenera: *Pseudononis* and *Viciastrum*, four sections, *Monocicer*, *Chamaecicer*, *Polycicer* and *Acanthocicer*, and 14 series (van der Maesen, 1987).

The subgenus *Pseudononis* is characterized by small flowers (normally 5–10 mm), subregular calyx, hardly gibbous base, with sublinear, nearly equal teeth. It comprises two sections, *Monocicer* (annuals, with firm erect or horizontal stems branched from the base or at middle) and *Chamaecicer* (annuals or perennials, with thin, creeping, branched stem, and small flowers). The section *Monocicer* is the most important section for chickpea improvement and includes eight annual species, namely *C. arietinum, C. reticulatum, C. echinospermum, C. judaicum, C. bijugum, C. pinnatifidum, C. cuneatum* and *C. yamashitae*. This section is further subdivided into three series, *Arietina* (characterized by imparipinnate leaves, with none to small arista), *Cirrhifera* (leaves ending in a tendril, with short arista) and *Macro-aristae* (leaves imparipinnate, long arista). The second section, *Chamaecicer*, includes one annual species, *C. chorassanicum,* and one perennial species, *C. incisum*, and is divided into two series, *Annua* and *Perennia* (Kazan & Muehlbauer, 1991; Muehlbauer, Kaiser, & Simon 1994).

The subgenus *Viciastrum* (perennials, characterized by medium large flowers, calyx strongly gibbous at the base, with unequal teeth) comprises two sections, *Polycicer* and *Acanthocicer*. *Polycicer* (leaf rachis ending in a tendril or a leaflet, never a spine) contains 23 perennial species and is divided into two subsections, *Nano-polycicer* (with creeping rhizome, short stem, imparipinnate leaves, weak and short arista) and *Macro-polycicer* (with short rhizome, non-creeping, stems ascending to 75 cm, firm arista longer than pedicel). *Macro-polycicer* is further divided into six series: (i) *Persica* (inflorescences 1–2 flowered, flowers 14–15 mm, calyx teeth 2–4 times the tube, stipules 14–15 mm, half as large as the leaflets, which are in 2–12 pairs); (ii) *Anatolo-persica* (inflorescences 1–2 flowered, flowers 20–27 mm, calyx teeth short, stipules smaller than the largest leaflets, which are in 4–9 pairs); (iii) *Europaeo-anatolica* (inflorescences 2–5 flowered, bracts foliolate, stipules small or up to half as large as the leaflets, which are in 4–8 pairs); (iv) *Flexuosa* (inflorescences 1–2 flowered, bracts minute, stipules much smaller than the leaflets, which are in 4–13 pairs); (v) *Songarica* (inflorescences 1–2 flowered, bracts minute,

Table 4.1 List of Various *Cicer* Species

S. No.	Species	S. No.	Species
Cultivated species			
1	*Cicer arietinum* (Chickpea)		
Annual wild *Cicer* species			
1	*Cicer reticulatum*	5	*Cicer pinnatifidum*
2	*Cicer echinospermum*	6	*Cicer chorassanicum*
3	*Cicer judaicum*	7	*Cicer cuneatum*
4	*Cicer bijugum*	8	*Cicer yamashitae*
Perennial wild *Cicer* species			
1	*Cicer acanthophyllum*	20	*Cicer macracanthum*
2	*Cicer anatolicum*	21	*Cicer microphyllum*
3	*Cicer atlanticum*	22	*Cicer mogolatvicum*
4	*Cicer balcaricum*	23	*Cicer montbretii*
5	*Cicer baldshuanicum*	24	*Cicer multijugum*
6	*Cicer canariense*	25	*Cicer nuristanicum*
7	*Cicer fedtschenkoi*	26	*Cicer oxyodon*
8	*Cicer flexuosum*	27	*Cicer paucijugum*
9	*Cicer floribundum*	28	*Cicer pungens*
10	*Cicer graecum*	29	*Cicer rassuloviae*
11	*Cicer grande*	30	*Cicer rechingeri*
12	*Cicer heterophyllum*	31	*Cicer songaricum*
13	*Cicer incanum*	32	*Cicer spiroceras*
14	*Cicer incisum*	33	*Cicer stapfianum*
15	*Cicer isauricum*	34	*Cicer subaphyllum*
16	*Cicer kermanense*	35	*Cicer tragacanthoides*
17	*Cicer korshinskyi*		*Cicer tragacanthoides* var. *tragacanthoides*
18	*Cicer laetum*		*Cicer tragacanthoides* var. *turcomanicum*
19	*Cicer luteum*		

stipules more or less equal to the largest leaflets, which are in 2–18 pairs) and (vi) *Microphylla* (inflorescences 1–2 flowered, bracts minute, stipules smaller than or equal to the largest leaflets, which are in 7–10 pairs). Section *Acanthocicer* (perennials, with branched stems with woody base, persistent spiny leaf rachis, spiny calyx teeth, and large flowers) encompasses nine perennial species and is divided into three series: *Pungentia* (foliate or small spiny stipules), *Macrocantha* (long spiny stipules) and *Tragacanthoidea* (small, triangular, incised stipules).

4.2.1 Gene Pool

In the genus *Cicer*, 43 wild species are classified into three gene pools based on their crossability status, with the cultivated chickpea following the Harlan and de Wet (1971) gene pool concept. The primary gene pool consists of cultivated chickpea, its landraces and the progenitor species, *C. reticulatum*, the species that are freely crossable with cultivated chickpea with regular gene exchange. The secondary gene pool consists of *C. echinospermum*, a species that is crossable with cultivated chickpea, but with reduced fertility of the resulting hybrids and progenies. The tertiary gene pool consists of remaining six annual and 35 perennial species that are not readily crossable with cultivated chickpea and require specialized techniques for gene transfer into the cultivated background.

4.3 Erosion of Genetic Diversity from the Traditional Areas

The major factors responsible for genetic erosion include replacement of the traditional varieties, indigenous species and landraces with genetically uniform, high-yielding, modern cultivars resulting in loss of about three-quarters of the genetic diversity of agricultural crops, climate change posing serious threats on crop germplasm, intensive recent development activities, habitat destruction by modern agriculture and poor knowledge of germplasm and of its scientific, social, cultural and economic importance, resulting in the loss of this treasure. In most of the crops including chickpea, only a fraction of the diversity of wild species is stored in the existing collections. In gene banks also, many accessions have been lost because of improper storage, poor seed viability following introduction and short storage viability even in good facilities. Further, much of this diversity is threatened by decades of underfunding and neglect as well as by wars and natural disasters. In genus *Cicer* six species, namely *C. atlanticum*, *C. echinospermum*, *C. floribundum*, *C. graecum*, *C. isauricum* and *C. reticulatum*, were categorized as rare (R) and were included in the 1997 World Conservation Union (International Union for Conservation of Nature, IUCN) List of Threatened Plants (Walter & Gillett, 1998). The tertiary gene pool species, *C. bijugum*, has been considered a priority for collection. Due to the introduction of high-yielding varieties, a number of landraces carrying vast amount of genetic diversity are lost from farmers' fields in many countries (Berger, Abbo, & Turner, 2003). In Georgia, where chickpea is one of the traditional crops, local varieties are rarely cultivated today (Akhalkatsi, Ekhvaia, & Asanidze, 2012). Dekaprelevich and Menabde (1929) reported that three subspecies and 24 varieties were available in western Georgia – Racha-Lechkhumi, Svaneti and Imereti up to the 1920s, but in the 1970s the same three subspecies – *C. arietinum* subsp. *mediterraneum* G. Pop., *C. arietinum* subsp. *eurasiaticum* G. Pop., *C. arietinum* subsp. *orientalis* G. Pop. – and only 6 of 24 varieties – *C. arietinum* subsp. *mediterraneum* var. *ochroleucum* A. Kob., *C. arietinum* subsp. *mediterraneum* var. *rozeum* G. Pop., *C. arietinum* subsp. *eurasiaticum* var. *aurantiacum* G. Pop., *C. arietinum* subsp. *orientalis* var. *fulvum* G. Pop., *C. arietinum* subsp. *orientalis* var. *rufescens* G. Pop. and *C. arietinum* subsp. *orientalis* var. *rufescens brunneopunctatus* A. Kob. – were in cultivation (Kobakhidze, 1974). In Svaneti also, chickpea was traditionally available, but by the 1970s only one farmer was sowing it in the Kala community village Khe (Zhizhizlashvili & Berishvili, 1980). The genetic erosion of chickpea has also been noticed in the Mianwali district of Punjab along the Indus (Ahmad et al., 1984). Several *Cicer* species are found in eastern Anatolian deciduous forests in the centre of Southwest Asia (Turkey, Iran and Afghanistan), but the high level of habitat conversion and low level of protection in this region is posing a major threat to the chickpea genetic diversity and has warranted considerable conservation concerns in recent years (Stolton, Maxted, Ford-Lloyd, Kell, & Dudley, 2006).

4.4 Status of Germplasm Resources Conservation

Large-scale collection and conservation efforts have been initiated to protect the crop biodiversity, and *ex situ* gene banks have been established by the Food and Agriculture Organization (FAO) and the World Bank for the collection and conservation of plant genetic resources. Globally, about 7.4 million germplasm accessions of different crops have been collected and/or assembled and conserved in over 1750 gene banks (FAOSTAT, 2010). For chickpea, there are a large number of gene banks conserving over 98,000 germplasm accessions comprising of landraces, modern cultivars, genetic stocks, mutants and wild *Cicer* species (http://apps3.fao.org/wiews/germplasm_query.htm?i_l=EN). The major gene banks holding chickpea germplasm are given in Table 4.2. The RS Paroda gene bank at International Crops Research Institute for the Semi-Arid Tropics (ICRISAT) has the largest collection: 19,959 accessions of cultivated chickpea and 308 accessions of 18 wild *Cicer* species from 60 countries. These accessions were obtained from donations as well as from collection missions in different countries. Other major gene banks holding chickpea germplasm include the National Bureau of Plant Genetic Resources (NBPGR) (16,881 accessions), New Delhi, India; the International Centre for Agricultural Research in Dry Areas (ICARDA) (13,818 accessions), Aleppo, Syria; Australian Temperate Field Crops Collection (ATFCC) (8655 accessions), Horsham, Victoria; and Western Regional Plant Introduction Station (WRPIS), United States Department of Agriculture - Agricultural Research Service (USDA-ARS) (6789 accessions), Pullman (Table 4.2). Besides conserving germplasm accessions in these gene banks, duplication agreements have been negotiated for safety between gene banks within and outside the Consultative Group on International Agricultural Research (CGIAR) system for a majority of crops. At the global level, the Svalbard Global Seed Vault will definitely contribute to combating the loss of biological diversity, reducing vulnerability to climate change and securing future food production.

4.5 Germplasm Evaluation and Maintenance

The characterization, evaluation and maintenance of germplasm are essential for their effective utilization in crop improvement programmes and for efficient management of genetic resources. At ICRISAT chickpea germplasm accessions have been characterized and evaluated for various morpho-agronomic traits following the Chickpea Descriptors (IBPGR, ICRISAT, & ICARDA, 1993) since 1974. A multidisciplinary approach is followed for the characterization and evaluation of chickpea germplasm for various biotic and abiotic stresses and for agronomic and nutrition-related traits. Besides, germplasm sets are also evaluated jointly with National Agricultural Research Systems (NARS) scientists in different countries and more intensively with the

Table 4.2 Major Holdings of Chickpea Germplasm in Different Gene Banks of the World

Country	Institute	Wild Accessions	Wild Species	Cultivated Accessions	Total
Australia	Australian Temperate Field Crops Collection (ATFCC), Horsham, Victoria	246	18	8409	8655
Ethiopia	Institute of Biodiversity Conservation (IBC), Addis Ababa			1173	1173
Hungary	Institute for Agrobotany (RCA), Tápiószele	9	5	1161	1170
India	Indian Agricultural Research Institute (IARI), New Delhi			2000	2000
	International Crop Research Institute for the Semi-Arid Tropics (ICRISAT), Patancheru	308	18	19,959	20,267
	National Bureau of Plant Genetic Resources (NBPGR), New Delhi	69	10	16,812	16,881
Iran	College of Agriculture, Tehran University, Karaj			1200	1200
	National Plant Gene Bank of Iran, Seed and Plant Improvement Institute (NPGBI-SPII), Karaj			5700	5700
Mexico	Estación de Iguala, Instituto Nacional de Investigaciones Agrícolas (IA-Iguala), Iguala			1600	1600
Pakistan	Plant Genetic Resources Institute (PGRP), Islamabad	89	3 (1)	2057	2146
Russian Federation	N.I. Vavilov All-Russian Scientific Research Institute of Plant Industry (VIR), St. Petersburg			2091	2091
Syria	International Centre for Agricultural Research in Dry Areas (ICARDA), Aleppo	270	11 (1)	13,548	13,818
Turkey	Plant Genetic Resources Department, Aegean Agricultural Research Institute (AARI), Izmir	21	4	2054	2075
Ukraine	Institute of Plant Production n.a. V.Y. Yurjev of UAAS, Kharkiv			1021	1021

(Continued)

Table 4.2 (Continued)

Country	Institute	Wild Accessions	Wild Species	Cultivated Accessions	Total
USA	Western Regional Plant Introduction Station, USDA-ARS, Pullman	205	22	6584	6789
Uzbekistan	Uzbek Research Institute of Plant Industry (UzRIPI), Botanica			1055	1055

Source: http://apps3.fao.org/wiews/germplasm_query.htm?i_l=EN.

NBPGR, New Delhi. About 99% of chickpea germplasm accessions have been characterized for agronomic and morphological traits at ICRISAT. Chickpea has orthodox seeds that can be dried to low seed moisture content (about 5–7%) for efficient conservation. For conservation of germplasm, a two-tier system is being followed in the ICRISAT gene bank. Seeds are dried in cool and dry conditions to reduce the moisture content to a desired level (5%±1%) and then stored as active collections in medium-term storage (at 4°C, 20–30% relative humidity) in aluminium cans and as base collection in long-term storage (at −20°C) after packing in vacuum-sealed aluminium foil pouches. The entire chickpea collection consisting of 20,267 accessions is stored as active and base collection in the ICRISAT gene bank. A recent monitoring of the health of seed conserved for 10–25 years under medium-term storage has indicated greater than 85% seed viability for the majority of the accessions. Regeneration is one of the most important gene-bank activities, which aims at seed multiplication by maintaining the genetic integrity of the original sample. Accessions with declining seed viability (less than 75% seed germination) and/or quantity (<100g) have high priority for regeneration. Further, the regeneration of accessions that have low viability is given the highest priority over accessions with low seed quantity. Besides, special requirements for seed multiplication may arise for accessions requiring safety duplication and repatriation. Breeding behaviour of the crop and the sample size are the two key factors affecting efficient regeneration. Since chickpea is a self-pollinated crop, regeneration is carried out in field without any control on pollination by using at least 80 plants for regenerating an accession. Regeneration of cultivated types is carried out in solarized fields during the post-rainy season. Solarization is the process of heating soil by covering it with polyethylene sheets during hot summer to control soilborne diseases like *Fusarium* wilt that represent a major limitation on chickpea growth during regeneration. Solarization is conducted for at least 6 weeks during the hottest part of the year. However, critical accessions of wild *Cicer* species that need long day length and cool weather to grow and produce seeds are regenerated under controlled greenhouse conditions (Figure 4.1). Newly acquired germplasm of foreign origin is first grown in the post-entry quarantine isolation area under the supervision of the National Plant Quarantine Services. Recently, the management practices of different gene banks were reviewed to develop the best practices and procedures for chickpea germplasm management (Upadhyaya et al., 2009; http://cropgenebank.sgrp.cgiar.org/).

Figure 4.1 Regeneration of wild *Cicer* species under controlled environmental conditions in the greenhouse at ICRISAT, Patancheru, India.

4.6 Use of Germplasm in Crop Improvement

4.6.1 Status of Germplasm in Chickpea Improvement

Since 1974, the ICRISAT gene bank has distributed about 321,251 chickpea seed samples to researchers in 88 countries. The evaluation of chickpea germplasm by national programmes has led to the release of 17 accessions directly as varieties in 15 countries. Studies have shown scanty use of germplasm (<1%) in chickpea improvement programmes. India has one of the largest chickpea improvement programmes and has released 126 chickpea cultivars in the past four decades. Surprisingly, 41% of cultivars have Pb 7 as one of the parents, with IP 58, F 8, S 26 and Rabat being the most extensively used parents (Kumar, Gupta, Chandra, & Singh, 2004). However, ICRISAT, has the largest chickpea germplasm collections; our chickpea breeding programme has used 12,887 (586 unique) parents including only 91 germplasm lines to develop the 3,548 advanced breeding lines; L 550 and K 850 being the most frequently used cultivars (Upadhyaya, Gowda, Buhariwalla, & Crouch, 2006). This shows the breeders' preference for selecting parental genotypes from their working collections. Working collections usually exhibit good agronomic performance and provide a quick way for the breeders to make steady progress in the shortest possible time. Further, the chances of diluting the agronomic performance become higher with the involvement of new germplasm lines (Kannenberg & Falk, 1995). Thus, the use of parental genotypes from working collections results in recirculation of the same germplasm, hence the narrow genetic base of the cultivars. This results in genetic vulnerability, which has already caused havoc in the past, such as the southern corn leaf blight epidemic in United States of America during 1969–1970, due to the large-scale use of genetically uniform male sterile lines.

4.6.2 Small Subsets for Enhancing the Utilization of Germplasm

Frankel and Brown (1984) suggested that greater use of germplasm in crop improvement is possible if a small collection representing the diversity of the entire large collection is made available to researchers for meaningful evaluation and utilization. Frankel (1984) coined the term "core collection" to sample representative variability from the entire collection. A core collection contains 10% of the accessions from the entire collection that capture most of the available diversity in the species (Brown, 1989a). Thus, a core collection has a reduced size containing a diverse set of germplasm and is representative of the entire collection. Such core collections can be evaluated extensively and the information derived could be used to guide the more efficient utilization of the entire collection (Brown, 1989b).

4.6.2.1 Core Collection

Using passport information and characterization and evaluation data generated over a period of time, a chickpea core collection consisting of 1956 accessions has been developed from the global collection of 16,991 accessions from 44 countries at ICRISAT (Upadhyaya, Bramel, & Singh, 2001). Similarly, a core collection of 505 accessions was developed from 3350 chickpea accessions by the scientists at the USDA in Pullman, Washington (Hannan, Kaiser, & Muehlbauer, 1994). A kabuli chickpea core collection consisting of 103 accessions has been developed at the Seed and Plant Improvement Institute (SPII), Karaj, Iran (Pouresmael, Akbari, Vaezi, & Shahmoradi, 2009). Recently, a core collection consisting of 158 germplasm accessions has been developed for the Ethiopian chickpea germplasm collection at ICRISAT (Kibret, 2011) (Table 4.3).

4.6.2.2 Mini-Core Collection

The germplasm collections at the International Agricultural Research Center (IARC) gene banks are very large in size such as the International Maize and Wheat Improvement Center (CIMMYT) gene bank holding more than 100,000 wheat accessions and the International Rice Research Institute (IRRI) gene bank with over 110,000 rice accessions; hence, the core collections with about 10,000 accessions could be unmanageably large and unwieldy, which would restrict its proper evaluation and use by crop breeders. Even at ICRISAT, the chickpea core collection of 1956 accessions is too large for its meaningful multilocation evaluation. This forced the scientists to develop a new strategy to further reduce the size of the core collection without losing the spectrum of diversity. Upadhyaya and Ortiz (2001) postulated

Table 4.3 Small-Sized Subsets for Chickpea Germplasm

Crop	Accessions	Subset Developed	Accessions in Subset	Reference
Chickpea	16,991	Core collection	1956	Upadhyaya et al. (2001)
	3350	Core collection	505	Hannan et al. (1994)
	1002	Core collection	158	Kibret (2011)
	N/A	Kabuli chickpea core collection	103	Pouresmael et al. (2009)
	1956	Mini-core collection	211	Upadhyaya and Ortiz (2001)
	482	Mini-core collection	39	Biabani et al. (2011)
	N/A	Composite collection	3000	Upadhyaya et al. (2006)
	3000	Reference set	300	Upadhyaya et al. (2008a)

the mini-core concept following a seminal two-stage strategy for sampling the entire and core collections to develop a mini-core collection, which consists of roughly 10% of the accessions of the core collection (about 1% of the entire collection) representing the diversity of the entire collection with minimum loss of diversity. They suggested using the core collection as a basis for developing a mini-core collection. The first stage in constituting a mini-core collection thus involves developing a representative core collection (about 10%) from the entire collection using the available information on origin, geographical distribution, characterization and evaluation data. The second stage involves evaluation of the core collection for various morphological, agronomic and grain quality traits, and selecting a further set of about 10% accessions from the core collection. At both the stages, standard clustering procedures are used to create groups of similar accessions and various statistical tests are used to evaluate and validate core and mini-core collections. Following this strategy, a mini-core collection was constituted in chickpea (Upadhyaya & Ortiz, 2001), which consists of 211 accessions representing the diversity of over 16,000 accessions (Table 4.3). Validation studies of this mini-core collection with the core collection and of the core collection with the entire collection revealed that the mini-core and core collections represented adequate diversity for most of the traits detected in the entire collection and will improve the efficiency of identifying valuable genes in the entire large collections for their effective utilization in chickpea improvement programmes. Another chickpea mini-core collection consisting of 39 accessions has been developed at the WRPIS at Pullman, Washington, USA (Biabani et al., 2011).

4.6.2.3 Composite Collection and Reference Set

Large collections of chickpea germplasm are maintained by ICRISAT, India and ICARDA, Syria (Table 4.2). As a part of the Generation Challenge Programme

Table 4.4 Composition of Global Composite Collections of Chickpea Germplasm

Germplasm/Traits	No. of Accessions
Accessions from ICRISAT	
Core collection	1956
Cultivars/breeding lines	39
Ascochyta blight	13
Botrytis gray mold	8
Stunt	8
Fusarium wilt	50
Collar rot	9
Black root rot	8
Dry root rot	6
Helicoverpa	16
Leaf miner	5
Nematode	8
Low temperature	12
High temperature	4
Drought	10
Salinity	4
Early maturity	25
High protein	10
Multiseeded pods	7
Seed size	18
High-input responsive	4
Twin pods	8
Nodulation	8
Morphological diversity	35
Accessions from ICARDA	
Based on characterization and evaluation data	599
Based on agro-climatological data	110
Cicer echinospermum	7 (1 from ICRISAT)
Cicer reticulatum	13 (2 from ICRISAT)

(GCP; http://www.generationcp.org), ICRISAT and ICARDA jointly developed a global composite collection of 3000 accessions to capture the global diversity available in these two gene banks and other materials such as released cultivars, sources of resistance/tolerance to various biotic/abiotic stresses including wild species (Tables 4.3 and 4.4) (Upadhyaya et al., 2006). The composite collection, which includes core and mini-core collections (Table 4.4), was molecularly profiled using 48 Simple Sequence Repeat (SSR) markers to study its genetic structure. A total of 1683 alleles were detected, of which 935 were rare, 720 common and 28 most frequent. The alleles per locus ranged from 14 to 67 and averaged 35; the polymorphic information content was from 0.467 to 0.974, averaging 0.854; and the gene diversity ranged from 0.533 to 0.974 with an average of 0.869. Kabuli chickpea as a group were genetically more diverse than other seed types. Desi and kabuli shared

436 alleles, while wild *Cicer* shared 17 and 16 alleles with desi and kabuli types, respectively. Desi chickpea contained a higher proportion of rare alleles (53%) than kabuli (46%), while wild *Cicer* accessions were devoid of rare alleles. Several group-specific unique alleles were also detected as 104 in kabuli, 297 in desi, and 69 in wild *Cicer*. Geographically, 114 unique alleles were found each in West Asia (WA) and Mediterranean, 117 in SSEA, and 10 in African accessions. The accessions from SSEA and WA shared 74 alleles, while those from Mediterranean shared 38 and 33 alleles with WA and SSEA, respectively (Upadhyaya et al., 2008a). The composite collection was also characterized for qualitative and quantitative traits at ICRISAT. A reference set consisting of the 300 genetically most diverse accessions was selected based on SSR markers, qualitative and quantitative traits, and their combinations. The reference set based on 48 SSR markers (78.1% alleles) was similar to the reference set based on seven qualitative traits (73.5%), whereas the reference set based on both captured 80.5% of the alleles of the composite collection (1683 alleles) (Upadhyaya et al., 2008b). This demonstrated that both SSR markers and qualitative traits were equally effective in sampling allelic diversity.

4.6.3 Trait-Specific Germplasm for Use in Chickpea Improvement

Evaluation of germplasm accessions, especially the small subsets, has resulted in the identification of new sources of resistance/tolerance to important biotic/abiotic stresses as well as promising accessions for important agronomic traits as follows.

4.6.3.1 Biotic Stresses

Resistance to Diseases

Evaluation of the chickpea mini-core collection resulted in the identification of three accessions (ICC 1915, ICC 6306 and ICC 11284) moderately resistant to *Ascochyta* blight, 55 accessions (ICC 1180, ICC 2990, ICC 4533, ICC 4841, ICC 4872 and others) to *Botrytis* gray mold, six accessions (ICC 1710, ICC 2242, ICC 2277, ICC 11764, ICC 12328 and ICC 13441) to dry root rot, 21 asymptomatic (ICC 637, ICC 1205, ICC 1356, ICC 1396, ICC 2065 and others) and 24 resistant (ICC 67, ICC 95, ICC 791, ICC 867, ICC 1164 and others) to *Fusarium* wilt (Pande, Kishore, Upadhyaya, & Rao, 2006). Combined resistance to *Ascochyta* blight and *Botrytis* gray mold was identified only in one accession, ICC 11284; for *Botrytis* gray mold and dry root rot in two accessions (ICC 11764 and ICC 12328); for *Botrytis* gray mold and *Fusarium* wilt in 11 accessions (ICC 2990, ICC 4533, ICC 6279, ICC 7554, ICC 7819 and others); and for dry root rot and *Fusarium* wilt in four accessions (ICC 1710, ICC 2242, ICC 2277 and ICC 13441) (Pande et al., 2006).

Resistance to Insect Pests

The chickpea mini-core collection was evaluated for pod borer (*Helicoverpa armigera* L.) resistance. Five accessions (ICC 5878, ICC 6877, ICC 11764, ICC 16903 and ICC 18983) had very low leaf-feeding score under detached leaf assay screening; five accessions (ICC 12537, ICC 9590, ICC 7819, ICC 2482 and ICC 4533) had least larval survival rate and five accessions (ICC 16903, ICC 6877, ICC 3946, ICC 11746 and ICC 18983) were identified as the best accessions for lower larvae weight, when compared to resistant control cultivar ICC 506-EB (ICRISAT Archival Report, 2009). Similarly, evaluation of the chickpea reference set consisting of 300 accessions identified 13 accessions (ICC 1230, ICC 2263, ICC 3325, ICC 4567, ICC 5135, ICC 6874, ICC 10466, ICC 11198, ICC 12307, ICC 14831, ICC 15406, ICC 15606 and ICC 16524) with low *H. armigera* damage and plant mortality, which also exhibited high yield potential under unprotected conditions (ICRISAT Archival Report, 2010). Further, one mini-core accession, ICC 4969, has been identified as a resistant source for pulse beetle (*Callosobruchus maculatus* F.) in both free-choice and no-choice tests (Erler, Ceylan, Erdemir, & Toker, 2009).

4.6.3.2 Abiotic Stresses

Drought

Drought stress, especially terminal drought stress, is one of the major adverse factors affecting chickpea production. The importance of an extensive and deep root system is well recognized as a means to improve drought tolerance and hence crop productivity through enhanced water uptake. Evaluation of chickpea mini-core accessions for the root traits using a cylinder culture system revealed a large genetic variability among accessions and identified two accessions (ICC 8261 and ICC 10885) with high root length density (RLD), six accessions (ICC 13124, ICC 14506, ICCV 2, ICC 8261, ICC 15333, ICC 7315) with large shoot to root length density ratio (S/RLD) and several accessions having a deep root system in comparison to the then-known most drought-tolerant accession, ICC 4958. A kabuli type landrace ICC 8261, from Turkey, had the most prolific root system, the largest RLD, as well as larger biomass allocation into the root system, which could be of high importance under severe drought conditions (Kashiwagi et al., 2005). Similarly, evaluation of 50 large-seeded kabuli germplasm accessions with four control cultivars (KAK 2, JGK 1, ICCV 2 and ICC 4958) for drought-avoidance root traits identified one accession, ICC 17450 (EC 543583) with larger RLD than ICC 4958, which could be utilized for a larger-seeded kabuli chickpea improvement programme (Kashiwagi, Upadhyaya, Krishnamurthy, & Singh, 2007). Kashiwagi, Krishnamurthy, Upadhyaya, and Gaur (2008) also used canopy temperature as a simple screening method to screen for drought tolerance and identified ICC 14799 as having the highest relatively cool canopy temperature, followed by ICC 867, ICC 3325 and ICC 4958. Similarly, evaluation of 289 chickpea accessions for drought tolerance has identified several promising accessions (ICC 2580, ICC 7272, ICCV 92311, ICC 3362, ICCV 95311, ICC 506 and EC 583311) with high grain yield, high harvest index (HI) and/or pest resistance and was to be evaluated further in multilocation trails (Mulwa, Kimurto, & Towett, 2010). Following field screening techniques, the chickpea mini-core germplasm accession ICC 13124 had the highest drought tolerance efficiency, least drought susceptibility index, the highest HI and minimum reduction in seed yield under drought, and was identified as the most drought-tolerant accession for moisture stress conditions (Parameshwarappa & Salimath, 2008; Parameshwarappa et al., 2010). Similarly, evaluation of the chickpea mini-core for drought tolerance index over 3 years identified five accessions (ICC 867, ICC 1923, ICC 9586, ICC 12947 and ICC 14778) as highly drought tolerant (Krishnamurthy, Kashiwagi, Gaur, Upadhyaya, & Vadez, 2010). Of these five accessions, ICC 867 and ICC 14778 have also been found to maintain the coolest canopy temperatures (Kashiwagi et al., 2008).

Water Use Efficiency

The soil plant analysis development chlorophyll meter reading (SCMR) has been recognized as a useful measure to estimate leaf chlorophyll content for the plant's nitrogen acquisition capability and is a surrogate trait for selecting genotypes with improved nitrogen status leading to improved yield. Kashiwagi, Krishnamurthy, Singh, and Upadhyaya (2006) evaluated the chickpea mini-core collection and identified two accessions, ICC 16374 and ICC 4958, with high and stable SCMR values. Similarly, based on transpiration efficiency (TE) and carbon isotope discrimination ($\delta^{13}C$), promising accessions were identified such as ICC 5337 and ICC 4958 are having high $\delta^{13}C$ under stress condition, and ICC 5337 having the highest TE under stress and well-watered conditions. Later, evaluation of the chickpea mini-core collection for SCMR identified ICC 4958 as having the best SCMR performance. The same genotype, ICC 4958, has also been identified to possess the most prolific and deep root systems as well as the largest relatively cool canopy temperature (Kashiwagi et al., 2008), which makes it a unique breeding material for improving the acquisition of both soil water and soil nitrogen. Additional accessions with high SCMR values, such as ICC 1422, ICC 10945, ICC 16374 and ICC 16903, were also identified (Kashiwagi, Upadhyaya, & Krishnamurthy, 2010).

Salinity

Two hundred and eleven chickpea mini-core germplasm accessions and 41 popular varieties and breeding lines were evaluated under saline conditions (100 mM NaCl; pot screening) and 10 highly tolerant accessions (ICC 10755, ICC 13124, ICC 13357, ICC 15406, ICC 15697 and others) were identified (Serraj, Krishnamurthy, & Upadhyaya, 2004). Similarly, 263 chickpea accessions comprising 211 mini-core accessions and some lines reported as tolerant to sodicity, popular cultivars and breeding lines, and one cultivar released by the Central Soil Salinity Research Institute (CSSRI) for salinity tolerance (CSG 8962) were evaluated under saline conditions (80 mM NaCl; pot screening) to identify salinity-tolerant chickpea genotypes based on their seed yield under salinity (Vadez et al., 2007). Sixteen salinity-tolerant accessions yielding more than the previously identified salt-tolerant genotype CSG 8962 were identified. Of these, three accessions, ICC 5003, ICC 15610 and ICC 1431, had about 20% higher yield than the tolerant control, CSG 8962. Vadez et al. (2007) also reported that the desi genotypes had more salinity tolerance than the kabuli genotypes. Recently, Krishnamurthy, Turner, et al. (2011) also evaluated chickpea germplasm accessions including 211 mini-core accessions for salinity tolerance and identified 12 accessions (ICC 9942, ICC 6279, ICC 11121, ICC 456, ICC 12155 and others), which were highly tolerant in both a Vertisol and an Alfisol soil. Of these, one accession, ICC 9942, had the highest and most consistent seed yield performance in both soil types.

Heat Tolerance

Evaluation of 35 chickpea germplasm accessions selected from the core collection along with a control cultivar, ICCV 92944, for tolerance to heat stress identified ICC 14346 as the most heat-tolerant germplasm accession, followed by ICC 5597, ICC 5829, ICC 6121, ICC 7410, ICC 111916, ICC 13124, ICC 14284, ICC 14368 and ICC 14653. These accessions were consistently high yielding (>1400 kg/ha) as compared with the control, ICCV 92944 (1333 kg/ha) (Upadhyaya, Dronavalli, Gowda, & Singh, 2011). Similarly, Krishnamurthy, Gaur, et al. (2011) evaluated the chickpea reference set collection for heat tolerance at two locations (Patancheru and Kanpur) in India and identified 18 stable heat-tolerant accessions (ICC 456, ICC 637, ICC 1205, ICC 3362, ICC 3761 and others).

4.6.3.3 Agronomic Traits

Early Maturity

Chickpea breeding programmes aim at developing early-maturing cultivars especially to increase crop adaptation by avoiding terminal drought and high temperature stress in the sub-tropics. Twenty-eight early-maturing chickpea germplasm accessions (ICC 16641, ICC 16644, IC 11040, ICC 11180, ICC 12424 and others), which were similar or earlier than control cultivars Harigantars and ICCV 2 and produced about 23% more seed yield as compared to the average of four control cultivars (ICCV 2, Harigantars, ICCV 96029 and Annigeri) have been identified (Upadhyaya, Dwivedi, Gowda, & Singh, 2007).

Large Seed Size

In chickpea, seed size and colour are important traits for trade purposes. Large-seeded kabuli cultivars with a 100-seed weight of >40 g have higher consumer preference and fetch about three times higher price in the market. Evaluation of 65 large-seeded kabuli germplasm lines in three sets and across environments identified the six best large-seeded kabuli chickpea genotypes in three sets having high stability. One accession, ICC 14190, a *Fusarium* wilt–resistant large-seeded (37.4 g 100-seed weight) landrace from India, ranked first with average yield of 1430 kg/ha and high productivity (13.64 kg/ha/day). Three accessions, ICC 14194, ICC 7344 and ICC 7345, were early-flowering, extra-large-seeded types (48.2–54.1 g 100-seed weight), with grain yields similar to the best control, L 550. The other two superior lines were ICC 17452 (54.0 g 100-seed weight) and ICC 19189 (50.7 g 100-seed weight), both early-flowering, extra-large-seeded types with grain yield similar to the control KAK 2. All these accessions exhibited high stability with regression value near unity and deviation near zero (Gowda, Upadhyaya, Dronavalli, & Singh, 2011). Kaul, Kumar, and Gurha (2007) evaluated 150 kabuli chickpea germplasm accessions belonging to diverse geographical regions for phenological and morpho-agronomic traits at Kanpur, India, and identified four large-seeded kabuli accessions, ICC 12033, ICC 14199, ICC 14197 and ICC 14203 (46.2–60.2 g 100-seed weight and originating from Mexico) having high yield potential of >18 q/ha. In a similar study, nine large-seeded accessions (ICC 7345, ICC 11883, ICC 17450, ICC 17452,

ICC 17456, ICC 17457, ICC 18591, ICC 19189 and ICC 19195) having 100-seed weight ranging from 50.0 to 61.6 g and high yield (1154.4–1708.3 kg/ha) comparable to the control cultivar, KAK 2 (35.4 g 100-seed weight and 1359.5 kg/ha yield) have been identified for their use in developing new large-seeded kabuli cultivars with a broad genetic base (Kashiwagi et al., 2007).

Yield and Component Traits

Evaluation of the chickpea core collection for 14 agronomic traits identified 39 accessions (19 desi, 15 kabuli and 5 intermediate) performing better for a combination of agronomic traits such as early maturity, seed size and grain yield (Upadhyaya et al., 2007). The most desirable accessions having high seed yield and greater 100-seed weight than controls are ICC 1836 among the desi type and ICC 5644, ICC 7200, ICC 8042, ICC 10783 and ICC 11904 among the kabuli type; for early maturity and greater 100-seed weight than controls are ICC 6122, ICC 8474 and ICC 12197 in desi, ICC 8155, ICC 12034, ICC 14190 and ICC 14203 among kabuli type, and ICC 4871 among intermediate type (Upadhyaya et al., 2007). These accessions represent new and diverse sources of germplasm for use in breeding programmes to develop new chickpea cultivars. Meena et al. (2010) identified six promising and diverse accessions, ICC 14778, ICC 6279, ICC 4567, ICC 4533, ICC 1397 and ICC 12328, for more than one trait for use in chickpea improvement. Further, evaluation of the chickpea mini-core collection under three environments identified one accession, ICC 13124, promising for earliness, large seed size, and high yield per plant in all the three environments, and concluded that this accession is best suited for cultivation under both rain-fed and irrigated conditions during the post-rainy season (Parameshwarappa, Salimath, Upadhyaya, Patil, & Kajjidoni, 2011).

4.7 Limitations in Germplasm Use

Although plant breeders recognize the limitations of working with collections and the importance of crop genetic resources, yet they are often reluctant to use these resources for several reasons. The main reason for the low utilization of germplasm in crop improvement programmes is the lack of information on the large number of accessions, particularly for traits of economic importance such as yield, stable resistance/tolerance to biotic/abiotic stresses and nutrition-related traits, which often show high genotype × environment interactions and require replicated multilocational evaluation. However, the large size of germplasm collections makes it a costly and resource-demanding task. Another major reason for the low use of germplasm is the apprehensions among breeders about poor adaptability of germplasm and a linkage load of many undesirable genes associated especially with utilizing exotic germplasm and wild relatives in crop improvement programmes. While using unknown and wild germplasm, comparatively more effort and time is needed to generate breeding materials. Further, inadequate linkages between gene banks and germplasm users, lack of an informative and user-friendly gene bank database management system, restricted access to germplasm collections due to limited seed availability and regulations governing germplasm exchange are the important factors responsible for the low use of germplasm in chickpea improvement programmes.

4.8 Germplasm Enhancement Through Wide Crosses

The narrow genetic base of cultivated chickpea is one of the major limitations in improving chickpea production and productivity. Further, the global production is affected drastically by several biotic and abiotic constraints. Limited genetic variation present in the cultivated type of chickpea germplasm necessitates the utilization of wild *Cicer* species for germplasm enhancement. Wild *Cicer* species have been extensively screened and several of them have been reported to have very high levels of resistance/tolerance to many biotic and abiotic stresses, which includes resistance to *Ascochyta* blight (Collard, Ades, Pang, Brouwer, & Taylor, 2001; Croser, Ahmad, Clarke, & Siddique, 2003; Pande, Ramgopal, et al., 2006; Rao, Reddy, & Bramel, 2003; Singh, Hawtin, Nene, & Reddy, 1981; Singh & Reddy, 1993; Stamigna, Crino, & Saccardo, 2000), *Botrytis* gray mold (Pande, Ramgopal, et al., 2006; Rao et al., 2003; Stevenson & Haware, 1999), *Fusarium* wilt (Croser et al., 2003; Infantino, Porta-Puglia, & Singh, 1996; Rao et al., 2003), *Helicoverpa* pod borer (Sharma, Chen, & Muehlbauer, 2005), drought (Croser et al., 2003; Kashiwagi et al., 2005; Toker, Canci, & Yildirim, 2007), cold (Berger et al., 2012; Croser et al., 2003; Singh, Malhotra, & Saxena, 1990; Singh, Malhotra, & Saxena, 1995; Toker, 2005) and drought and heat (Canci and Toker, 2009). Besides resistant/tolerant sources, wild *Cicer* species harbour beneficial alleles/genes for high seed protein (Rao et al., 2003; Singh & Pundir, 1991) and improvement of agronomic traits in cultivated chickpea. Keeping in view the importance of wild *Cicer* species, most of the chickpea improvement programmes emphasize utilizing wild species to develop new cultivars with a broad genetic base. Of the eight annual wild *Cicer* species, only *C. reticulatum* is readily crossable with cultivated chickpea resulting in a fertile hybrid, whereas for exploitation of the remaining seven annual wild *Cicer* species for chickpea improvement, specialized techniques such as application of growth hormones, embryo rescue, ovule culture and other tissue culture techniques have been suggested by various researchers (Badami, Mallikarjuna, & Moss, 1997; Lulsdorf, Mallikarjuna, Clarke, & Tar'an, 2005; Mallikarjuna, 1999; Mallikarjuna & Jadhav, 2008). Utilization of the *C. reticulatum* accession ILWC 119 in a crossing programme has resulted in the development of two cyst–nematode-resistant chickpea germplasm lines: ILC 10765 and ILC 10766 (Malhotra, Singh, Vito, Greco, & Saxena, 2002). Promising high-yielding lines with good agronomic and seed traits, such as early flowering and high 100-seed weight, have also been obtained from crosses involving *C. reticulatum* and *C. echinospermum* with cultivated chickpea (Jaiswal, Singh, Singh, & Singh, 1986; Malhotra et al., 2003; Singh, Gumber, Joshi, & Singh, 2005; Singh, Jaiswal, Singh, & Singh, 1984; Singh & Ocampo, 1997; Upadhyaya, 2008). High-yielding cold-tolerant lines with high biomass have been obtained from *C. arietinum* × *C. echinospermum* crosses (ICARDA, 1995). Using various techniques, interspecific hybrids have been produced between *C. arietinum* and *C. judaicum* (Singh, Singh, Asthana, & Singh, 1999; Verma, Ravi, & Sandhu, 1995; Verma, Sandhu, Rrar, & Brar, 1990), *C. arietinum* × *C. pinnatifidum* (Badami et al., 1997; Mallikarjuna, 1999; Mallikarjuna & Jadhav, 2008; Verma et al., 1990), *C. arietinum* × *C. cuneatum* (Singh & Singh, 1989), and *C. arietinum* × *C. bijugum* (Singh et al., 1999; Verma et al., 1990) to exploit the possibility of introgression of desirable alien genes from these wild *Cicer* species into the cultivated chickpea. These interspecific hybrids have contributed significantly towards the development of genomic resources for chickpea improvement. From *C. arietinum* × *C. judaicum* cross, a pre-breeding line IPC 71 having a high number of primary branches, more pods per plant and green seeds has been developed for use in chickpea improvement programmes (Chaturvedi & Nadarajan, 2010).

4.9 Chickpea Genomic Resources

Average chickpea productivity is less than 1 t ha^{-1}, which is much less than its potential, 6 t ha^{-1} (Singh, 1985). Biotechnological tools can help to increase chickpea productivity by using the marker-assisted selection (MAS) approach in breeding programmes (Varshney, Graner, & Sorrells, 2005; Varshney, Nayak, May, & Jackson, 2009). Trait mapping provides the first step to employ MAS in breeding programmes. Recent developments in genomics technology have helped to explain the mechanism of complex traits controlling chickpea productivity and the genetic architecture of traits of economic importance to accelerate breeding programmes. A number of marker-trait associations have been identified in chickpea along with the dense genetic maps which have allowed MAS to become a routine in breeding programmes (Kulwal, Thudi, & Varshney, 2011; Varshney, Hoisington, & Tyagi, 2006). A huge amount of genomic and genetic resources developed by ICRISAT in collaboration with partners have regularly been used in accelerating the genomic and breeding application to increase chickpea productivity. Since 2005, ICRISAT has regularly been focussing on the development of molecular markers, construction of comprehensive genetic and consensus maps, identification of marker-trait associations and Quantitative Trait Loci (QTLs), and initiation of molecular breeding for various disease resistance and drought tolerance in chickpea.

4.9.1 Molecular Markers and Genotyping Platforms

A number of marker systems have been introduced recently, such as hybridization-based diversity arrays technology (DArT) and sequence-based single nucleotide polymorphism (SNP) markers. These marker systems can easily be automated and provide medium- to high-throughput genotyping. Still, microsatellite (SSR) markers are the marker of choice for geneticist and breeders. SSRs are highly polymorphic, multi-allelic and codominant in nature; therefore suitable for genotyping the germplasm with a narrow genetic base and for segregating populations (Gupta & Varshney, 2000). Development of SSRs was mainly dependent on the screening of size-selected genomic and cDNA libraries, but recently *in silico* approaches of mining the expressed sequence tags (ESTs) and Bacterial Artificial Chromosome (BAC)-end sequences have also become popular for the identification of genic SSRs (Varshney, Glaszmann, Leung, & Ribaut, 2010). To supplement the chickpea genomics, more than 2000 SSR markers (Table 4.5) have been developed in the past few years using various approaches including genomic DNA libraries (Gaur et al., 2011; Nayak et al., 2010), cDNA libraries (Varshney, Hiremath, et al., 2009) and 454/FLX transcript reads (Garg, Patel, Tyagi, & Jain, 2011; Garg, Patel, Jhanwar, et al., 2011; Hiremath et al., 2011). On the other hand, a new set of 487 functional markers including EST-SSRs, Intron-targeted primers (ITP), expressed sequence tag polymorphisms and SNPs have been developed by the National Institute of Plant Genome Research (NIPGR), New Delhi, India (Choudhary, Gaur, Gupta, & Bhatia, 2012).

ICRISAT in collaboration with DArT Pty Ltd., Australia, has also developed another marker resource namely DArT arrays representing 15,360 features (Table 4.5) for chickpea (Varshney et al., 2010). This set has regularly been used for diversity studies and saturating linkage maps (Thudi et al., 2011). These arrays showed very little polymorphism when screened on the elite chickpea germplasm (Thudi et al., 2011), and the parental genotypes of mapping populations showed only 35% polymorphism when screened with these DArT arrays. This suggests that DArT arrays are not cost-effective to screen the cultivated chickpea germplasm. Another type of marker system, SNP, is gaining popularity in several crop species due to its genome-wide distribution, abundance, flexibility of automation and amenability to high throughput. For identification of SNP, three different approaches were used. First, RNA sequencing approach was used to sequence the parents of mapping population. Alignment of these short reads led to identification of thousands of SNPs. The second approach focussed on the allele-specific sequencing of parental genotypes using conserved orthologous sequence markers and led to identification of 768 SNPs (Table 4.5). In the third approach, 220 candidate genes were sequenced on 2–20 genotypes and 1893 SNP were identified based on allele-specific sequencing (Gujaria et al., 2011). In total, a large number of SNPs were identified and made available for use in chickpea improvement. To use these SNPs in breeding programmes and other applications, selection of an appropriate genotyping platform is very important. University of California – Davis in collaboration with its partners has developed Illumina GoldenGate assays for 768 SNPs. These GoldenGate assays are cost-effective only when dealing with large number of SNPs to genotype a large number of samples. However, where fewer markers are required for genotyping, another genotyping platform, BeadXpress based on VeraCode technology, suits well. Therefore, VeraCode assay for 96-plex SNP (Table 4.5) has been developed at ICRISAT to be used on Illumina's BeadXpress system (R. K. Varshney, unpublished data). Another SNP genotyping platform, KASPar, developed by KBiosciences (www.kbioscience.co.uk), provides a flexible and cost-effective assay for SNP genotyping. ICRISAT has developed 2068 KASPar assays (Table 4.5) in chickpea (Hiremath et al., 2012).

In recent years, next-generation sequencing (NGS) technologies have been adapted by researchers to produce a huge amount of sequencing data at very low cost and in less time. In chickpea, two NGS approaches 454 and Illumina were used for

Table 4.5 Summary of Genomic Resources in Chickpea Developed at ICRISAT, India

Resource	Number
SSRs	Approx. 2000
SNPs	9000
DArTs	15,360
GoldenGate assays	768 SNPs
KASPar assays	2068 SNPs
VeraCode assays	96 SNPs
Sanger ESTs	Approx. 30,000
454/FLX reads	435,018
TUSs	103,215
Illumina reads (million reads)	>108 (4 parents)

characterization of the chickpea transcriptome. Sanger sequencing was used to generate the EST from drought- and salinity-stress-challenged cDNA libraries. 454/FLX sequencing was undertaken to generate 435,018 transcript reads (Table 4.5), which were used along with the Sanger ESTs to improve the chickpea transcript assembly (Hiremath et al., 2011). In a similar study, National Institute of Plant Genome Research (NIPGR) generated a hybrid assembly with 34,760 tentative consensus sequences (Garg, Patel, Jhanwar, et al., 2011). Recently, a transcriptome of a wild chickpea, *C. reticulatum* (genotype PI 489777) with 37,265 *C. reticulatum* tentative consensus (CrTC) was reported using GS-FLX Roche 454 NGS technology (Jhanwar et al., 2012). Previously, the higher cost and need for time and expertise were the main constraints in whole-genome sequencing, but recent advancements in NGS technologies have allowed initiating genome sequencing at very low cost and less time. Very recently, ICRISAT in collaboration with Beijing Genomics Institute (BGI), Shenzhen, China and other international collaborators reported the draft whole-genome shotgun sequence of CDC Frontier, a kabuli chickpea variety (Varshney et al., 2013). Along with the genome sequence, resequencing of 90 cultivated and wild chickpea accessions has also been reported. An effort to sequence ICC 4958, a desi landrace, has also been initiated at NIPGR, New Delhi. These resources can be used for chickpea improvement through molecular breeding and to explain chickpea genome diversity and domestication events.

4.9.2 Genetic Maps and Trait Mapping

A first step in crop improvement using molecular breeding/genomics-assisted breeding is the discovery of marker-trait association between the trait of interest and a genetic marker. However, QTL analysis has suffered severely from the lack of saturated genetic maps. Large-scale genomic resources developed by ICRISAT and partners during the last 5 years have been used for the construction of comprehensive/consensus genetic maps in chickpea. An interspecific reference mapping population has been developed from a cross, ICC 4958×PI 489777 and used for generating genetic maps (Upadhyaya, Thudi, et al., 2011). The first genetic map in chickpea was developed on this reference population (ICC 4958×PI 489777) using markers like Random Amplified Polymorphic DNA (RAPD), Amplified Fragment Length Polymorphism (AFLP) and very few SSR markers. To saturate this map, a high-density chickpea genetic map with 1291 loci has been developed by Thudi et al. (2011). This map comprises a range of markers starting from BES-SSRs (157), genic molecular markers (145), DArT (621) and earlier published legacy markers (368), spanning a total of 845.56 cM across eight linkage groups (LG) with an average marker distance of 0.65 cM. The number of markers on each LG ranged from 68 (LG 8) to 219 (LG 3). Genetic maps constructed using the gene-based markers are referred to as *transcript maps*. In chickpea, a transcript map with 126 genic molecular markers, including 53 CAPS-SNPs, 55 EST-SSRs and 18 CISR loci has been developed (Gujaria et al., 2011). In another study using the same reference population, an advanced linkage map spanning 1497.7 cM with 406 loci including 177 gene-based markers and 126 genomic SSRs (gSSRs) has been developed (Choudhary et al., 2012). Recently, KASPar assays have been adopted for SNP genotyping and been used to develop a second-generation genetic map with 1328 loci including 625 Chickpea KASPar Assay Markers (CKAMs), 314 TOG-SNPs and 389 already published markers with an average intermarker distance of 0.59 cM (Hiremath et al., 2012).

Besides interspecific mapping populations, several intraspecific mapping populations have also been developed to identify the markers associated with *Fusarium* wilt (Sharma, Winter, Kahl, & Muehlbauer, 2004; Sharma et al., 2005), *Ascochyta* blight (Anbessa, Taran, Warkentin, Tullu, & Vandenberg, 2009; Iruela et al., 2007) and drought. For drought tolerance in chickpea, ICRISAT has developed two intraspecific mapping populations (ICC 4958×ICC 1882 and ICC 283×ICC 8261) (Chamarthi et al., 2011). Both populations were used for the construction of SSR-based genetic maps comprising 240 and 170 loci, respectively. QTL analysis using the extensive phenotyping data revealed a genomic region that harbours QTLs for several root-related and other drought tolerance–related traits contributing approximately 35% of the phenotyping variation. Therefore, this genomic region has been targeted for introgression in elite chickpea lines to enhance drought tolerance using the marker-assisted backcross (MABC) approach.

4.9.3 Molecular Breeding

Once the QTLs for trait of interest are identified, the next step is to use this information in a crop improvement programme using genomic-assisted breeding for developing superior lines with better response to stress and high yield. With the recent development in NGS technology, it has become common practice to use molecular markers for phenotype prediction and selection of progenies for the next generation in breeding (Varshney et al., 2012). Several genomics-assisted breeding approaches, namely MABC, marker-assisted recurrent selection (MARS) and genomic selection have regularly been used in crop improvement programmes. MABC focusses on the introgression of the QTL and/or genomic region associated with the trait(s) of interest from a donor parent into an elite recurrent parent using molecular markers

(Hospital, 2005). This approach leads to the generation of near-isogenic lines (NILs) containing only the major gene/QTL from the donor parent, while retaining the whole genome of the recurrent parent (Gupta, Kumar, Mir, & Kumar, 2010). MABC can also be used for gene pyramiding, where different genes for the same trait or for different traits are accumulated in one background.

In chickpea, ICRISAT has been working on two MABC programmes. The first initiative, supported by the CGIAR GCP and the Bill & Melinda Gates Foundation, focusses on improved drought response in elite chickpea lines. Efforts have been made to introgress the genomic region harbouring QTLs for several drought-related traits into JG 11 genetic background from the germplasm accession, ICC 4958. BC_3F_4 lines have been generated and were evaluated under both rain-fed and irrigated conditions in India, Ethiopia and Kenya in the main crop season during 2011–2012. Results of the first-year field trial were very encouraging: the BC lines possessed the RLD of the donor parent with the seed quality and yield of the recipient parents. BC lines showed 6–11% higher yield in the rain-fed condition, while in the irrigated condition, the gains were up to 24%. The success story of JG 11 inspired several institutes, such as Indian Institute of Pulses Research (IIPR), Kanpur and Indian Agricultural Research Institute (IARI), New Delhi from India, Egerton University, Kenya, and Ethiopian Institute of Agricultural Research (EIAR), Ethiopia, to start MABC programmes for introgressing this genomic region from ICC 4958 into the leading varieties of different regions.

In an another initiative, sponsored by the Department of Biotechnology (DBT), Government of India, ICRISAT in collaboration with Jawaharlal Nehru Krishi Vishwavidyalaya (JNKVV) of Jabalpur, Mahatma Phule Krishi Vidyapeeth (MPKV) of Rahuri and ARS-Gulbarga has been working on gene pyramiding of resistance to two races (foc1 and foc3) for *Fusarium* wilt (FW) and two QTLs conferring resistance to *Ascochyta* blight (AB). Efforts have been initiated for introgression of resistance to FW from WR 315 and resistance to AB from ILC 3279 into elite chickpea cultivars (C 214, JG 74, Pusa 256, Phule G12 and Annigeri-1) from different agro-climatic zones through MABC. Presently, homozygous BC_3F_4 lines are available for preliminary evaluation for resistance to FW and AB.

4.10 Conclusions

The presence of enormous genetic variation and the means to exploit such variability is the key to success of crop improvement programmes. Large collections of chickpea germplasm comprising landraces and wild *Cicer* species have been conserved in various gene banks worldwide, representing a large spectrum of diversity in the genus *Cicer*. Development and evaluation of small subsets such as core and mini-core collections have resulted in the identification of trait-specific germplasm accessions for important abiotic and biotic stresses as well as for agronomic and nutrition-related traits, which results in the enhanced utilization of genetic resources for developing broad-based climate-resilient chickpea cultivars. Besides cultivated type germplasm, new sources of variability for traits of interest exists in wild *Cicer* gene pools, which can be exploited using widespread hybridization techniques. Promising lines having resistance genes and good agronomic performance have been developed from crosses involving cultivated and wild *Cicer* species. Further, recent advances in plant biotechnology in combination with the traditional breeding approaches, coupled with genomics and transgenic technologies, provide new tools to exploit the genes locked up in cross-incompatible secondary and tertiary gene pools. The availability of genomic resources such as the development of molecular markers, genetic and physical maps and the generation of expressed sequenced tags (ESTs), genome sequencing and association studies revealing marker-trait associations has facilitated the identification of QTLs and discovery of genes associated with tolerance/resistance to abiotic and biotic stresses including agronomic traits. These advancements in chickpea genomic resources can assist in identifying and tracking allelic variants associated with beneficial traits and identifying desirable recombinant plants with the markers of interest, which will accelerate the chickpea improvement programmes.

References

Abbo, S., Berger, J., & Turner, N. C. (2003). Evolution of cultivated chickpea: Four bottlenecks limit diversity and constrain adaptation. *Functional Plant Biology*, *30*, 1081–1087.

Ahmad, Z., Haqqani, A. M., Mohmand, A. S., Hashmi, N. I., Bhatti, M. S., & Malik, B. A. (1984). Collection of natural genetic variability of chickpea, lentil and other crops along the Indus. *Pakistan Journal of Agriculture Research*, *5*, 47–50.

Akhalkatsi, M., Ekhvaia, J., & Asanidze, Z. (2012). Diversity and genetic erosion of ancient crops and wild relatives of agricultural cultivars for food: Implications for nature conservation in Georgia (Caucasus), perspectives on nature conservation – patterns, pressures and prospects, J. Tiefenbacher (Ed.), ISBN: 978-953-51-0033-1, InTech. Available from <http://www.intechopen.com/books/perspectives-on-nature-conservation-patterns-pressures-andprospects/diversity-and-genetic-erosion-of-ancient-crops-and-wild-relatives-of-agricultural-cultivars-for-food/>.

Anbessa, Y., Taran, B., Warkentin, T. D., Tullu, A., & Vandenberg, A. (2009). Genetic analysis and conservation of QTL for *Ascochyta* blight resistance in chickpea. *Theoretical and Applied Genetics*, *119*, 757–765.

Badami, P. S., Mallikarjuna, N., & Moss, J. P. (1997). Interspecific hybridization between *Cicer arietinum* and *C. Pinnatifidum*. *Plant Breeding*, *116*, 393–395.

Berger, J., Abbo, S., & Turner, N. C. (2003). Ecogeography of annual wild *Cicer* species: The poor state of the world collection. *Crop Science*, *43*, 1076–1090.

Berger, J. D., Kumar, S., Nayyar, H., Street, K. A., Sandhu, J. S., Henzell, J. M., et al. (2012). Temperature-stratified screening of chickpea (*Cicer arietinum* L.) genetic resource collections reveals very limited reproductive chilling tolerance compared to its annual wild relatives. *Field Crops Research*, *126*, 119–129.

Biabani, A., Carpenter-Boggs, L., Coyne, C. J., Taylor, L., Smith, J. L., & Higgins, S. (2011). Nitrogen fixation potential in global chickpea mini-core collection. *Biology and Fertility of Soils*, *47*, 679–685.

Brown, A. H. D. (1989a). Core collections: A practical approach to genetic resources management. *Genome*, *31*, 818–824.

Brown, A. H. D. (1989b). The case for core collections. In A. H. D. Brown, O. H. Frankel, D. R. Marshall, & J. T. Williams (Eds.), *The use of plant genetic resources* (pp. 136–155). Cambridge, UK: Cambridge University Press.

Canci, H., & Toker, C. (2009). Evaluation of annual wild *Cicer* species for drought and heat resistance under field conditions. *Genetic Resources and Crop Evolution, 56*, 1–6.

Chamarthi, S. K., Kumar, A., Vuong, T. D., Blair, M. W., Gaur, P. M., Nguyen, H. T., et al. (2011). Trait mapping and molecular breeding. In A. Pratap & J. Kumar (Eds.), *Biology and breeding of food legumes* (pp. 296–313). CABI International.

Chandel, K. P. S. (1984). A note on the occurrence of wild *Cicer microphyllum* Benth. and its nutrient status. *International Chickpea Newsletter, 10*, 4–5.

Chaturvedi, S. K., & Nadarajan, N. (2010). Genetic enhancement for grain yield in chickpea – accomplishments and resetting research agenda. *Electronic Journal of Plant Breeding, 14*, 611–615.

Choudhary, S., Gaur, R., Gupta, S., & Bhatia, S. (2012). EST-derived genic molecular markers: Development and utilization for generating an advanced transcript map of chickpea. *Theoretical and Applied Genetics, 124*, 1449–1462.

Collard, B. C. Y., Ades, P. K., Pang, E. C. K., Brouwer, J. B., & Taylor, P. W. J. (2001). Prospecting for sources of resistance to *Ascochyta* blight in wild *Cicer* species. *Australasian Plant Pathology, 30*, 271–276.

Croser, J. S., Ahmad, F., Clarke, H. J., & Siddique, K. H. M. (2003). Utilization of wild *Cicer* in chickpea improvement – progress, constraints and prospects. *Australian Journal of Agricultural Research, 54*, 429–444.

Cubero, J.I. (1975). The research on chickpea (*Cicer arietinum*) in Spain. In *Proceedings of the international workshop on grain legumes, 13–16 January 1975* (pp. 17–122). Hyderabad, India: International Crops Research Institute for the Semi-Arid Tropics.

Dekaprelevich, L., & Menabde, V. (1929). K izucheniu polevykh kultur zapadnoi Gruzii. Racha, I (Study of cereal cultivars in Georgia. I. Racha). *Scientific Papers of the Applied Sections of the Tbilisi Botanical Garden, 6*, 219–252. (in Russian).

Erler, F., Ceylan, F., Erdemir, T., & Toker, C. (2009). Preliminary results on evaluation of chickpea, *Cicer arietinum*, genotypes for resistance to the pulse beetle, Callosobruchus maculates. *Journal of Insect Science, 9*, 58–61.

FAOSTAT, (2010). The second report on the state of the world's plant genetic resources for food and agriculture: *Commission on genetic resources for food and agriculture (CGRFA), food and agriculture organization of the United Nations*. Rome, Italy: FAO. ISBN: 978-92-5-106534-172.

FAOSTAT. (2011). <http://faostat.fao.org/site/567/desktopdefault.aspx?pageid=567/>.

Frankel, O. H. (1984). Genetic perspective of germplasm conservation. In W. Arber, K. Limensee, W. J. Peacock, & P. Stralinger (Eds.), *Genetic manipulations: Impact on man and society* (pp. 161–170). Cambridge, UK: Cambridge University Press.

Frankel, O. H., & Brown, A. H. D. (1984). Plant genetic resources today: A critical appraisal. In J. H. W. Holden & J. T. Williams (Eds.), *Crop genetic resources: Conservation and evaluation* (pp. 249–257). London: George Allen and Unwin Ltd.

Garg, R., Patel, R. K., Jhanwar, S., Priya, P., Bhattacharjee, A., Yadav, G., et al. (2011). Gene discovery and tissue-specific transcriptome analysis in chickpea with massively parallel pyrosequencing and web resource development. *Plant Physiology, 156*, 1661–1678.

Garg, R., Patel, R. K., Tyagi, A. K., & Jain, M. (2011). De novo assembly of chickpea transcriptome using short reads for gene discovery and marker identification. *DNA Research, 18*, 53–63.

Gaur, R., Sethy, N. K., Choudhary, S., Shokeen, B., Gupta, V., & Bhatia, S. (2011). Advancing the STMS genomic resources for defining new locations on the intraspecific genetic linkage map of chickpea (*Cicer arietinum* L.). *BMC Genomics*, *12*, 117.

Gowda, C. L. L., Upadhyaya, H. D., Dronavalli, N., & Singh, S. (2011). Identification of large-seeded high-yielding stable Kabuli Chickpea germplasm lines for use in crop improvement. *Crop Science*, *51*, 198–209.

Gujaria, N., Kumar, A., Dauthal, P., Dubey, A., Hiremath, P., Bhanu Prakash, A., et al. (2011). Development and use of genic molecular markers (GMMs) for construction of a transcript map of chickpea (*Cicer arietinum* L.). *Theoretical and Applied Genetics*, *122*, 1577–1589.

Gupta, P. K., Kumar, J., Mir, R. R., & Kumar, A. (2010). Marker-assisted selection as a component of conventional plant breeding. *Plant Breeding Revolution*, *33*, 145–217.

Gupta, P. K., & Varshney, R. K. (2000). The development and use of microsatellite markers for genetic analysis and plant breeding with emphasis on bread wheat. *Euphytica*, *13*, 163–185.

Hannan, R. M., Kaiser, W. J., & Muehlbauer, F. J. (1994). Development and utilization of the USDA chickpea germplasm core collection. *Agronomy Abstracts*, 219.

Harlan, J. R., & de Wet, J. M. T. (1971). Toward a rational classification of cultivated plants. *Taxonomy*, *20*, 509–517.

Hiremath, P. J., Farmer, A., Cannon, S. B., Woodward, J., Kudapa, H., Tuteja, R., et al. (2011). Large-scale transcriptome analysis in chickpea (*Cicer arietinum* L.), an orphan legume crop of the semi-arid tropics of Asia and Africa. *Plant Biotechnology Journal*, *9*, 922–931.

Hiremath, P. J., Kumar, A., Penmetsa, R. V., Farmer, A., Schlueter, J. A., Chamarthi, S. K., et al. (2012). Large-scale development of cost-effective SNP marker assays for diversity assessment and genetic mapping in chickpea and comparative mapping in legumes. *Plant Biotechnology Journal*, *10*, 716–732.

Hospital, F. (2005). Selection in backcross programme. *Philippines Trans Research Society*, *360*, 1503–1511.

IBPGR, ICRISAT, & ICARDA. (1993). *Descriptors for chickpea (Cicer arietinum L.)* (p. 31). Rome, Italy: International Board for Plant Genetic Resources (IBPGR); Patancheru, Andhra Pradesh, India: International Crops Research Institute for the Semi-Arid Tropics (ICRISAT); Aleppo, Syria: International Center for Agricultural Research in the Dry Areas.

ICARDA. (1995). *Annual report for 1995. Germplasm program legumes* (p. 210). Aleppo, Syria: ICARDA.

ICRISAT Archival Report. (2009). <http://intranet/ddg/Admin%20Pages2009/Archival_Report_2009.aspx/>.

ICRISAT Archival Report. (2010). <http://intranet/ddg/Admin%20Pages2009/Archival_Report_2010.aspx/>.

Infantino, A., Porta-Puglia, A., & Singh, K. B. (1996). Screening of wild *Cicer* species for resistance to *Fusarium* wilt. *Plant Disease*, *80*, 42–44.

Iruela, M., Castro, P., Rubio, J., Cubero, J. I., Jacinto, C., Milla, T., et al. (2007). Validation of a QTL for resistance to *Ascochyta* blight linked to resistance to *Fusarium* wilt race 5 in chickpea (*Cicer arietinum* L.). *European Journal of Plant Pathology*, *119*, 29–37.

Jaiswal, H. K., Singh, B. D., Singh, A. K., & Singh, R. M. (1986). Introgression of genes for yield and yield traits from *C. reticulatum* into *C. arietinum*. *International Chickpea Newsletter*, *14*, 5–8.

Jhanwar, S., Priya, P., Garg, R., Parida, S. K., Tyagi, A. K., & Jain, M. (2012). Transcriptome sequencing of wild chickpea as a rich resource for marker development. *Plant Biotechnology Journal*, *10*, 690–702.

Jukanti, A. K., Gaur, P. M., Gowda, C. L. L., & Chibbar, R. N. (2012). Nutritional quality and health benefits of chickpea (*Cicer arietinum* L.): A review. *Breeding Journal of Nutrition, 108*(Suppl. 1), S11–S26.

Kannenberg, L. W., & Falk, D. E. (1995). Models for activation of plant genetic resources for crop breeding programs. *Canadian Journal of Plant Science, 75*, 45–53.

Kashiwagi, J., Krishnamurthy, L., Singh, S., & Upadhyaya, H. D. (2006). Variation of SPAD chlorophyll meter readings (SCMR) in the mini-core germplasm collection of chickpea. *Journal of SAT Agricultural Research, 2*, 1–3.

Kashiwagi, J., Krishnamurthy, L., Upadhyaya, H. D., & Gaur, P. M. (2008). Rapid screening technique for canopy temperature status and its relevance to drought tolerance improvement in chickpea. *Journal of SAT Agricultural Research, 6*, 4–105. ISSN 0973-3094.

Kashiwagi, J., Krishnamurthy, L., Upadhyaya, H. D., Krishna, H., Chandra, S., Vadez, V., et al. (2005). Genetic variability of drought-avoidance root traits in the mini-core germplasm collection of chickpea (*Cicer arietinum* L.). *Euphytica, 146*, 213–222.

Kashiwagi, J., Upadhyaya, H. D., Krishnamurthy, L., & Singh, S. (2007). Identification of high-yielding large-seeded kabuli chickpeas with drought avoidance root traits. *Journal of SAT Agricultural Research, 3*, 1–2. ISSN 0973-3094.

Kashiwagi, J., Upadhyaya, H. D., & Krishnamurthy, L. (2010). Significance and genetic diversity of SPAD chlorophyll meter reading in chickpea germplasm in the semi-arid environments. *Journal of Food Legumes, 23*, 99–105. ISSN 0970-6380.

Kaul, J., Kumar, S., & Gurha, S. N. (2007). Evaluation of exotic germplasm of kabuli chickpea. *Indian Journal of Plant Genetic Resources, 20*, 160–164.

Kazan, K., & Muehlbauer, F. J. (1991). Allozyme variation and phylogeny in annual species of *Cicer* (*Leguminosae*). *Plant Systematics Evolution, 175*, 11–21.

Kibret, K. T. (2011). Development and utilization of genetic diversity based ethiopian chickpea (*Cicer arietinum* L.) germplasm core collection for association mapping. PhD thesis, Hyderabad, India, International Crops Research Institute for the Semi-Arid Tropics.

Kobakhidze, A. (1974). Sakartvelos samartsvle-parkosan mtsenareta botanikur-sistematikuri shestsavlisatvis (Botanical–systematic study of cereals in Georgia). In N. Ketskhoveli (Ed.), *Botanika (Botany)* (pp. 58–190). Metsniereba, Tbilisi (in Georgian).

Krishnamurthy, L., Gaur, P. M., Basu, P. S., Chaturvedi, S. K., Tripathi, S., Vadez, V., et al. (2011). Large genetic variation for heat tolerance in the reference collection of chickpea (*Cicer arietinum* L.) germplasm. *Plant Genetic Resources, 9*, 59–69.

Krishnamurthy, L., Kashiwagi, J., Gaur, P. M., Upadhyaya, H. D., & Vadez, V. (2010). Sources of tolerance to terminal drought in the chickpea (*Cicer* arietinum L.) minicore germplasm. *Field Crops Research, 119*, 322–330.

Krishnamurthy, L., Turner, N. C., Gaur, P. M., Upadhyaya, H. D., Varshney, R. K., Siddique, K. H. M., et al. (2011). Consistent Variation across Soil Types in Salinity Resistance of a Diverse Range of Chickpea (*Cicer* arietinum L.) Genotypes. *Journal of Agronomy and Crop Science, 197*(3), 214–227.

Kulwal, P. L., Thudi, M., & Varshney, R. K. (2011). Genomics interventions in crop breeding for sustainable agriculture. In *Encyclopaedia of sustainability science and technology*. New York: Springer. http://dx.doi.org/10.1007/978-1-4419-0851-3.

Kumar, S., Gupta, S., Chandra, S., & Singh, B. B. (2004). How wide is genetic base of pulse crops?. In M. Ali, B. B. Singh, S. Kumar, & V. Dhar (Eds.), *Pulses in new perspective* (pp. 211–221). Kanpur: Indian Society of Pulses Research and Development, Indian Institute of Pulses Research. Proceedings of the national symposium on Crop diversification and natural resources management. 20–22 December 2003.

Kumar, J., & van Rheenen, H. A. (2000). A major gene for time of flowering in chickpea. *Journal of Heredity*, *91*, 67–68.

Kupicha, F. K. (1981). Vicieae. In R. M. Polhill & P. H. Raven (Eds.), *Advances in legume systematics* (pp. 377–381). Kew: *Royal Botanic Gardens*.

Lulsdorf, M., Mallikarjuna, N., Clarke, H., & Tar'an, B. (2005). Finding solutions for interspecific hybridization problems in chickpea (*Cicer arietinum* L.). In *4th international food legumes research conference 18–22 October* (p. 44). New Delhi, India.

Malhotra, R. S., Baum, M., Udupa, S. M., Bayaa, B., Kabbabe, S., & Khalaf, G. (2003). *Ascochyta* blight resistance in chickpea: Present status and future prospects. In R. N. Sharma, G. K. Shrivastava, A. L. Rathore, M. L. Sharma, & M. A. Khan (Eds.), *Chickpea research for the millennium* (pp. 108–117). Raipur, Chhattisgarh, India: Indira Gandhi Agriculture University. International chickpea conference, 20–22 January 2003.

Malhotra, R. S., Singh, K. B., Vito, M., Greco, N., & Saxena, M. C. (2002). Registration of ILC 10765 and ILC 10766 chickpea germplasm lines resistant to cyst nematode. *Crop Science*, *42*, 1756.

Mallikarjuna, N. (1999). Ovule and embryo culture to obtain hybrids from interspecific incompatible pollinations in chickpea. *Euphytica*, *110*, 1–6.

Mallikarjuna, N., & Jadhav, D. R. (2008). Techniques to produce hybrids between *Cicer arietinum* L. × *Cicer pinnatifidum* Jaub. *Indian Journal of Genetics*, *68*, 398–405.

McIntosh, M., & Miller, C. (2001). A diet containing food rich in soluble and insoluble fiber improves glycemic control and reduces hyperlipidemia among patients with type 2 diabetes mellitus. *Nutrition Reviews*, *59*, 52–55.

Meena, H. P., Kumar, J., Upadhyaya, H. D., Bharadwaj, C., Chauhan, S. K., Verma, A. K., et al. (2010). Chickpea mini core germplasm collection as rich sources of diversity for crop improvement. *Journal of SAT Agricultural Research*, *8*, 1–5.

Muehlbauer, F. J. (1993). Use of wild species as a source of resistance in cool-season food legume crops. In K. B. Singh & M. C. Saxena (Eds.), *Breeding for stress tolerance in cool-season food legumes* (pp. 359–372). Chichester, UK: John Wiley & Sons.

Muehlbauer, F. J., Kaiser, W. J., & Simon, C. J. (1994). Potential for wild species in cold season food legume breeding. *Euphytica*, *73*, 109–114.

Mulwa, R. M. S., Kimurto, P. K., & Towett, B. K. (2010). Evaluation and selection of drought and pod borer (*Helicoverpa armigera*) tolerant chickpea genotypes for introduction in semi-arid areas of Kenya. In *Second RUFORUM biennial meeting, 20–24 September 2010*. Entebbe, Uganda.

Nayak, S. N., Zhu, H., Varghese, N., Datta, S., Choi, H. K., Horres, R., et al. (2010). Integration of novel SSR and gene-based SNP marker loci in the chickpea genetic map and establishment of new anchor points with *Medicago truncatula* genome. *Theoretical and Applied Genetics*, *120*, 1415–1441.

Pande, S., Kishore, G. K., Upadhyaya, H. D., & Rao, J. N. (2006). Identification of multiple diseases resistance in mini-core collection of chickpea. *Plant Disease*, *90*, 1214–1218.

Pande, S., Ramgopal, D., Kishore, G. K., Mallikarjuna, N., Sharma, M., Pathak, M., et al. (2006). Evaluation of wild *Cicer* species for resistance to *Ascochyta* blight and *Botrytis* gray mold in controlled environment at ICRISAT, Patancheru, India. *SAT eJournal*, ICRISAT

Parameshwarappa, S. G., & Salimath, P. M. (2008). Field screening of chickpea genotypes for drought resistance. *Karnataka Journal of Agricultural Science*, *21*, 113–114.

Parameshwarappa, S. G., Salimath, P. M., Upadhyaya, H. D., Patil, S. S., Kajjidoni, S. T., & Patil, B. C. (2010). Characterization of drought tolerant accessions identified from the minicore of chickpea (*Cicer arietinum* L.). *Indian Journal of Genetics and Plant Breeding*, *70*, 125–131. ISSN 0019-5200.

Parameshwarappa, S. G., Salimath, P. M., Upadhyaya, H. D., Patil, S. S., & Kajjidoni, S. T. (2011). Genetic divergence under three environments in a minicore collection of chickpea (*Cicer* arietinum L.). *Indian Journal of Plant Genetic Resources*, *24*, 177–185. ISSN 0971-8184.

Pittaway, J. K., Ahuja, K. D., Cehun, M., Chronopoulos, A., Robertson, I. K., Nestel, P. J., et al. (2006). Dietary supplementation with chickpeas for at least 5 weeks results in small but significant reductions in serum total and low-density lipoprotein cholesterols in adult women and men. *Annals of Nutrition and Metabolism*, *50*, 512–518.

Pouresmael, M., Akbari, M., Vaezi, S., & Shahmoradi, S. (2009). Effects of drought stress gradient on agronomic traits in Kabuli chickpea core collection. *Iranian Journal of Crop Sciences*, *11*, 308–324.

Queiroz, K. S., de Oliveira, A. C., & Helbig, E. (2002). Soaking the common bean in a domestic preparation reduced the contents of raffinose-type oligosaccharides but did not interfere with nutritive value. *Journal of Nutritional Science and Vitaminology*, *48*, 283–289.

Rao, N. K., Reddy, L. J., & Bramel, P. J. (2003). Potential of wild species for genetic enhancement of some semi-arid food crops. *Genetic Resources and Crop Evolution*, *50*, 707–721.

Serraj, R., Krishnamurthy, L., & Upadhyaya, H. D. (2004). Screening chickpea minicore germplasm for tolerance to soil salinity. *International Chickpea and Pigeonpea Newsletter*, *11*, 29–32.

Sharma, K. D., Chen, W., & Muehlbauer, F. J. (2005). Genetics of chickpea resistance to five races of *Fusarium* wilt and a concise set of race differentials for *Fusarium oxysporum* f. sp. *ciceris*. *Plant Disease*, *89*, 385–390.

Sharma, K. D., Winter, P., Kahl, G., & Muehlbauer, F. J. (2004). Molecular mapping of *Fusarium oxysporum* f. sp. *Ciceris* race 3 resistance gene in chickpea. *Theoretical and Applied Genetics*, *108*, 1243–1248.

Singh, B. D., Jaiswal, H. K., Singh, R. M., & Singh, A. K. (1984). Isolation of early flowering recombinants from the interspecific cross between *Cicer arietinum* and *C. reticulatum*. *International Chickpea Newsletter*, *11*, 14.

Singh, K. B., Hawtin, G. C., Nene, Y. L., & Reddy, M. V. (1981). Resistance in chickpeas to *Ascochyta rabiei*. *Plant Disease*, *65*, 586–587.

Singh, K. B., Malhotra, R. S., & Saxena, M. C. (1990). Sources for tolerance to cold in *Cicer* species. *Crop Science*, *30*, 1136–1138.

Singh, K. B., Malhotra, R. S., & Saxena, M. C. (1995). Additional sources of tolerance to cold in cultivated and wild *Cicer* species. *Crop Science*, *35*, 1491–1497.

Singh, K. B., & Ocampo, B. (1997). Exploitation of wild *Cicer* species for yield improvement in chickpea. *Theoretical and Applied Genetics*, *95*, 418–423.

Singh, K. B., Pundir, R. P. S., Robertson, L. D., van Rheene, H. A., Singh, U., Kelley, T. J., et al. (1997). Chickpea. In D. Fuccillo, L. Sears, & P. Stapleton (Eds.), *Biodiversity in trust* (pp. 100–113). Cambridge University Press.

Singh, K. B., & Reddy, M. V. (1993). Sources of resistance to *Ascochyta* blight in wild *Cicer* species. *Netherland Journal of Plant Pathology*, *99*, 163–167.

Singh, N. P., Singh, A., Asthana, A. N., & Singh, A. (1999). Studies on inter-specific crossability barriers in chickpea. *Indian Journal of Pulses Research*, *12*, 13–19.

Singh, R. P., & Singh, B. D. (1989). Recovery of rare interspecific hybrids of gram *Cicer arietinum* × *C. cuneatum* L. through tissue culture. *Current Science*, *58*, 874–876.

Singh, S., Gumber, R. K., Joshi, N., & Singh, K. (2005). Introgression from wild *Cicer reticulatum* to cultivated chickpea for productivity and disease resistance. *Plant Breeding*, *124*, 477–480.

Singh, U. (1985). Nutritional quality of chickpea (*Cicer* arietinum L.): Current status and future research needs. *Plant Foods Human Nutrition*, *35*, 339–351.

Singh, U., & Pundir, R. P. S. (1991). Amino acid composition and protein content of chickpea and its wild relatives. *International Chickpea Newsletter*, *25*, 19–20.

Stamigna, C., Crino, P., & Saccardo, F. (2000). Wild relatives of chickpea: Multiple disease resistance and problems to introgression in the cultigen. *Journal of Genetics and Breeding*, *54*, 213–219.

Stevenson, P. C., & Haware, M. P. (1999). Maackiain in *Cicer bijugum* Rech. F. associated with resistance to *Botrytis* gray mold. *Biochemical System Ecology*, *27*, 761–767.

Stolton, S., Maxted, N., Ford-Lloyd, Kell, S., & Dudley, N.. (2006). *Food stores: Using protected areas to secure crop genetic diversity*. Available at <http://assets.panda.org/downloads/food_stores.pdf/>.

Tanno, K., & Willcox, G. (2006). The origins of cultivation of *Cicer arietinum* L. and *Vicia faba* L.:Early finds from north west Syria (Tell el-Kerkh, late 10th millennium BP). *Vegetation History and Archaeobotany*, *15*, 197–204.

Thudi, M., Bohra, A., Nayak, S. N., Varghese, N., Shah, T. M., Penmetsa, R. V., et al. (2011). Novel SSR markers from BAC-end sequences, DArT arrays and a comprehensive genetic map with 1291 marker loci for chickpea (*Cicer* arietinum L.). *PLoS ONE*, *6*, e27275.

Toker, C. (2005). Preliminary screening and selection for cold tolerance in annual wild *Cicer* species. *Genetic Resources and Crop Evolution*, *52*, 1–5.

Toker, C., Canci, H., & Yildirim, T. (2007). Evaluation of perennial wild *Cicer* species for drought resistance. *Genetic Resources and Crop Evolution*, *54*, 1781–1786.

Upadhyaya, H. D. (2008). Crop Germplasm and wild relatives: A source of novel variation for crop improvement. *Korean Journal of Crop Science*, *53*, 12–17. ISSN 0252-9777.

Upadhyaya, H. D., Bramel, P. J., & Singh, S. (2001). Development of a chickpea core collection using geographic distribution and quantitative traits. *Crop Science*, *41*, 206–210.

Upadhyaya, H. D., Dronavalli, N., Gowda, C. L. L., & Singh, S. (2011). Identification and evaluation of chickpea germplasm for tolerance to heat stress. *Crop Science*, *51*, 2079–2094.

Upadhyaya, H. D., Dwivedi, S. L., Baum, M., Varshney, R. K., Udupa, S. M., Gowda, C. L. L., et al. (2008a). Genetic structure, diversity, and allelic richness in composite collection and reference set in chickpea (*Cicer arietinum* L.). *BMC Plant Biology*, *8*, 106. doi:10.1186/1471-2229-8-106.

Upadhyaya, H. D., Dwivedi, S. L., Baum, M., Varshney, R. K., Udupa, S. M., Gowda, C. L. L., et al. (2008b). Allelic richness and diversity in global composite collection and reference sets in chickpea (*Cicer arietinum* L.). In *5th international crop science congress & exhibition, Jeju, Korea, 13–18 April 2008*.

Upadhyaya, H. D., Dwivedi, S. L., Gowda, C. L. L., & Singh, S. (2007). Identification of diverse germplasm lines for agronomic traits in a chickpea (*Cicer arietinum* L.) core collection for use in crop improvement. *Field Crops Research*, *100*, 320–326.

Upadhyaya, H. D., Furman, B. J., Dwivedi, S. L., Udupa, S. M., Gowda, C. L. L., Baum, M., et al. (2006). Development of a composite collection for mining germplasm possessing allelic variation for beneficial traits in chickpea. *Plant Genetic Resources*, *4*, 13–19.

Upadhyaya, H. D., Gowda, C. L. L., Buhariwalla, H. K., & Crouch, J. H. (2006). Efficient use of crop germplasm resources: Identifying useful germplasm for crop improvement through core and mini core collections and molecular marker approaches. *Plant Genetic Resources*, *4*, 25–35.

Upadhyaya, H. D., & Ortiz, R. (2001). A mini core collection for capturing diversity and promoting utilization of chickpea genetic resources in crop improvement. *Theoretical and Applied Genetics*, *102*, 1292–1298.

Upadhyaya, H. D., Sharma S., Gowda, C. L. L., Sastry, D. V. S. S. R., Singh S., Sharma, S. K., et al. (2009). *Best practices for chickpea genebank management*. Available at <http://cropgenebank.sgrp.cgiar.org/>.

Upadhyaya, H. D., Thudi, M., Dronavalli, N., Gujaria, N., Singh, S., Sharma, S., et al. (2011). Genomic tools and germplasm diversity for chickpea improvement. *Plant Genetic Resources*, *9*, 45–58.

Vadez, V., Krishnamurthy, L., Serraj, R., Gaur, P. M., Upadhyaya, H. D., Hoisington, D. A., et al. (2007). Large variation in salinity tolerance in chickpea is explained by differences in sensitivity at reproductive stage. *Field Crops Research*, *104*, 123–129.

van der Maesen, L. J. G. (1972). *A monograph of the genus, with special references to the chickpea (Cicer arietinum L.) its ecology and cultivation*. Wageningen, The Netherlands: Mendelingen Landbouwhoge School.

van der Maesen, L. J. G. (1987). Origin, history and taxonomy of chickpea. In M. C. Saxena & K. B. Singh (Eds.), *The chickpea* (pp. 11–34). Wallingford, UK: CAB International.

Varshney, R. K., Glaszmann, J. C., Leung, H., & Ribaut, J. M. (2010). More genomic resources for less-studied crops. *Trends in Biotechnology*, *28*, 452–456.

Varshney, R. K., Graner, A., & Sorrells, M. E. (2005). Genomics-assisted breeding for crop improvement. *Trends in Plant Science*, *10*, 621–630.

Varshney, R. K., Hiremath, P. J., Lekha, P., Kashiwagi, J., Balaji, J., Deokar, A. A., et al. (2009). A comprehensive resource of drought- and salinity-responsive ESTs for gene discovery and marker development in chickpea (*Cicer arietinum* L.). *BMC Genomics*, *10*, 523.

Varshney, R. K., Hoisington, D. A., & Tyagi, A. K. (2006). Advances in cereal genomics and applications in crop breeding. *Trends in Biotechnology*, *24*, 490–499.

Varshney, R. K., Kudapa, H., Roorkiwal, M., Thudi, M., Pandey, M. K., Saxena, R. K., et al. (2012). Advances in genetics and molecular breeding of three legume crops of semi-arid tropics using next-generation sequencing and high-throughput genotyping technologies. *Journal of Bioscience*, *37*, 811–820.

Varshney, R. K., Nayak, S. N., May, G. D., & Jackson, S. A. (2009). Next-generation sequencing technologies and their implications for crop genetics and breeding. *Trends in Biotechnology*, *27*, 522–530.

Varshney, R. K., Song, C., Saxena, R. K., Azam, S., Yu, S., Sharpe, A. G., et al. (2013). Draft genome sequence of chickpea (*Cicer* arietinum) provides a resource for trait improvement. *Nature of Biotechnology*, doi:10.1038/nbt.2491.

Vavilov, N. I. (1951). The origin, variation, immunity and breeding of cultivated plants. *Chronica Botanica*, *13*, 1–366.

Verma, M. M., Ravi, , & Sandhu, J. S. (1995). Characterization of the interspecific cross *Cicerarietinum* L. × *C. judaicum* (Boiss.). *Plant Breeding*, *114*, 549–551.

Verma, M. M., Sandhu, J. S., Rrar, H. S., & Brar, J. S. (1990). Crossability studies in different species of *Cicer*(L.). *Crop Improvement*, *17*, 179–181.

Walter, K.S., & Gillett, H.J. (Eds.). (1998). IUCN red list of threatened plants. Compiled by the world conservation monitoring centre (pp. lxiv+862). Switzerland and Cambridge, UK: IUCN – The World Conservation Union, Gland.

Williams, P. C., & Singh, U. (1987). Nutritional quality and the evaluation of quality in breeding programmes. In M. C. Saxena & K. B. Singh (Eds.), *The chickpea* (pp. 329–356). Wallingford, UK: CAB International.

Zhizhizlashvili, K., & Berishvili, T. (1980). Zemo Svanetis kulturul mtsenareta shestsavlisatvis (Study of cultivated plants in Upper Svaneti). *Bulletin of Georgian Academy of Sciences*, *100*, 417–419. ISSN 0132-1447 (in Georgian).

5 Faba Bean

Maalouf Fouad[1], *Nawar Mohammed*[1], *Hamwieh Aladdin*[1], *Amri Ahmed*[1], *Zong Xuxiao*[2], *Bao Shiying*[3] *and Yang Tao*[2]

[1]International Center for Agricultural Research in the Dry Areas (ICARDA), Aleppo, Syria, [2]Institute of Crop Science/National Key Facility for Crop Gene Resources and Genetic Improvement, Chinese Academy of Agricultural Sciences, Beijing, China, [3]Institute of Grain Crops, Yunnan Academy of Agricultural Sciences, Kunming, China

5.1 Introduction

Faba bean (*Vicia faba* L.) is grown worldwide under different cropping systems as a dry grain (pulse), green grains/pods and a green-manure legume. Faba bean contributes to the sustainability of cropping systems through

- its ability to contribute nitrogen (N) to the system by biologically fixing N_2;
- diversification of production systems leading to decreased diseases, pests and weed build-up, and potentially increased biodiversity;
- its capacity to reduce fossil energy consumption;
- providing food and feed rich in protein (Jensen, Peoples, & Hauggaard-Nielsen, 2010).

Faba bean is cultivated under rainfed and irrigated conditions and is distributed in more than 55 countries. The harvested area is 2.56 million ha and 4.56 million tons of dry grains are produced. Asia and Africa accounted for 72% of the area and 80% of the production of dry faba bean grains (FAOSTAT, 2012). Faba bean remains in short supply in some countries. For example, Morocco imports around 9% of its annual needs to supplement its present production of 153,000 tons. Egypt imports around 43% of its annual needs to add to the present production of 297,620 tons. Globally, faba bean production showed a decline of 41%, from 5.4 million tons in 1961–62 to 3.2 million tons in the period of 1991–1993. This was followed by an increase of 33%, to 4.25 million tons, in the period of 2008–2010. However, up to today, the overall production is dominated by landraces, despite a number of improved varieties having been released by various national breeding programs. The major reasons for the decline in production were the susceptibility of landraces and cultivars to different biotic and abiotic stresses. Among biotic stresses, *Orobanche crenata* is a major factor in the declining production in North African countries like Morocco. Faba bean necrotic yellow virus (FBNYV) was the major cause of disappearance of faba bean from middle Egypt. Additionally, 3.5 million ha sown to faba bean in the period 1961–1963 in China declined to 0.95 million ha in the period

Genetic and Genomic Resources of Grain Legume Improvement. DOI: http://dx.doi.org/10.1016/B978-0-12-397935-3.00005-0

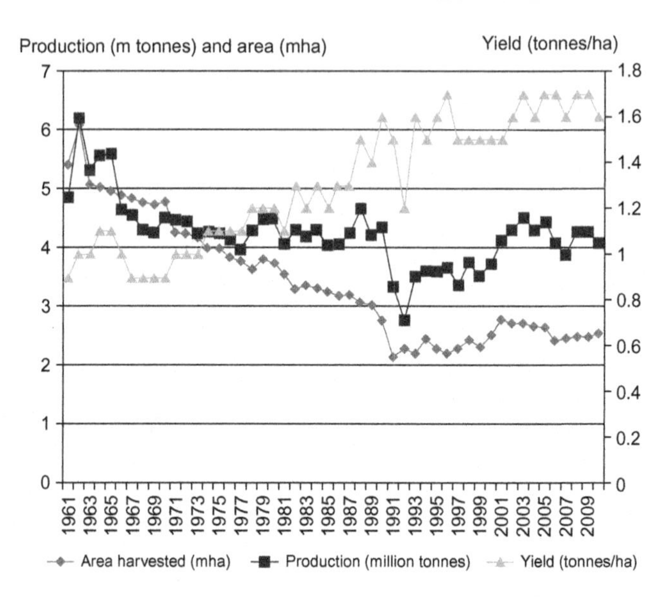

Figure 5.1 Trends in grain yield, area harvested and production over the last 40 years.

of 2008–2010. This reflects a general trend, observed since the 1960s, of increasing reliance by farmers on N fertilizers rather than legumes as a source of N input (Crews & Peoples, 2004) and the effects of recurrent and severe droughts. In many countries, faba bean has been subjected continuously to various biotic and abiotic stresses, which have led to genetic erosion among the landraces grown around the world. Despite the 0.3 million ha gain observed since the period of 1991–1993 and the overall decline of cultivated area from its peak value, the grain yield increased from 980 kg/ha in the period of 1961–1963 to 1700 kg/ha in the period of 2008–2010 (FAOSTAT, 2012), a yield gain of 15.4 kg/ha/year (Figure 5.1). This clear increase in yield is a result of the replacement of old cultivars with new improved varieties. Therefore, the clear fluctuation in area and the drastic effects of different biotic and abiotic stresses have resulted in a reduction in genetic diversity among traditional landraces.

5.2 Origin, Distribution, Diversity and Taxonomy

Faba bean was domesticated with the beginning of agriculture in the Fertile Crescent of the Near East following the Neolithic era around 9000–10,000 BC. Subsequently, its cultivation has spread around the world (Cole, 1970; Tanno & Willcox, 2006). The centre of diversity area includes Iraq, Iran, Georgia, Armenia, Azerbaijan, Syria and Turkey (Maxted, 1995). Cubero (1973, 1974) postulated that there are different routes radiating from the Near East to Europe and other parts of the world. The first could be across Anatolia to Greece and other Mediterranean regions towards Europe. The second could begin at the Nile delta and move towards the coastal areas to the

Maghreb and Iberian lands. The third could be along the River Nile to Abyssinia, now known as Ethiopia. The last could be from Mesopotamia to India. Secondary centres of diversity are postulated to have occurred in Afghanistan and Ethiopia. However, Ladizinsky (1975) reported the origin to be in Central Asia. According to Muratova (1931) and Maxted (1995), the centre of origin for the genus *Vicia* is southeastern Europe and southwestern Asia. Trait analyses have distinguished two groups: the small-seeded forms in southwestern Asia, including India, Afghanistan and adjoining regions of Bukhara and Kashmir, and large-seeded forms in the west. The Eastern group is very ancient and can be traced back to Neolithic agriculture. This group has the greatest number of endemic forms and the greatest diversity of characteristics, having many specific traits that are lacking in other groups (few pairs and many pairs of leaflets and grey-green colour, presence of tender and of course pod valves, a wide range of variation in maturity period, size, colour and shape of seeds, dimensions of leaflets, height and branching of stem, etc.). This group is found over a large area (from Spain to the Himalayas) (Muratova, 1931). Recent archeological findings at Tell El-Kerkh, northwest Syria, indicate a date of origin for faba bean domestication during the late 10th millennium BC (Tanno & Willcox, 2006). All these data point to southwestern Asia as the principal centre of origin of *V. faba*. The earliest archaeological findings of major types come from Iraq and are dated at around 1000 AD (Schultze-Motel, 1972). The migration of faba bean towards South America, especially the Andean region, probably occurred in the fifteenth century and was helped by Spanish and Portuguese travellers. This resulted in development of Peruvian and Bolivian landraces displaying a large variability in seed size, colour and shape (Duc et al., 2010). According to Zheng, Wang, and Zong (1997), faba bean (*V. faba* var. major) was first introduced to the northern part of China from the Middle East 2100 years ago through the Silk Road. However, a faba bean seed image on ancient pottery was found in a historical site in the Guanghe county of Gansu province in northern China (spring sowing area) in 1973, which was dated to between 4000 and 5000 years ago (Ye, Lang, Xia, & Tu, 2003). Faba bean grain fossils indicated that faba bean has been grown in southern China (winter sowing area) for more than 4000–5000 years (Ye et al., 2003). China is likely to be another secondary centre of diversity for faba bean, especially the Chinese winter gene pool, which has been reproductively isolated from the European and West Asian gene pools (Zong et al., 2009). Bond and Crofton (1999) described the development of winter faba beans in the nineteenth century in Europe. These were bred from Russian and French small-seeded, winter-hardy populations. The major geographical regions for faba bean cultivation are East Asia (34%), East Africa (20%), Central and West Asia and North Africa (CWANA; 18.8%), Europe (12.7%), Australia (6.3%) and Latin America (7.3%). In East Africa, Ethiopia is the major producer of summer-sown faba bean, cultivating 0.52 million ha in the highlands and producing 0.62 million tons. In the CWANA region, Morocco, Egypt, Sudan and Tunisia are the main producers, growing winter-sown faba bean on 0.48 million ha and producing 0.76 million tons (FAOSTAT, 2012). Faba bean consumption is primarily in East Asia, East Africa and West Asia and North Africa (WANA), where 6 of the 10 top producing countries are found. The temperate and herbaceous genus,

Table 5.1 Seed and Pod Characteristics of the Four Botanical Groups of *V. faba*

Botanical Group	Seed Weight and Shape	Pod Characteristics
Major	SW[a] ≥ 100 mg Very plate	Small to large (from 2 to 10 seeds) Plate, thick, nondehiscent pods
Equina	$50<SW<100$ mg Plate	Medium size, 3–5 seeds Plate
Minor	$30<SW<50$ mg Cylindrical to rounded form	Small with 3–4 seeds, cylindrical form
Paucijuga	$20<SW<30$ mg Rounded to elliptical form	Very small, dehiscent or nondehiscent types

[a]SW, average seed weight.

Vicia L. is a member of the legume tribe Vicieae of the Papilionoideae (Frediani, Maggini, Gelati, & Cremonini, 2004). *Vicia* comprises 166 annual or perennial species (Allkin, Goyder, Bisby, & White, 1986) distributed mainly in Europe, Asia and North America but also extending to the temperate regions of South America and tropical Africa (Maxted, 1993). Maxted, Callimassia, and Bennett (1991) divided the genus into two subgenera, *Vicilla* and *Vicia* (Kupicha, 1976). The two subgenera can be distinguished using the following characters: stipule nectary, peduncle length, style type, keel shape, legume and canavanine (Kupicha, 1976). Maxted (1993) classified the subgenus *Vicia* into 9 sections, 9 series, 38 species, 14 subspecies and 22 varieties. *V. faba* has suffered very little intraspecific differentiation as substantiated by the studies showing the presence of a partial incompatibility system; this is stronger in the central European populations studied, weak (to various degrees) in the Spanish ones, and absent in at least one population of the *paucijuga* group (Cubero, 1974). Cubero (1973) postulated four botanical groups of faba bean: *major*, *minor*, *equina* and *paucijuga* (Table 5.1).

5.2.1 Genetic Diversity in Faba Bean

The morphological and agronomic characterization of 900 accessions of faba bean held in the ICARDA gene bank at Tel Hadya experimental station, Syria, during the 2010–11 season (Table 5.2) showed limited degrees of variation for most of the qualitative and quantitative traits. The highest variation was recorded for first (lowest) pod length, the number of seeds per plant and 100-seed weight, which could be confounding effects of different botanic groups and would indicate a low genetic diversity within the cultivated faba bean groups. However, the use of amplified fragment length polymorphism (AFLP) (Zong et al., 2009) and simple sequence repeat (SSR) markers (Wang et al., 2012) have allowed genetic resources to be distinguished according to their geographic origin and the structuring of collections. Combined genotyping and phenotyping activities must continue on *V. faba* so that core collections can be defined. These will help in the discovery of new genes and alleles of interest for breeders. The AFLP markers were used to study the genetic

Table 5.2 Mean, Range and Coefficient of Variation for Morphologic and Agronomic Traits Measured on 900 Accessions Evaluated at Tel Hadya Station During the 2010–11 Season

Trait	Mean	Range	Coefficient of Variation
Leaflet size	6.07	Min. 1–Max. 9	21.96
Leaflet shape	2.31	Min. 2–Max. 6	20.62
Number of leaflets per leaf	5.10	Min. 4–Max. 6	14.63
Stem thickness	6.14	Min. 1–Max. 10	18.17
Branching from basal node	3.87	Min. 1–Max. 6	22.22
Stem pigmentation at flowering	4.09	Min. 1–Max. 7	33.52
Number of flowers per inflorescence	4.09	Min. 3–Max. 6	14.67
Flower ground colour	1.09	Min. 1–Max. 5	48.04
Wing petal colour	2.99	Min. 0–Max. 3	5.76
Pod surface reflectance	1.64	Min. 1–Max. 2	29.42
Pod distribution on stem	1.77	Min. 1–Max. 2	23.53
Days to 50% flowering	110.35	Min. 97–Max. 135	7.82
Days to 90% maturity	163.89	Min. 148–Max. 187	5.74
First (lowest) pod height (cm)	18.20	Min. 2–Max. 150	54.06
Number of nodes with pods	1.89	Min. 1–Max. 2	16.82
Number of pods per plant	13.68	Min. 0–Max. 48	48.67
Pod length (cm)	78.33	Min. 0–Max. 151	26.25
Pod width	10.93	Min. 1–Max. 21	24.09
Pod shape	1.27	Min. 1–Max. 12	44.11
Number of seeds per pod	2.15	Min. 0–Max. 4.4	23.17
Number of seeds per plant	29.28	Min. 4–Max. 141	52.31
Hundred-seed weight	84.00	Min. 0–Max. 205.48	55.04

diversity among a large set (n=79) of inbred lines of recent elite faba bean cultivars of Asian, European (northern and southern) and North African origin. These inbred lines were analysed using 8 selected AFLP primer combinations and produced 477 polymorphic fragments (Zeid, Schoen, & Link, 2003). The genetic diversity of 1000 faba bean accessions, comprising 505 accessions from the ICARDA global collection, 250 accessions from Instituto de Agricultura Sostenible and 245 accessions from Institut National de la Recherche Agronomique (INRA), was assessed using 16 SSR markers. Pozarkova et al. (2002) developed 25 SSRs in faba bean from a nonenriched library VffJF01, which was screened with a mix of (CTTT)n, (ACT)n, (AAG)n, and (AAC)n probes. Further, the development of 41 novel EST-SSR markers for *Pisum sativum* showed 53.7% of these markers could be transferred to the related species, *V. faba* (Xu et al., 2012). ICARDA, under the Generation Challenge Program (GCP), has also developed a new set of 100 SSRs, which are being used to characterize the faba bean collections representing genetic variation of the species. The primary results using 18 SSRs showed 10.6% heterozygosity (unpublished data, Table 5.3).

Table 5.3 Summary of Genotyping 1000 Faba Bean Accessions with 18 Microsatellite Primers

Primer Name	Max (bp)	Min (bp)	Range (bp)	Heterozygocity (%)
A110-1	245	117	129	8.28
F112-1	308	250	57	2.12
E115-1	300	211	89	8.92
E114-1	306	219	86	14.12
C7-1	250	204	46	12.10
O25-JF1-AG2	217	145	71	28.13
A105-1	329	248	81	16.77
G114-1	137	92	44	11.15
A102-1	254	146	108	27.28
A9	301	250	60	17.3
O23-GA1154	252	176	76	12.21
O13-GA3	237	150	87	11.57
F117-1	250	197	53	9.98
F11-1	307	266	40	2.34
E109-1	282	194	88	7.01
A117-1	214	171	44	6.37
A116-1	300	239	61	2.76
O3-GATA2	198	128	70	6.48
A109-1	240	176	64	2.76
Average	256.9	185.0	72.0	10.6

5.3 Erosion of Genetic Diversity from the Traditional Areas

The following information could indicate past and ongoing erosion of faba bean landraces in their various locations.

- Worldwide reduction of the cultivated area of faba bean as shown by the data compiled from FAO. Figure 5.1 lists the global annual harvested area, yield and production of faba bean and shows a reduction of 50% of the overall area since 1961. This reduction in area could be accompanied by a loss of some landraces, which in turn could be reflected in the change or loss of alleles because of a reduced population size and shrinking in number of distinct habitats or environments (Figure 5.1). In Morocco, the area allocated to faba bean has been reduced by 50% following infestation by the *Orobanche* parasitic weed, which has compelled farmers to abandon faba bean cultivation and replace the prevailing susceptible landraces with newly developed cultivars. In middle Egypt, FBNYV devastated the crop in 1992, which has led to the complete disappearance of all types of faba bean landraces and cultivars (Katul et al., 1993; Makkouk et al., 1994)
- Replacement of old landraces with new resistant/tolerant cultivars or by other species. In addition to improved agricultural practices, the observed increase in average yield could result from the increased adoption of modern varieties, replacing traditional landraces; this could be another indicator of the genetic erosion of this crop. In Egypt, 20 varieties have

Table 5.4 Gene Banks with More Than 500 Faba Bean Accessions

Country/City	Organization	No. of Accessions
Australia/Victoria	DPI	2445
Bulgaria/Sadovo	IIPGR	692
China/Beijing	CAAS	5200
Ethiopia/Addis Ababa	PGRC	1118
France/Dijon	INRA	1900
Germany/Gatersleben	Genebank IPK/	1920
Italy/Bari	Genebank	1876
Morocco/Rabat	INRA	1715
Netherlands/Wageningen	DLO	726
Poland/Poznam	IOPG-PAS	1258
Poland/Radzikow	PBAI	856
Portugal/Oeiras	INRB-IP	788
Russia/St Petersburg	VIR	1881
Spain/Córdoba	IFAPA	1091
Spain/Madrid	CNR	1622
Syria/Aleppo	ICARDA	10,045
USA/Pullman	USDA	750

been released since 1980 with a 30% adoption rate. In China, the cultivar Yundou 147, released from a K0285 × ILB8047 cross, is estimated to account for more than 30% of the faba bean acreage in Yunnan province. Several varieties replacing the old landraces in different regions have been released by various Chinese academies (Bao Shiying, personal communication).

- Surveys undertaken by ICARDA within the dry-land agrobiodiversity project, including four countries of the Fertile Crescent – Jordan, Lebanon, Syria and the Palestinian Authority – showed that the landraces of several field crops (cereals and food legumes) were replaced by introduced fruit tree species, such as apples, cherries and olive (Mazid, Shideed, & Amri, 2006).

5.4 Status of Germplasm Resources Conservation

ICARDA safeguards the largest collection of faba bean worldwide (32% of the total world collection). This global collection conserves materials from 71 countries with a high percentage of unique accessions. A total of 8628 of these accessions comprise the international collection held in trust for the global community. The collection held at ICARDA also conserves over 6000 accessions of other *Vicia* species, including about 3000 accessions of wild species of *Vicia*. The accession type and source data in Table 5.4 provide an indication of the uniqueness of the collections. Collections with a high percentage of wild relatives, landraces and materials originally collected by ICARDA are most likely to encompass unique accessions prioritized in a rational global system.

5.5 Germplasm Maintenance

Maintenance and evaluation of any species depends on its reproductive system. Faba bean is an entomophilic and partially cross-pollinated legume. The reproductive system follows a mixed mating model in the major populations. The outcrossing rate varies widely among cultivars and locations (Gasim, Abel, & Link, 2004; Suso, Pierre, Moreno, Esnault, & Le Guen, 2001). Much of its pollination depends on wild vectors (Bond & Kirby, 1999; Pierre, Suso, Moreno, Esnault, & Le Guen, 1999). Most of the data on faba bean gene flow were from experiments by Bond and Pope (1974) and Link and von Kittlitz (1989). The methods used for faba bean germplasm maintenance and genetic resources multiplication and regeneration are based on preventing the effect of insect pollinators. The use of insect-proof cages is one efficient technique applied in faba bean germplasm maintenance. However, in addition to being an expensive system to prevent intercrossing in this crop, this technique has limited capacity and is advisable only for small sets (Hawtin & Omar, 1980). It also increases the inbreeding depression of faba bean, affecting the yield potential of different cultivars (Drayner, 1959). When breeding programs are managing a large number of samples with large seed numbers per sample, it is not advisable to use the isolation cages, as the cost will be very high, their use will be very difficult to manage and there is yield reduction through inbreeding depression. The techniques developed for the maintenance of germplasm are based on an adequate gene flow between different faba bean plots and the isolated crop used. Link and von Kittlitz (1989) used seed and hilum colour marker genes to measure gene flow. Allozyme and isozyme markers and different experimental genotypes have been used to measure the patterns of variation of gene flow in small plots of a field of germplasm multiplication (Suso, Gilsanz, Duc, Marget, & Moreno, 2006). In order to reduce gene flow among plots, a combination of isolation by a distance of 3 m and pollination barriers using *Brassica napus* L. and *Triticosecale* reduced intercrossing between adjacent plots by more than 95% (Robertson & Cardona, 1986). Suso, Nadal, Román, and Gilsanz (2008) assumed that planting a border surrounding the faba bean plots is more efficient than using a noncultivated area between two adjacent plots. At ICARDA, collections and improved germplasm were maintained in two different ways. A small sample derived from single plant selection is maintained in insect-proof cages and large-seed germplasm is maintained in isolation in the open field using *Brassica* or *Vicia narbonesis* as border crops. For large seed multiplication, a faba bean field has to be far away – at least 50–100 m – from any other faba bean plot or farmer's field or experimental site to ensure the seed purity.

5.6 Use of Genetic Diversity in Faba Bean Breeding

Faba bean breeding is carried out by only a few research institutes; the main operational breeding programs are found at:

- International Center for Agricultural Research in the Dry Areas (ICARDA), Aleppo, Syria
- The Instituto de Investigación y Formación Agraria y Pesquera de Andalucía (IFAPA), Spain

- Institut National de la Recherche Agronomique (INRA), Rennes, France
- University of Adelaide, Australia
- Yunnan Academy for Agricultural Science (YAAS), China
- Field Crop Research Institute (FCRI), Egypt
- Institut National de la Recherche Agronomique de Tunisie (INRAT), Tunisia
- Ethiopian Institute for Agricultural Research (EIAR), Ethiopia
- Institut National de la Recherche Agronomique (INRA), Morocco
- Field Crop Research Institute (FCRI), Sudan.

The regional research program on faba bean was started officially in 1972 at Arid Land Agriculture Development (ALAD), Lebanon, to fulfil the needs of the WANA region. The program, based at Tel Amara in the Bekaa Valley, got underway in 1973 with significant financial support provided by the International Development Research Centre (IDRC), Canada, in cooperation with the Agricultural Research Institute (ARI) of Lebanon. The identification of lines resistant to major diseases was made in collaboration with the University of Manitoba. When the civil war broke out in Lebanon in 1975, the program continued for a while and then moved to Syria. In 1976, the program was developed in Egypt through the Ford Foundation office in Cairo. By 1977, ALAD had transmuted into ICARDA, the base had moved from the Bekaa to Aleppo (Geoffrey Hawtin, personal communication).

5.6.1 Breeding for Abiotic Stresses

The major abiotic stresses affecting faba bean production are terminal drought, frost and heat. Drought, an interval of water deficiency leading to a significant reduction in yield, is widely considered to be the most important environmental constraint to crop productivity (Borlaug & Dowswell, 2005, chap. 2; Fischer & Turner, 1978). Faba bean is reputed to be more sensitive to water deficits than other grain legumes (Amede & Schubert, 2003; McDonald & Paulsen, 1997). In many production regions in the Mediterranean basin, the crop is seldom if ever irrigated and generally relies on stored soil moisture and current rainfall for its growth and development (Sau & Mínguez, 2000). Variation in the amount and distribution of rainfall is generally considered the major reason for variability in the grain yield of faba bean (Abdelmula, Link, Kittlitz, & von Stelling, 1999; Bond et al., 1994; Siddique, Regan, Tennant, & Tomson, 2001). In drought-prone regions of North and East Africa, a shortage of water, especially during the flowering period, can cause a drastic reduction in yield. Terminal drought is one of the important constraints to faba bean production in regions like Ethiopia and Morocco, where the crop is largely grown under rainfed conditions. Several elite lines were identified as drought tolerant, like ILB 938/2 in the ICARDA germplasm collection (Khan, Link, Hocking, & Stoddard, 2007; Khan, Paull, Siddique, & Stoddard, 2010).

Extreme low temperature is one of the abiotic constraints for growing autumn-sown faba beans in cool temperate climates. Winter hardiness is a complex trait which depends not only on frost tolerance but also on tolerance to other abiotic stresses (e.g. saturation level of water in the soil, frost-drought) and biotic constraints (e.g. snow mold). To overcome this constraint, experiments under controlled conditions have been conducted for several crop species, revealing that frost tolerance is a major component of winter hardiness (Arbaoui, Balko, & Link, 2008; Link, Balko, & Stoddard, 2010). Sources of resistance to cold are Cote d'Or 1 (an inbred line derived from the winter-hardy French landrace Cote d'Or) and BPL4628 (an inbred line derived from the Chinese line in the ICARDA germplasm collection) (Arbaoui et al., 2008). The breeding program at ICARDA identified different lines with tolerance to frost damage. In addition, the screening for winter hardiness of more than 5200 entries from the Chinese gene banks led to the identification of a few sources for cold tolerance. Likewise, extreme heat is the major threat to faba bean production in south Egypt, Sudan, and the Ethiopian lowlands. Artificially induced terminal heat stress can significantly reduce yield and the yield components of faba bean genotypes (Ahmed, 1989; Abdelmula & Abuanja, 2007). This adverse effect could be attributed mainly to high temperature during the vegetation period, which checked growth and led to the development of a small, short-stemmed crop with few branches and pods. Abdelmula and Abuanja (2007) concluded that the genotype C.52/1/1/1 could be used to improve heat tolerance in faba bean and make it possible to extend production to the nontraditional areas of Sudan.

5.6.2 Breeding for Biotic Stresses

More than five foliar diseases caused by *Ascochyta* blight, chocolate spot, rust, powdery mildew, *Cercospora* leaf spot, different root rot complexes, nematodes, *Orobanche* and a large number of viruses affect the production and productivity of faba bean (Sillero et al., 2010). In North and East Africa, the major biotic stresses are, *Ascochyta* blight, black root rot, bruchids, chocolate spot and rust (Bayaa, Kabakebji, Khalil, Kabbabeh, & Street, 2004). Other biotic stresses include *Orobanche* (Khalil, Kharrat, Malhotra, Saxena, & Erskine, 2004; Maalouf et al., 2011) and different types of viruses, like bean yellow mosaic virus (BYMV), pea enation mosaic virus, bean leaf roll virus (BLRV), FBNYV, true broad bean mosaic virus, broad bean mottle virus, and broad bean stain virus (Bond et al., 1994; Saxena, 1991; van Leur, Kumari, Makkouk, & Rose, 2006). *O. crenata* can reduce the yield of faba bean in infested areas by up to 90%. The estimated average yield losses due to *O. crenata* in Morocco ranged from 7% to 80% depending on the level of infestation (Gressel et al., 2004). Around 78% of the Moroccan faba bean fields were infested by *Orobanche* (Mesa-García & García-Torres, 1991). *Orobanche*-tolerant lines have been developed in faba bean (Khalil & Erskine, 1999; Khalil, Kharrat, et al., 2004; Maalouf et al., 2011). Efforts have been focused on identifying sources of resistance/tolerance to *Ascochyta* blight, chocolate spot, rust and *Orobanche* (Bayaa et al., 2004; Hanounik & Roberston, 1989; Khalil, Bayaa, Malhotra, Erskine, & Saxena, 2004; Khalil, Kharrat, et al., 2004; Maalouf, Ahmed, Kabakebji, Kabbabeh, & Street, 2009; Maalouf, Ahmed, Nawar, Khalil, & Bayaa, 2012; Maalouf et al., 2010, 2011), at ICARDA and in other advanced research institutes (Bernier & Conner, 1982; Bond et al., 1994; Rashid & Bernier, 1984, 1986). Among the breeding lines resistant to rust developed at ICARDA are ILB403, ILB411, ILB420, ILB 431, ILB 479, ILB 490, ILB 866, ILB 919, ILB 938, Reina Blanca ILB 249/803/80, ILB 249/804/40, ILB 938, ILB 159-1, ILB 159-4, BPL 710, BPL 1179,

BPL 7, BPL 8, BPL 260, BPL 261, BPL 263, BPL 309, BPL 406, BPL 417, BPL 427, BPL 490, BPL 484, BPL 524,BPL 533, BPL 539, BPL 552, BPL 554, BPL 567, BPL 571, BPL 573, BPL 576, BPL 588, BPL 604, BPL 610, BPL 627, BPL 649, BPL 663, BPL 665, BPL 667, BPL 680, BPL 640, BPL 643 and BPL 702 (Bernier & Conner, 1982; Bond et al., 1994; Khalil, Nassib, & Mohammed, 1985; ICARDA, 1987; Rashid and Bernier, 1984, 1986). As regards pathogenic diversity, several races of *U. viciae-fabae* have been identified. Using established reference sets (Conner and Bernier, 1982; Emeran, Sillero, & Rubiales, 2001) the highest virulence was identified in the Egyptian populations. The evidence of the physiologic specialization in *U. viciae-fabae* described above suggests that the use of single resistance genes in cultivars would not likely result in long-term rust control. So it is a major need to search for strategies to prolong durability. Complete resistance is common (Khalil et al., 1985; Rashid & Bernier, 1984, 1991; Sillero, Moreno, & Rubiales, 2000).

Ascochyta blight is caused by the fungus *Ascochyta fabae*. It is a common disease that causes yield losses of up to 90% in susceptible cultivars when environmental conditions are favourable for disease development (Hanounik & Roberston, 1989). The fungus infects all the above-ground plant parts including the seeds. Sexual reproduction allows new virulence combinations and, as a consequence, the pathogen may respond over time to selection exerted by the introduction of host resistance genes. Physiological specialization between pathogen isolates and host genotypes has been described in the *A. fabae* – faba bean pathosystem (Ali & Bernier, 1985; Avila et al., 2004; Hanounik & Roberston, 1989; Kharbanda & Bernier, 1980; Kohpina, Knight, & Stoddard, 1999; Rashid, Bernier, & Conner, 1991), which is problematic in breeding, making it necessary to evaluate segregating breeding materials against a range of isolates to ensure good success. Among the faba bean lines identified as resistant to *Ascochyta* blight are BPL 74, BPL 460, BPL 471, BPL 472, BPL 646, BPL 818, BPL 2485, ILB 1814, 14434-2, 14434-3, 15025-2, 15035-1, 15041-2, BPL 2485-1, BPL 2485-2, ERF-3-14, BPL 230, BPL 266, BPL 365, BPL 465, ILB 752, L83118, L83120, L83124, L83125, L83127, L83129, L83136, L83142, L83149, L83151, L83155, L83156, L82001, L831818-1, Line 224, ILB 757 Ascot, V-46, V-47, V-165, V-175, V-494, V-1122, V-1220, ILB 1414 and ILB 6561 (Bond et al., 1994; Hanounik & Roberston, 1989; Lawsawadsiri, 1995; Maurin & Tivoli, 1992; Ramsey, Knight, & Paull, 1995; Rashid et al., 1991; Sillero, Avila, Moreno, & Rubiales, 2001).

Chocolate spot is especially severe in humid areas and reported to be the cause of heavy reductions in yields in places, such as Morocco, Tunisia, Egypt, Ethiopia, China, and United Kingdom. The faba bean lines resistant to chocolate spot are BPL 74, BPL 460, BPL 471, BPL 472, BPL 646, BPL 818, BPL 248, 14434-2, 14434-3, 15025-2, 15035-1, 15041-2, BPL 2485-1, BPL 2485-2, BPL 230, BPL 266, BPL 365, BPL 465, ILB 752, L83118, L83120, L83124, L83125, L83127, L83129, L83136, L83142, L83149, L83151, L83155, L83156, L82001, L831818-1, Sel.97Lat.97 132-1, Sel.97Lat.97 132-3 (Bayaa et al., 2004; Kharrat, Le Guen, & Tivoli, 2006). Little is known about the mechanism of resistance to *Botrytis*. There is a need to establish differential lines and then use these to evaluate the virulence of a collection of isolates of diverse origin, under the same environmental conditions, for the major diseases and broomrapes that attack faba bean.

In addition, more than 180 new sources for resistance to chocolate spot, *Ascochyta* blight and rust were identified at ICARDA under heavy soils infested by a mixture of the most virulent pathogens collected in Syria. Lines with combined resistance have been developed at ICARDA and sent to different national agricultural research systems to observe the response of the resistant lines to different races in varying environments. In the last 5 years, 70 lines with resistance to chocolate spot and 70 lines with resistance to *Botrytis* were sent to different national agricultural institutes to evaluate their resistance under their specific races and environments. National breeding programs, mainly in Morocco, Sudan and Syria, selected 28 promising lines (Maalouf et al., 2012). In Ethiopia, the major disease problems were chocolate spot, rust and root rots. Several varieties with a high-level resistance to chocolate spot, derived directly or indirectly from the ICARDA breeding program, were released by EIAR. In addition, because of the high prices of faba bean in Ethiopia, farmers are expanding faba bean production on vertisols that are confronted with root rots favoured by stagnant water. Through extensive collaborative research, EIAR researchers have released several high-yielding faba bean varieties through direct selection from the germplasm supplied by ICARDA. Among the faba bean varieties released with good levels of disease resistance are 'Moti' (ILB 4432 x Kuse-2-27-33); 'Gebelcho' (ILB 4726 x 'Tesfa'); 'Obsie' (ILB 4427 x CS20DK) and 'Walki' (ILB 4615 z Bulga 70). The variety 'Walki' was developed for waterlogged areas and is gaining popularity in the central highlands of Ethiopia. Viruses that infect faba bean crop are not host species-specific; they can affect a range of food and pasture legumes as well as numerous weeds. A 'green bridge' between cropping seasons is apparently necessary for the transmission of viruses. The other means of virus survival is seed transmission, which is almost absent or not of economic importance for faba bean viruses (van Leur et al., 2006). Because of the uncertainty of virus epidemics and the lack of virus control options, growers can perceive viruses as a higher risk than fungal diseases. However, some inbred lines such as 2N23, 2N65, 2N85, 2N101, 2N138, 2N295 and 2N425 were reported in Canada some decades ago as sources of resistance to BYMV, but only one of them, line 2N138, was highly resistant to the necrotic strain of this virus (Gadh & Bernier, 1984). ICARDA has identified different accessions resistant to BLRV (BPL 756, BPL 757, BPL 758, BPL 769, BPL 5278 and BPL 5279), and resistant to BYMV (BPL 1351, BPL 1363, BPL 1366 and BPL 1371) (Bond et al., 1994; Kumari & Makkouk, 2003; Robertson, Singh, Erskine, & Abd El Moneim, 1996).

Efforts to breed faba bean resistant to *Orobanche* have resulted in the release of cultivars with useful levels of incomplete resistance combined with a degree of tolerance (Cubero, Moreno, & Hernandez, 1992; Cubero, Pieterse, Khalil, & Sauerborn, 1994; Kharrat, Abbes, & Amri, 2010; Khalil & Erskine, 1999; Khalil, Kharrat, et al., 2004; Maalouf et al., 2011). The resulting resistance, which might be based on a combination of resistance mechanisms, is more likely to last longer than resistance based on a single gene (Perez-de-Luque, Lozano, Moreno, Testillano, & Rubiales, 2007; Rubiales et al., 2006). Little resistance to *O. crenata* was available in faba bean until the appearance of the Egyptian line F402 (Nassib, Ibrahim, & Khalil, 1982). Some accessions with moderate to low levels of resistance and/or tolerance have been reported (Table 5.4), but the first significant finding of resistance was the identification of family 402 derived from a 3-year cycle of individual plant selection in an F7 from the cross (Rebaya 40 x F216) made at ICARDA (Cubero et al., 1994). Different cultivars have been developed from this cross (Giza 402, BPL 2210, Baraca, Lines 18009, 18015, 1835, Cairo 241, Cairo 348, Cairo 2, Line 402/294, Lines 402/29/84, 674/154/85, L3-4, Line X-843, Giza 429, Giza 674, Giza 843, ILB 4347, ILB 4357, ILB 4360, Bader, XBJ 90.03-16-1-1-1, Misr1 and Misr2 (Abbes, Kharrat, Delavault, Simier, & Chaibi, 2007; Abdalla & Darwish, 1994, 1996; Cubero et al., 1992; Hanounik, Jellis, & Hussein, 1993; Khalil & Erskine, 1999; Khalil, Kharrat, et al., 2004; Kharrat & Halila, 1994; Nassib et al., 1982; Saber et al., 1999; ter Borg et al., 1994)).

5.6.3 Breeding for Antinutritional and Nutritional Components

The nutritional value of faba bean has been traditionally attributed to its high protein content, which ranges from 20% to 37%, (Crépon et al., 2010; Santidrián, Sobrini, & Larralde, 1981). Most of these proteins are globulins (60%), albumins (20%), glutelins (15%) and prolamins (Cubero, 1984). Additionally, faba bean is also a good source of sugars, minerals (Ca, Mg, Fe and Zn), vitamins (B-complex, vitamin C, and vitamin A) (Sobrini, Santidrian, & Larralde, 1982). Thus, the chemical analysis of this legume reveals a 50–60% carbohydrate content, which is mainly starch. Faba bean is rich in tannins and two glucosidic aminopyrimidine derivatives, V and C, which generate the redox aglycones divicine (D) (2,6-diamino-4,5-dihydroxypyrimidine) and isouramil (I) (6-amino-2,4,5-trihydroxypyrimidine), respectively, upon hydrolysis of the beta-glucosidic bond between the glucose and hydroxyl group at C-5 on the pyrimidine ring. Faba bean also contains high amount of ascorbate and varying amounts of L-DOPA glucoside (Arese & De Flora, 1990). Small children and old people are at high risk because their gastric juice is less acidic and the beta-glycosidase of the bean is not inactivated. In normal red blood cells, oxidized glutathione (GSH) is rapidly regenerated by a metabolic cycle in which glucose-6-phosphate dehydrogenase (G6PD) is an essential component. G6PD deficiency is widespread in humans.

5.7 Germplasm Enhancement Through Wide Crosses

The wild ancestor of faba bean remains unknown and no successful interspecific crosses with other *Vicia* species have been made (Hanelt, Schäfer, & Schultze-Motel, 1972; Muratova, 1931). The closest species to *V. faba* is considered to be *V. pliniana* (Trabut) Murat from Algeria (Muratova, 1931). Differences from *V. faba* in morphological characters, such as broad arillus, the anatomical structure of seed coat and its weak swelling, allowed Muratova to classify it as an independent species, *V. pliniana*. Pods of this wild form, which has slightly different morphology from that of *V. faba*, were used for cooking (Trabut, 1911). Another presumed ancestor is the *paucijuga* type, which was found by the traveller Slagintwein in Tibet and Pendjub (Alefeld, 1866).

Hopf (1973) proposed that *V. narbonensis* L. is a probable wild ancestor of *V. faba*. These two species have many morphological similarities and coincide in their distribution. However, Ladizinsky (1975) argued against considering *V. narbonensis* and other wild species as immediate ancestors of the cultivated *V. faba*. Although *V. narbonensis*, *V. johannis* and *V. bithynica* all cross well with each other and many attempts to cross *V. faba* with any of its relatives have failed to produce viable hybrids (Bond, Lawes, Hawtin, Saxena, & Stephens, 1985; Cubero, 1982; Hanelt & Mettin, 1989).

Hybridization between *V. faba* bean and *V. narbonesis* was tried by Roupakias (1986). Fertilized embryo sac development and pod growth were studied in one *V. faba* cultivar, one *V. narbonensis* population, and their reciprocal crosses. The initial development of endosperm and embryo was at least 4 days faster in *V. narbonensis* than in *V. faba*. Pods and ovules also developed faster in *V. narbonensis* than in *V. faba*. The growth rate of the hybrid pods followed the growth rate of the mother species, but was slower than that of the pods from selfed flowers. In the cross *V. narbonensis* × *V. faba*, the ovules stopped growing 9 days after pollination, while in the reciprocal cross they stopped growing 15 days after pollination. Hybrid embryo sacs from *V. faba* × *V. narbonensis* were aborted before they reached the stage of 256 endosperm nuclei or 200 embryo cells. Selfed *V. faba* embryo sacs reached this stage <9 days after pollination. In the reciprocal cross, the embryo sacs were aborted before they reached the stage of 128 endosperm nuclei or 80 embryo cells. Selfed *V. narbonensis* embryo sacs reached this stage at the fourth day after pollination. Given that at these stages the embryo has <200 cells, it was concluded that an in-ovule embryo culture technique should be developed to obtain viable hybrid plants. Molecular investigations have indicated the independence of *V. faba* and its large genetic difference from the *V. narbonensis* complex (Przybylska & Zimniak-Przybylska, 1997; Raina & Ogihara, 1994; van de Ven et al., 1993). Restriction fragment length polymorphism (RFLP) data has divided the *Vicia* gene pool into the species *narbonensis*, *peregrinae* and *faba*, which is in good agreement with the classification by Maxted et al. (1991). However, it has also been suggested that *V. faba* is more closely aligned to species from the genus *Hypechusa* and the genus *Peregrinae* than to those in the *V. narbonensis* complex (van de Ven et al., 1993).

5.8 Faba Bean Genomic Resources

A composite map of the *V. faba* genome based on morphological markers, isozymes, seed protein genes and microsatellites was constructed by Román et al. (2004). The map incorporates data from 11F_2 families for a total of 654 individuals all sharing the common female parent *Vf* 6. The integrated map is arranged in 14 major linkage groups (LGs; 5 of which were located in specific chromosomes). These LGs included 192 loci and cover 1559 cM with an overall average marker interval of 8 cM. By joining data of a new F_2 population segregating for resistance to *Ascochyta*, and broomrape, other traits of agronomic interest were revealed. The combination of trisomic segregation, linkage analysis among loci from different families with a recurrent parent and the analysis of new physically located markers has allowed the establishment of a *V. faba* map with wide coverage. This map provides an efficient tool in breeding applications, such as disease-resistance mapping, quantitative trait loci (QTL) analyses and marker-assisted selection (MAS). Comparative genomics and synteny analysis with closely related legumes will reveal new candidate genes and selectable markers for use in MAS. Ellwood et al. (2008) used 151 intron-targeted amplified polymorphic (ITAP) markers to construct a comparative genetic map of the faba bean. Linkage analysis revealed 7 major and 5 small LGs, 1 pair and 12 unlinked markers. Each LG was composed of 3–30 markers and varied in length from 23.6cM to 324.8cM. However, the high number of LGs compared to the number of chromosomes may be because faba bean possesses one of the largest genomes among cultivated legumes (~13,000Mbp). The map spanned a total length of 1685.8cM (Ellwood et al., 2008).

One hundred and four of the 127 mapped markers in the 12 LGs, which were previously assigned to *Medicago truncatula* genetic and physical maps, were found in regions syntenic between the faba bean and *M. truncatula* genomes. However, chromosomal rearrangements were observed that could explain the difference in chromosome numbers between faba bean, lentil and *M. truncatula*. Multiple polymerase chain reaction (PCR) amplicons and comparative mapping were suggestive of small-scale duplication events in faba bean. They provided a preliminary indication of finer scale macro-synteny between *M. truncatula*, lentil and faba bean. Markers originally designed from genes on the same *M. truncatula* bacterial artificial chromosomes (BACs) were found to be grouped together in corresponding syntenic areas in lentil and faba bean (Ellwood et al., 2008), which may facilitate a more efficient selection of new cultivars free of antinutritional compounds.

5.8.1 Current QTLs Available in Faba Bean

Díaz-Ruiz et al. (2010) used 165 F_6 recombinant inbred lines (RILs) to identify genetic regions associated with broomrape resistance in three environments across two locations in 2003–2004. Two hundred and seventy-seven molecular markers were assigned to 21 LGs (9 of them assigned to specific chromosomes) that covered 2856.7cM of *V. faba* genome. The composite interval mapping (CIM) on the F_6 map detected four QTLs controlling *O. crenata* resistance (Oc2–Oc5) in three different environments. Oc2 and Oc3 were found to be associated with *O. crenata* resistance in at least two of the three environments, while the remaining two, Oc4 and Oc5, were only detected in Córdoba-04 and Mengíbar-04 and seemed to be environment dependent. Six QTLs for *Ascochyta* blight resistance in faba bean were identified by Avila et al. (2004) by using an F_2 population from the cross between the inbred lines 29H (resistant) and Vf136 (susceptible). The six QTLs detected were named Af3–Af8. Af3 and Af4 were effective against *Ascochyta* isolates. Af5 was the only effective against isolate CO99-01, while Af6, Af7 and Af8 were only effective against isolate L098-01. Af3, Af4, Af5 and Af7 were revealed in both leaves and stems. In contrast, Af6 was only effective in leaves and Af8 only in stems.

Genetic improvement by MAS has been carried out with success in several legume crops, such as soybean, common bean and pea. However, in other species, such as faba bean, it is still in its early stages. Use of molecular markers in faba bean breeding for resistance to broomrape, *Ascochyta* blight, rust and chocolate spot is underway, and promising results have been obtained. Gutierrez et al. (2006) identified markers linked to the nutritional value of seed tannins and V&C content. Three F_2 populations, involving lines with zero tannin genes (*zt-1* and *zt-2*) and with the zero vicine–convicine mutant (*vc*−=line1268), have been analysed by the group at IFAPA. Bulked segregant analysis (BSA) was used to identify random amplified polymorphic DNA (RAPD) markers linked to these genes and the RAPD fragments associated with tannin and V&C content have been transformed into more consistent sequence-characterized amplified regions (SCARs) (Gutierrez, Avila, Moreno, & Torres, 2008; Gutierrez, Avila, Rodriguez-Suarez, Moreno, & Torres, 2007). The cleaved amplified polymorphic sequence (CAPS) and SCAR markers linked in the coupling and repulsion phase to zero tannin and low V&C content can be used to introgress the appropriate alleles and help in developing cultivars with low V&C content and improved nutritional value, avoiding the cost and difficulties of the chemical determination of these products.

5.9 Conclusions

Faba bean is one of the oldest crops grown by man and is used as a source of protein in human diets, as fodder and a forage crop for animals, and for available nitrogen for the biosphere. Despite its importance in food, feed and farming systems, the area under cultivation has declined drastically and useful genetic variation has been lost. However, the available genetic materials conserved at various gene banks need to be maintained and critically evaluated for their use in breeding programs. The useful genetic variations identified for key stresses should be used to develop cultivars with multiple resistances, in order to attain stable yields. Advanced biotechnical tools accelerate the process of selection for resistance to major traits of interest; ICARDA is developing appropriate RILs for this purpose. In addition, a number of mapping studies have identified QTLs controlling different traits for the major biotic and abiotic stresses in faba bean as well as for quality and determinate types (Torres et al., 2010). These advanced studies should lead to promising results, but are still insufficient for MAS because of the limited saturation of the genomic regions bearing putative QTLs. This fact makes it difficult to identify the most tightly linked markers and to accurately determine the position of the QTLs (Torres et al., 2010). More efforts are needed to better understand the complexity of resistance mechanisms in pests and the broad adoption of new improvements in marker technology integrated with comparative mapping and functional genomics (Dita, Rispail, Parts, Rubiales, & Singh, 2006; Rispail et al., 2010).

References

Abbes, Z., Kharrat, M., Delavault, P., Simier, P., & Chaibi, W. (2007). Field evaluation of the resistance of some faba bean (*Vicia faba* L.) genotypes to the parasitic weed *Orobanche foetida* Poiret. *Crop Protection*, *26*, 1777–1784.

Abdalla, M. M. F., & Darwish, D. S. (1994). Breeding faba bean for orobanche tolerance at Cairo University. In A. H. Pieterse, J. A. C. Verkleij, & S. J. ter Borg (Eds.), *Biology and management of orobanche: Proceedings of the 3rd International Workshop on Orobanche and Related Striga Research* (pp. 450–454). Amsterdam, Netherlands: Royal Tropical Institute.

Abdalla, M.M.F. & Darwish, D.S. (1996). Investigations of faba beans, *Vicia faba*, L.: Cairo 2 and Cairo 241, two new Orobanche tolerant varieties. In *Proceedings of the 7th conference of agronomy*, Mansoura University, Mansoura, Egypt, 9–10 September 1996, pp. 187–201.

Abdelmula, A.A. & Abuanja, I.K. (2007). Genotypic responses, yield stability, and association between characters among some of Sudanese faba bean (*Vicia faba* L.) genotypes under heat stress. In *Proceedings of the conference utilization of diversity in land use systems: Sustainable and organic approaches to meet human needs*, University of Kassel-Witzenhausen and University of Göttingen, Witzenhausen, Germany, 9–11 October 2007, pp. 1–7.

Abdelmula, A. A., Link, W., Kittlitz, E., & von Stelling, D. (1999). Heterosis and inheritance of drought tolerance in faba bean, *Vicia faba* L. *Plant Breeding*, *118*, 458–490.

Ahmed, M.E.M. (1989). Responses of faba bean to water and heat stress. *M.Sc. Thesis*. Khartoum, Sudan: University of Khartoum.

Arbaoui, M., Balko, C., & Link, W. (2008). Study of faba bean (*Vicia faba* L.) winter hardiness and development of screening methods. *Field Crops Research*, *106*, 60–67.

Alefeld, F. (1866). *Genus Vicia. Landwirtschaftliche flora*. Berlin, Germany: Wiegandt and Hempel.

Ali, F. H., & Bernier, C. (1985). Evaluation of components of resistance of *Ascochyta fabae* on faba beans. *Phytopathology*, *75*, 962.

Allkin, R., Goyder, D. J., Bisby, F. A., & White, R. J. (1986). *Names and synonyms of species and subspecies in the Vicieae [Issue 3]*. Southampton, UK: Biology Department, University of Southampton.

Amede, T., & Schubert, S. (2003). Mechanisms of drought resistance in grain legumes. I. Osmotic adjustment. *Ethiopian Journal of Science*, *26*, 37–46.

Arese, P., & De Flora, A. (1990). Pathophysiology of hemolysis in glucose-6-phosphate dehydrogenase deficiency. *Seminars in Hematology*, *27*, 1–40.

Avila, C. M., Satovic, Z., Sillero, J. C., Rubiales, D., Moreno, M. T., & Torres, A. M. (2004). Isolate and organ-specific QTLs for Ascochyta blight resistance in faba bean (*Vicia faba* L.). *Theoretical and Applied Genetics*, *108*(6), 1071–1078.

Bayaa, B., Kabakebji, M., Khalil, S., Kabbabeh, S., & Street, K. (2004). Pathogenicity of Syrian isolates of *Ascochyta fabae* Speg. and *Botrytis fabae* Sard. and sources of resistance to both pathogens in a germplasm collection from central Asia and Caucasia European Association for Grain Legume Research-AEP (Ed.), *Legumes for the benefit of agriculture, nutrition and the environment: Their genomics, their products, and their improvement* (p. 308). Dijon, France: INRA. Proceedings of the 5th European conference on grain legumes.

Bernier, C. C., & Conner, R. L. (1982). Breeding for resistance to faba bean rust. In G. Hawtin & C. Webb (Eds.), *Faba bean improvement* (pp. 251–257). The Hague, Netherlands: Martinus Nijhoft Publishers.

Bond, D. A., & Crofton, G. R. A. (1999). History of winter beans in the UK. *Journal of the Royal Agricultural Society of England*, *160*, 200–209.

Bond, D. A., Jellis, G. J., Rowland, G. G., Le Guen, J., Robertson, L. D., Khalil, S. A., et al. (1994). Present status and future strategy in breeding faba beans (*Vicia faba* L.) for resistance to biotic and abiotic stresses. *Euphytica*, *73*, 151–166.

Bond, D. A., & Kirby, E. J. M. (1999). *Anthophora plumipes* (Hymenoptera: Anthophoridae) as a pollinator of broad bean (*Vicia faba* major). *Journal of Agricultural Research, 38*, 199–203.

Bond, D. A., Lawes, D. A., Hawtin, G. C., Saxena, M. C., & Stephens, J. S. (1985). Faba bean (*Vicia faba* L.). In R. J. Summerfield & E. H. Roberts (Eds.), *Grain legume crops* (pp. 199–265). London, UK: William Collins Sons Co. Ltd.

Bond, D. A., & Pope, M. (1974). Factors affecting the proportions of cross-bred and selfed seed obtained from field bean (*Vicia faba* L.) crops. *Journal of Agricultural Science, 83*, 343–351.

Borlaug, N. E., & Dowswell, C. R. (2005). Feeding a world of ten billion people: a 21st century challenge. In R. Tuberosa, R. L. Phillips, & M. D. Gale (Eds.), *The wake of the double helix: From the green revolution to the gene revolution: Proceedings of an international congress, University of Bologna, Italy, 27–31 May 2003*. Bologna, Italy: Avenue Media.

Cole, S. (1970). *The neolithic revolution*. London, UK: Trustees of the British Museum of Natural History.

Crépon, K., Marget, P., Peyronnet, C., Carrouee, B., Arese, P., & Duc, G. (2010). Nutritional value of faba bean (*Vicia faba* L.) seeds for feed and food. *Field Crops Research, 115*, 329–339.

Crews, T. E., & Peoples, M. B. (2004). Legume versus fertilizer sources of nitrogen: ecological tradeoffs and human needs. *Agriculture, Ecosystems and Environment, 102*, 279–297.

Cubero, J. I. (1973). Evolutionary trends in *Vicia faba*. *Theoretical and Applied Genetics, 43*(2), 59–65.

Cubero, J. I. (1974). On the evolution of *Vicia faba* L. *Theoretical and Applied Genetics, 45*, 47–51.

Cubero, J. I. (1982). Interspecific hybridization in *Vicia*. In G. Hawtin & C. Webb (Eds.), *Faba bean improvements* (pp. 91–108). The Hague, Netherlands: Martinus Nijhoff Publishers.

Cubero, J. I. (1984). Problems and perspectives in breeding for protein content in *Vicia faba*. *FABIS Newsletter – Faba Bean Information Service, 9*, 1–9.

Cubero, J. I., Moreno, M. T., & Hernandez, L. (1992). A faba bean (*Vicia faba* L.) cultivar resistant to broomrape (*Orobanche crenata* Forsk.). In P. Plancquaert (Ed.), *Proceedings of the 1st European conference on grain legumes, Angers, France, 1–3 June 1992* (pp. 41–42). Paris, France: L'association Europeenne des Proteagineux.

Cubero, J. I., Pieterse, A. H., Khalil, S. A., & Sauerborn, J. (1994). Screening techniques and sources of resistance to parasitic angiosperms. *Euphytica, 74*, 51–58.

Conner, R. L., & Bernier, C. C. (1982). Host range of *Uromyces viciae-fabae*. *Phytopathology, 72*, 687–689.

Díaz-Ruiz, R., Torres, A. M., Satovic, Z., Gutierrez, M. V., Cubero, J. I., & Román, B. (2010). Validation of QTLs for *Orobanche crenata* resistance in faba bean (*Vicia faba* L.) across environments and generations. *Theoretical and Applied Genetics, 120*(5), 909–919.

Dita, M. A., Rispail, N., Parts, E., Rubiales, D., & Singh, K. B. (2006). Biotechnology approaches to overcome biotic and abiotic stress constraints in legumes. *Euphytica, 147*, 1–24.

Drayner, J. M. (1959). Self- and cross-fertility in field beans (*Vicia faba* L.). *Journal of Agricultural Science, 53*, 387–403.

Duc, G., Bao, S. Y., Baum, M., Redden, B., Sadiki, M., Suso, M. J., et al. (2010). Diversity maintenance and use of *Vicia faba* L. genetic resources. *Field Crops Research, 115*, 270–278.

Ellwood, S. R., Phan, H. T., Jordan, M., Hane, J., Torres, A. M., Avila, C. M., et al. (2008). Construction of a comparative genetic map in faba bean (*Vicia faba* L.); conservation of genome structure with *Lens culinaris*. *BMC Genomics, 9*, 380.

Emeran, A. A., Sillero, J. C., & Rubiales, D. (2001). Physiological specialisation of *Uromyces viciae-fabae* European Association for Grain Legume Research-AEP (Ed.), *Towards the sustainable production of healthy food, feed and novel products* (p. 263). Paris, France: European Association for Grain Legume Research-AEP. Proceedings of the 4th European conference on grain legumes, 8–12 July 2001, Cracow, Poland.

FAOSTAT, (2012). *World statistics on faba bean.* Rome, Italy: Food and Agriculture Organization of the United Nations. Available at: <http://faostat.fao.org/>.

Fischer, R. A., & Turner, N. C. (1978). Plant productivity in the arid and semiarid zone. *Annual Review of Plant Physiology, 29*, 277–317.

Frediani, M., Maggini, F., Gelati, M. T., & Cremonini, R. (2004). Repetitive DNA sequences as probes for phylogenetic analysis in *Vicia* genus. *Caryologia, 57*, 379–386.

Gadh, I. P. S., & Bernier, C. C. (1984). Resistance in faba bean (*Vicia faba*) to bean yellow mosaic virus. *Plant Diseases, 68*, 109–111.

Gasim, S., Abel, S., & Link, W. (2004). Extent, variation and breeding impact of natural cross-fertilization in German winter faba beans using hilum colour as marker. *Euphytica, 136*, 193–200.

Gressel, J., Hanafi, A., Head, G., Marasas, W., Obilana, B., Ochanda, J., et al. (2004). Major heretofore intractable biotic constraints to African food security that may be amenable to novel biotechnological solutions. *Crop Protection, 23*, 661–689.

Gutierrez, N., Avila, C. M., Duc, G., Marget, P., Suso, M. J., Moreno, M. T., et al. (2006).). CAPs markers to assist selection for low vicine and convicine contents in faba bean (*Vicia faba* L.). *Theoretical and Applied Genetics, 114*, 59–66.

Gutierrez, N., Avila, C. M., Moreno, M. T., & Torres, A. M. (2008). Development of SCAR markers linked to zt-2, one of the genes controlling absence of tannins in faba bean. *Australian Journal of Agricultural Research, 59*, 62–68.

Gutierrez, N., Avila, C. M., Rodriguez-Suarez, C., Moreno, M. T., & Torres, A. M. (2007). Development of SCAR markers linked to a gene controlling absence of tannins in faba bean. *Molecular Breeding, 19*, 305–314.

Hanelt, P., & Mettin, D. (1989). Biosystematics of the genus *Vicia* L. (Leguminosae). *Annual Review of Ecological Systems, 20*, 199–223.

Hanelt, P., Schäfer, H., & Schultze-Motel, J. (1972). Die Stellung von *Vicia faba* L. in der Gattung *Vicia* L. und Betrachtungen zur Entstehung dieser Kulturart. *Kulturpflanze, 20*, 263–275.

Hanounik, S. B., Jellis, G. J., & Hussein, M. M. (1993). Screening for resistance in faba bean. In K. B. Singh & M. C. Saxena (Eds.), *Breeding for stress tolerance in cool-season food legumes* (pp. 97–106). Aleppo, Syria: ICARDA.

Hanounik, S. B., & Roberston, L. D. (1989). Resistance of faba bean germplasm to blight caused by *Ascochtya fabae. Plant Disease, 73*, 202–205.

Hawtin, G., & Omar, M. (1980). Estimation of out-crossing between isolation plots of faba beans. *FABIS Newsletter – Faba Bean Information Service, 25*, 36–39.

Hopf, M. (1973). Fruhe Kulturpflanzen aus bulgarien. *Jahrbuch des Romisch-Germanischer Zentralmuseums Mainz, 20*, 1–47.

International Center for Agricultural Research in the Dry Areas (ICARDA), (1987). *Faba bean pathology progress report 1986–1987: Food legume improvement program.* Aleppo, Syria: ICARDA.

Jensen, E. S., Peoples, M. B., & Hauggaard-Nielsen, H. (2010). Faba bean in cropping systems. *Field Crops Research, 115*, 203–216.

Katul, L., Vetten, H. J., Maiss, E., Makkouk, K. M., Lesemann, D. E., & Casper, R. (1993). Characterization and serology of virus-like particles associated with faba bean necrotic yellows. *Annals of Applied Biology, 123*, 629–647.

Khalil, S., Bayaa, B., Malhotra, R. S., Erskine, W., & Saxena, M. C. (2004). Breeding for combined resistance to chocolate spot (*Botrytis fabae* Sard.) and Ascochyta blight (*Ascochyta fabae* Speg.) diseases in faba bean (*Vicia faba* L.) European Association for Grain Legume Research-AEP (Ed.), *Legumes for the benefit of agriculture, nutrition and the environment: Their genomics, their products, and their improvement.* Paris, France: Dijon, France. INRA. Proceeding of the 5th European conference on grain legumes/2nd international conference on legume genomics and genetics, 7–11 June 2004.

Khalil, S., & Erskine, W. (1999). Breeding for Orobanche resistance in faba bean and lentil. In J. I. Cubero, M. T. Moreno, D. Rubiales, & J. C. Sillero (Eds.), *Resistance to broomrape: The state of the art* (pp. 63–76). Seville, Spain: Junta de Andalucia.

Khalil, S., Kharrat, M., Malhotra, R., Saxena, M., & Erskine, W. (2004). Breeding faba bean for Orobanche resistance. In R. Dahan & M. El-Mourid (Eds.), *Integrated management of orobanche in food legumes in the near east and North Africa* (pp. 1–18). Aleppo, Syria: ICARDA. Proceedings of the expert consultation on IPM for orobanche in food legume systems in the near East and North Africa, Rabat Morocco, 7–9 April 2003.

Khalil, S. A., Nassib, A. M., & Mohammed, H. A. (1985). Identification of some sources of resistance to diseases in faba beans. II – Rust (*Uromyces fabae*). *FABIS Newsletter – Faba Bean Information Service, 11*, 18–20.

Khan, H. R., Link, W., Hocking, T. J., & Stoddard, F. L. (2007). Evaluation of physiological traits for improving drought tolerance in faba bean (*Vicia faba* L.). *Plant and Soil, 292*, 205–217.

Khan, H. R., Paull, J. G., Siddique, K. H. M., & Stoddard, F. L. (2010). Faba bean breeding for drought-affected environments: a physiological and agronomic perspective. *Field Crops Research, 115*, 279–286.

Kharbanda, P. D., & Bernier, C. C. (1980). Cultural and pathogenic variability among isolates of *Ascochyta fabae*. *Canadian Journal of Plant Pathology, 2*, 139–142.

Kharrat, M., Abbes, Z., & Amri, M. (2010). A new faba bean small seeded variety Najeh tolerant to Orobanche registered in the Tunisian catalogue. *Tunisian Journal of Plant Protection, 5*, 125–130.

Kharrat, M., & Halila, M. H. (1994). Orobanche species on faba beans (*Vicia faba* L.) in Tunisia: problems and management. In A. H. Pieterse, J. A. C. Verkleij, & S. J. ter Borg (Eds.), *Biology and management of orobanche* (pp. 639–643). Amsterdam, Netherlands: Royal Tropical Institute. Proceedings of the third international workshop on orobanche and related Striga research.

Kharrat, M., Le Guen, J., & Tivoli, B. (2006). Genetics of resistance to 3 isolates of *Ascochyta fabae* on faba bean (*Vicia faba* L.) in controlled conditions. *Euphytica, 151*, 49–61.

Kohpina, S., Knight, R., & Stoddard, F. L. (1999). Variability of *Ascochyta fabae* in South Australia. *Australian Journal of Agricultural Research, 50*, 1475–1481.

Kumari, S. G., & Makkouk, K. M. (2003). Differentiation among bean leaf roll virus susceptible and resistant lentil and faba bean genotypes on the basis of virus movement and multiplication. *Journal of Phytopathology, 151*, 19–25.

Kupicha, F. K. (1976). The infrageneric structure of *Vicia*. *Notes from the Royal Botanic Garden. Edinburgh, 34*, 287–326.

Ladizinsky, G. (1975). Seed protein electrophoresis of the wild and cultivated species of section *faba* of *Vicia*. *Euphytica, 24*, 785–788.

Lawsawadsiri, S. (1995). *Ascochyta fabae*. *PhD Thesis*. Adelaide, Australia: University of Adelaide.

Link, W., Balko, C., & Stoddard, F. L. (2010). Winter hardiness in faba bean: Physiology and breeding. *Field Crops Research, 115*, 287–296.

Link, W., & von Kittlitz, E. (1989). Rate of cross-fertilization between single plants and between plots. *FABIS Newsletter – Faba Bean Information Service*, *25*, 36–39.

Maalouf, F., Ahmed, S., Kabakebji, M., Kabbabeh, S., & Street, K. (2009). Breeding faba bean for resistance to chocolate spot: *Oral presentation to fungal diseases session*. Beirut, Lebanon: Arab Society for Plant Protection and National Council for Scientific Research. 10th Arab congress of plant protection, 26–30 October 2009.

Maalouf, F. S., Ahmed, S., Kabakebji, M., Khalil, S., Abang, M., Kabbabeh, S., et al. (2010). Sources of multiple resistances for key foliar disease of faba bean: *Poster presented at the 5th international conference on food legumes/7th European conference on grain legumes*. Antalya, Turkey: European Plant Science Organization. 26–30 May 2010.

Maalouf, F., Ahmed, S., Nawar, M., Khalil, S., & Bayaa, B. (2012). Breeding faba bean for Ascochyta blight resistance: *Proceedings of the 3rd international ascochyta workshop, 22–26 April 2012*. Córdoba, Spain: University of Córdoba, Institute of Sustainable Agriculture, Institute of Agricultural and Fishery Research and Training. p. 43.

Maalouf, F., Khalil, S., Ahmed, S., Akinnola, N., Kharrat, M., Hajjar, S., et al. (2011). Yield stability of faba bean lines under diverse broomrape prone production environments. *Field Crop Research*, *124*, 288–294.

Makkouk, K. M., Rizkallah, L., Madkour, M., El-Sherbeiny, M., Kumari, S. G., Amriti, A. W., et al. (1994). Survey of faba bean (*Vicia faba* L.) for viruses in Egypt. *Phytopathologia Mediterranea*, *33*, 207–211.

Maurin, N., & Tivoli, B. (1992). Variation in the resistance of *Vicia faba* to *Ascochyta fabae* in relation to disease development in field trials. *Plant Pathology*, *41*, 737–744.

Maxted, N. (1993). A phenetic investigation of *Vicia* L. subgenus *Vicia* (Leguminosae-Vicieae). *Botanical Journal of the Linnean Society*, *111*, 155–182.

Maxted, N. (1995). An ecogeographical study of *Vicia* subgenus *Vicia*: *Systematic and ecogeographic studies on crop genepools 8*. Rome, Italy: International Plant Genetic Resources Institute. Available at: <http://pdf.usaid.gov/pdf_docs/PNABU773.pdf/>.

Maxted, N., Callimassia, M. A., & Bennett, M. D. (1991). Cytotaxonomic studies of eastern Mediterranean *Vicia* species (Leguminosae). *Plant Systematics and Evolution*, *177*, 221–234.

Mazid, A., Shideed, K., & Amri, A. (2006). Status of and threats to on-farm agrobiodiversity and its impact on rural livelihoods in dry areas of West Asia: *Dryland agrobiodiversity report*. Aleppo, Syria: ICARDA.

McDonald, G. K., & Paulsen, G. M. (1997). High temperature effects on photosynthesis and water relations of grain legumes. *Plant and Soil*, *196*, 47–58.

Mesa-García, J., & García-Torres, L. (1991). Status of *Orobanche crenata* in faba bean in the Mediterranean region and its control. *Options Méditerranéennes Série Séminaires*, *10*, 75–78.

Muratova, V. S. (1931). Common beans (*Vicia faba* L.). *Bulletin of Applied Botany, Genetics and Plant Breeding Supplement*, *50*, 1–298.

Nassib, A. M., Ibrahim, A. A., & Khalil, S. A. (1982). Breeding for resistance to Orobanche. In G. Hawtin & C Webb (Eds.), *Faba bean improvement* (pp. 199–206). The Hague, Netherlands: Martinus Nijhoft Publishers.

Perez-de-Luque, A., Lozano, M. D., Moreno, M. T., Testillano, P. S., & Rubiales, D. (2007). Resistance to broomrape (*Orobanche crenata*) in faba bean (*Vicia faba*): cell wall changes associated with pre haustorial defensive mechanisms. *Annals of Applied Biology*, *151*, 89–98.

Pierre, J., Suso, M. J., Moreno, M. T., Esnault, R., & Le Guen, J. (1999). Diversite et efficacite de l'entomofaune pollinisatrice (Hymenoptera: Apidae) de la féverole (*Vicia faba* L.) sur deux sites, en France et en Espagne. *Annales de la Société entomologique de France, 35*(Supplément), 312–318.

Pozarkova, D., Koblizkova, A., Román, B., Torres, A. M., Lucretti, S., Lysak, M., et al. (2002). Development and characterization of microsatellite markers from chromosome 1-specific DNA libraries of *Vicia faba*. *Biologia Plantarum, 45*, 337–345.

Przybylska, J., & Zimniak-Przybylska, Z. (1997). Electrophoretic seed albumin patterns and species relationship in *Vicia* sect. *faba* (Fabaceae). *Plant Systematics and Evolution, 198*, 179–194.

Raina, S. N., & Ogihara, Y. (1994). Chloroplast DNA diversity in *Vicia faba* and its close wild relatives: implications for reassessment. *Theoretical and Applied Genetics, 88*, 261–266.

Ramsey, M., Knight, R., & Paull, J. (1995). Ascochyta and chocolate spot resistant faba beans (*Vicia faba* L.) for Australia: *Proceedings of the 2nd European conference on grain legumes.* (pp. 164–165). Copenhagen, Denmark: European Association for Grain Legume Research.

Rashid, K. Y., & Bernier, C. C. (1984). Evaluation of resistance in *Vicia faba* to two isolates of the rust fungus *Uromyces viciae-fabae* from Manitoba. *Plant Disease, 68*, 16–18.

Rashid, K. Y., & Bernier, C. C. (1986). The genetics of resistance in *Vicia faba* to two races of *Uromyces viciae-fabae* from Manitoba. *Canadian Journal of Plant Pathology, 8*, 317–322.

Rashid, K. Y., & Bernier, C. C. (1991). The effect of rust on yield of faba bean cultivars and slow-rusting populations. *Canadian Journal of Plant Science, 71*, 967–972.

Rashid, K. Y., Bernier, C. C., & Conner, R. L. (1991). Genetics of resistance in faba bean inbred lines to five isolates of *Ascochyta fabae*. *Canadian Journal of Plant Pathology, 13*, 218–225.

Rispail, N., Kalo, P., Kiss, G. B., Ellis, T. H. N., Gallardo, K., Thompson, R. D., et al. (2010). Model of legumes to contribute to faba bean breeding. *Field Crops Research, 115*, 253–269.

Robertson, L. D., & Cardona, C. (1986). Studies on bee activity and outcrossing in increase plots of *Vicia faba* L.. *Field Crops Research, 15*, 157–164.

Robertson, L. D., Singh, K. B., Erskine, W., & Abd El Moneim, A. M. (1996). Useful genetic diversity in germplasm collections of food and forage legumes from West Asia and North Africa. *Genetic Resources and Crop Evolution, 43*, 447–460.

Román, B., Satovic, Z., Pozarkova, D., Macas, J., Dolezel, J., Cubero, J. I., et al. (2004). Development of a composite map in *Vicia faba*, breeding applications and future prospects. *Theoretical and Applied Genetics, 108*(6), 1079–1088.

Roupakias, D. G. (1986). Interspecific hybridization between *Vicia faba* L. and *Vicia narbonesis* L.: Early pod growth and embryo-sac development. *Euphytica, 35*, 175–183.

Rubiales, D., Perez-de-Luque, A., Fernandez-Aparicio, M., Sillero, J. C., Román, B., Kharrat, M., et al. (2006). Screening techniques and sources of resistance against parasitic weeds in grain legumes. *Euphytica, 147*, 187–199.

Saber, H. A., Omer, M. A., El-Hady, M. M., Mohmoud, S. A., Abou-Zeid, N. M., & Radi, M. M. (1999). Performance of a newly-bred faba bean line (X-843) resistant to Orobanche in Egypt. In J. Kroschel, M. Abderahibi, & H. Betz (Eds.), *Advances in parasitic weed control at on-farm level. Vol. II: Joint action to control orobanche in the WANA region* (pp. 251–257). Weikersheim, Germany: Margraf Verlag.

Santidrián, S., Sobrini, F. J., & Larralde, J. (1981). Problemas que plantea la utilizaci6n de habas desde el punto de vista de la nutrici6n. *Revistas de la Institución Príncipe de Viana, 1*, 95–103.

Sau, F., & Mínguez, M. I. (2000). Adaptation of indeterminate faba beans to weather and management under a Mediterranean climate. *Field Crops Research, 66*, 81–99.

Saxena, M. C. (1991). Status and scope for production of faba bean in the Mediterranean countries. *Options Méditerranéennes Série Séminaires, 10*, 15–20.

Siddique, K. H. M., Regan, K. L., Tennant, D., & Tomson, B. D. (2001). Water use and water use efficiency of cool season grain legumes in low rainfall Mediterranean-type environments. *European Journal of Agronomy, 15*, 267–280.

Sillero, J., Avila, C. M., Moreno, M. T., & Rubiales, D. (2001). Identification of resistance to *Ascochyta fabae* in *Vicia faba* germplasm. *Plant Breeding, 120*, 529–531.

Sillero, J. C., Moreno, M. T., & Rubiales, D. (2000). Characterization of new sources of resistance to *Uromyces viciae-fabae* in a germplasm collection of *Vicia faba*. *Plant Pathology, 49*, 389–395.

Sillero, J. C., Villegas-Fernandez, A. M., Thomas, J., Rojas-Molina, M. M., Emeran, A. A., Fernandez-Aparicio, M., et al. (2010). Faba bean breeding for disease resistance. *Field Crops Research, 115*, 297–307.

Sobrini, F. J., Santidrian, S., & Larralde, J. (1982). Effect of tannin content of *Vicia faba* seeds on the growth and nutritive value. *FABIS Newsletter – Faba Bean Information Service, 5*, 32–35.

Suso, M. J., Pierre, J., Moreno, M. T., Esnault, R., & Le Guen, J. (2001). Variation in outcrossing levels in faba bean cultivars: role of ecological factors. *Journal of Agricultural Science, 136*, 399–405.

Suso, M. J., Gilsanz, S., Duc, G., Marget, P., & Moreno, M. T. (2006). Germplasm management of faba bean (*Vicia faba* L.): Monitoring intercrossing between accessions with inter-plot barriers. *Genetic Resources and Crop Evolution, 53*, 1427–1439.

Suso, M. J., Nadal, S., Román, B., & Gilsanz, S. (2008). *Vicia faba* germplasm multiplication floral traits associated with pollen-mediated gene flow under diverse between-plot isolation strategies. *Annals of Applied Biology, 152*, 201–208.

Schultze-Motel, J. (1972). Die archaologischen reste der ackerbohne, *Vicia faba* L. und die genese de art. *Kulturpflanze, 19*, 321–358.

Tanno, K., & Willcox, G. (2006). The origins of cultivation of *Cicer arietinum* L. and *Vicia faba* L.: Early finds from Tell el-Kerkh, north-west Syria, late 10th millennium B.P.. *Vegetation History and Archaeobotany, 15*, 197–204.

ter Borg, S. J., Willemsen, A., Khalil, S. A., Saber, H. A., Verkleij, J. A. C., & Pierterse, A. H. (1994). Field study of the interaction between *Orobanche crenata* Forsk. and some lines of *Vicia faba*. *Crop Protection, 13*, 611–616.

Trabut, L. (1911). L'indegenat de la Flore en Algeria. *Bulletin de la Société Nationale Africaine, 7*(15), 1–7.

Torres, A. M., Avila, C. M., Gutierrez, N., Palomino, C., Moreno, M. T., & Cubero, J. I. (2010).). Marker-assisted selection in faba bean (*Vicia faba* L.). *Field Crops Research, 115*, 243–252.

van de Ven, W. T. G., Duncan, N., Ramsay, G., Phillips, M., Powell, W., & Waugh, R. (1993). Taxonomic relationships between *V. faba* and its relatives based on nuclear and mitochondrial RFLPs and PCR analysis. *Theoretical and Applied Genetics, 86*, 71–80.

van Leur, J. A. G., Kumari, S. G., Makkouk, K. M., & Rose, I. A. (2006). Viruses on faba bean in north-east Australia and strategies for virus control. In C. Avila, J. I. Cubero, M. T. Moreno, M. J. Suso, & A. M. Torres (Eds.), *Proceedings of the international workshop on faba bean breeding and agronomy, Seville, Spain, 25–27 October 2006* (pp. 129–131). Córdoba, Spain: Junta de Andalucía.

Wang, H., Zong, X., Guan, J., Yang, T., Sun, X., Yu, M., et al. (2012). Genetic diversity and relationship of global faba bean (*Vicia faba* L.) germplasm revealed by ISSR markers. *Theoretical and Applied Genetics*, *124*, 789–797.

Xu, S. C., Gong, Y. M., Mao, W. H., Hu, Q. Z., Zhang, G. W., Fu, W., et al. (2012). Development and characterization of 41 novel EST-SSR markers for *Pisum sativum*. *American Journal of Botany*, *99*(4), 149–153.

Ye, Y., Lang, L., Xia, M., & Tu, J. (2003). *Faba beans in China*. Beijing, China: China Agriculture Press, in Chinese.

Zeid, M., Schoen, C., & Link, W. (2003). Genetic diversity in recent elite faba bean lines using AFLP markers. *Theoretical and Applied Genetics*, *107*, 1304–1314.

Zheng, Z., Wang, S., & Zong, X. (1997). *Food legume crops in China*. Beijing, China: China Agriculture Press, in Chinese.

Zong, X, Liu, X., Guan, J., Wang, S., Liu, Q., Paull, J. G., et al. (2009). Molecular variation among Chinese and global winter faba bean germplasm. *Theoretical and Applied Genetics*, *118*, 971–978.

6 Cowpea

Ousmane Boukar, Ranjana Bhattacharjee, Christian Fatokun, P. Lava Kumar and Badara Gueye

International Institute of Tropical Agriculture (IITA), Ibadan, Nigeria

6.1 Introduction

Cowpea is probably the most commonly grown and consumed legume in the dry savanna regions of sub-Saharan Africa (SSA). Because of its drought tolerance ability, it is well adapted to the dry savanna, where the bulk of the crop is produced successfully. It is mostly grown by small-scale farmers in their fields in association with cereals, such as millets, sorghum, maize and groundnut. The West African subregion contributes to about 95% of global cowpea production (Food and Agriculture Organization, 2012). Nigeria alone produced over 2.24 million metric tons in about 2.52 million ha followed by Niger with 1.77 million metric tons produced in 5.57 million ha (FAO, 2012). Brazil is another country where a high volume of cowpea is produced and consumed. In 2011 the country produced about 822,000 metric tons in 1.6 million ha at an average yield of 525 kg ha^{-1}, which is about 11% higher than the average yield in SSA farmers' fields. According to the FAO (http://www.faostat.org), the world cowpea grain production has increased from about 1.3 million metric tons in the 1970s to over 5 million metric tons in the 2000s. However, annual consumption of cowpea in Nigeria is over 3.0 million tons, whereas the country produces about 2.6 million tons. Baseline studies on cowpea in western and central Africa, which account for 75% of the total world production, have projected that demand will grow faster at the rate of 2.68% in each year than supply at 2.55% annually over the period of 2007 to 2030 in the subregion (Abate, 2012). The mean grain yield of cowpea in a typical SSA farmer's field is about 495 kg ha^{-1}, much lower than what is obtained under experimental conditions (FAO, 2012). The low grain yield is caused by a number of biotic and abiotic factors. Cowpea is susceptible to many insects and pests such as aphids in the seedling stage, flower bud thrips at flowering stage, maruca pod borer and a complex of pod-sucking bugs at flowering and podding stages. Bruchid (*Callosobruchus maculatus*) can cause significant loss to cowpea seeds in storage. Each of these insects is capable of causing significant reduction in the grain yield and thereby farmers' income. Apart from insects and pests, there are fungal, bacterial and viral diseases that afflict the crop in field and reduce yield (Allen, Thottappilly, Emechebe, & Singh, 1998; Emechebe & Lagoke, 2002). Through cowpea breeding activities, several improved cowpea lines and varieties have been developed and released to farmers in different countries. These lines

Genetic and Genomic Resources of Grain Legume Improvement. DOI: http://dx.doi.org/10.1016/B978-0-12-397935-3.00006-2

and varieties have been characterized by extra early, early or medium maturity, dual purpose, i.e. grain and fodder producing, *Striga* and *Alectra* resistance, drought tolerance and resistance to some diseases such as bacterial blight. Some other germplasm lines were identified with >30% protein in the grains (Boukar et al., 2011). As a result of this, cowpea is commonly referred to as 'poor man's meat', especially among the inhabitants of rural areas and urban slums of western and central Africa. The grains are processed into different types of food products, such as *kosai* or *akara*, *moi moi*. Green immature cowpea pods are harvested and sold in local markets for consumption as a vegetable. Cowpea leaves are also known to contain a high amount of protein and minerals, such as calcium, phosphorus and vitamin B (Maynard, 2008). Further, in the present global scenario with the regularly expanding need for varietal improvement, there is an urgent need for the systematic collection, conservation, characterization, evaluation and utilization of germplasm for both the present and posterity.

6.2 Origin, Distribution, Diversity and Taxonomy

Cowpea (*Vigna unguiculata* (L.) Walp.), a true diploid ($2n=2x=22$) species, belongs to the family *Leguminosae*, tribe *Phaseoleae*, genus *Vigna*, and section *Catiang* (Verdcourt, 1970). The genus *Vigna* comprises about 85 species, which Marechal, Mascherpa, and Stainier (1978) divided into seven subgenera, namely *Ceratotropis*, *Haydonia*, *Lasiocarpa*, *Macrorhycha*, *Plectotropis*, *Sigmoidotropis* and *Vigna*. The Asiatic *Vigna* includes green gram (*Vigna radiata*), black gram (*Vigna mungo*) and rice bean (*Vigna umbellata*) of the subgenus *Ceratotropis*, whereas cowpea along with its cross-compatible wild relatives are in a subgenus of *Vigna*. Taxonomic relationships between the members of *Vigna* species, based on restriction fragment length polymorphism (RFLP) markers, confirmed the distinctness of bambara groundnut (*V. subterranea*), cowpea along with members of section *Catiang*, Asiatic *Vigna* species and those belonging to subgenus *Plectotropis* (Fatokun, Danesh, Young, & Stewart, 1993). The study also revealed that members of subgenus *Plectotropis*, which include *V. vexillata*, are closer taxonomically to those belonging to section *Catiang*. According to Baudoin and Marechal (1988), *V. vexillata* is an intermediate type between the African and Asiatic *Vigna* species. Despite the phylogenetic proximity of *V. vexillata* and cowpea, there exists a strong barrier to cross compatibility between them (Fatokun, 2002). Most members of the *Vigna* species are true diploid with $2n=2x=22$ chromosome numbers (Marechal et al., 1978). However, some species, such as *V. ambacensis* Bak. f., *V. heterophylla*, *V. reticulata* Hook. f. and *V. wittei* Bak. f., have $2n=2x=20$ chromosome numbers, while *V. glabrescens* has $2n=4x=44$ chromosomes and is the only known amphidiploid in the subtribe *Phaseolinea* (Verdcourt, 1970). The progenitor of cowpea is *V. unguiculata* var. *spontanea* (formerly var. *dekindtiana*), whose habitat has been found in all lowland areas of SSA, outside the high rain forests and deserts. However, southern Africa has been suggested as the centre of origin for wild cowpea (Padulosi & Ng, 1997). According to these workers, the area from Namibia through Botswana,

Zambia, Zimbabwe, Mozambique, Republic of South Africa and Swaziland represents the highest genetic diversity and most primitive forms of wild *V. unguiculata*. The researchers further reported that some primitive wild cowpea relatives, such as *V. unguiculata* var. *rhomboidea*, var. *protracta*, var. *tenuis* and var. *stenophylla*, are found mainly in the Transvaal region of South Africa. The restricted distribution of these primitive forms of wild cross-compatible cowpea relatives in this part of southern Africa provides strong evidence that the region is probably the centre of origin of wild cowpea. The existence of substantial variation among traditional cowpea varieties grown by the farmers in western and central Africa confirms that the region is the possible centre of diversity for cowpea. The crop would have been growing in this area over a long period of time, during which a number of mutants and recombinants would have arisen and accumulated in germplasm lines and farmers' varieties. The oldest evidence that cowpea existed in West Africa was obtained from carbon dating of specimens from the Kimtampo rock shelter in central Ghana (Flight, 1976). Cultivated cowpea is divided into four cultivar groups, namely *Biflora*, *Sesquipedalis*, *Textilis* and *Unguiculata* (Ng & Marechal, 1985). Cowpea belongs to culti-group *unguiculata*, while the yard-long bean or asparagus bean belongs to *sesquipedalis*. Cowpea and yard-long bean cross readily and the progeny from these crosses are fertile and viable. While cowpea is grown mainly for its dry grains in SSA, South and Central America, southern USA and Europe, the yard-long bean is commonly grown in Southeast Asia for long green fleshy pods consumed as a vegetable. It is interesting to note that in SSA, where cowpea has its centre of origin, the pods are short with crowded seeds, while the yard-long bean found commonly in India and some southeast Asian countries have long pods that are fleshy and with seeds sparsely distributed. It has been further suggested that the yard-long bean has evolved in Asia from cowpea following deliberate selection by farmers in the region for plants with long pods that are consumed as a vegetable.

6.3 Erosion of Genetic Diversity from the Traditional Areas

The development of new crop varieties and their widespread adoption by farmers is a major factor responsible for genetic erosion. In addition, agricultural intensification, changes in land use planning, pests and pathogens, increased human population, land degradation and changes in the environment such as climate change may also contribute to genetic erosion. There has not been a concerted research effort aimed at understanding the population dynamics of cowpea and its wild relatives. However, a number of breeders are engaged in the development of new improved cowpea varieties, which generally perform better than farmers' own varieties/landraces. It is therefore reasonable to expect that with the passage of time, these improved varieties will replace farmers' varieties, which quite often are traditional varieties and have not undergone any breeding efforts. Farmers who have adopted improved varieties still plant, though in small areas, their traditional varieties, which seem to meet their culinary and some other special needs. Studies carried out in some parts of northern

Nigeria on adoption of new cowpea varieties showed that many farmers embrace the improved varieties because of their superior grain yield as compared with farmers' varieties, particularly in areas where *Striga gesnerioides* has become endemic. About 72% and 81% of cowpea farmers in Borno and Kano states of Nigeria have adopted one or more improved varieties, respectively (Amaza, 2011). These new varieties are mostly resistant to *Striga* and some other biotic stresses. Farmers are able to access the seeds of these new improved varieties through extension personnel, NGOs and researchers. However, farmer-to-farmer seed diffusion has also helped in some communities to spread improved cowpea varieties in the region. In addition, breeders now engage farmers in their breeding efforts by practising farmer participatory variety selection. This practice exposes farmers to the better performing new lines that are being selected, thereby enhancing their early and wider adoption. With current trends, most cowpea farmers in SSA may adopt planting of the new higher yielding varieties while discarding their traditional lines. This may result in the loss of farmers' traditional varieties if they are not collected soon and conserved in gene banks. Van de Wouw, van Hintum, Kik, van Treuren, and Visser (2010) have also stated, following a review of literature on the subject, that genetic erosion of crops has been associated with the introduction of modern varieties. These researchers have opined that it is not yet clear whether an active breeding programme with many new releases contributes to maintaining a certain level of diversity or is countereffective and hastens a potential process of genetic erosion. They concluded that the threat of genetic erosion due to modernization of agriculture is most probably highest for crops no breeders are interested in. The threat of genetic erosion to cowpea due to introduction of modern farming techniques may not be very serious at present since farmers still grow their traditional varieties in many SSA communities. Adoption of new varieties by farmers has so far not attained the level that calls for special attention. Besides widespread adoption of new improved varieties, pressure on available suitable farmland in the various communities may lead to loss of cowpea germplasm.

6.4 Status of Germplasm Resources Conservation

Given the importance of genetic resources conservation, IITA is committed to the collection and conservation of cowpea germplasm. The conservation activity started in the mid 1970s with the establishment of the IITA gene bank. Collection was carried out through several plant exploration missions in more than 30 countries, donations from or exchange with national programmes, individual scientists, IBPGR and the University of Gembloux. The IITA Genetic Resource Center (GRC) maintains in its *ex situ* collections more than 15,100 accessions of cultivated and more than 1900 accessions of wild relatives. The main cowpea wild species available in IITA collections include: *V. dekindtiana*, *V. vexillata*, *V. spontanea*, *V. tenuis*, *V. protracta*, *V. baoulensis* and *V. stenophylla*. The collection missions for cowpea and wild *Vigna* started in 1972 and 1976, respectively. The cowpea collections maintained at IITA have about 64% of their germplasm from Africa with 39% from Nigeria. In addition, the collection consists of 23% germplasm from India and 6% from the United States. More than 96% of the wild relatives were collected from Africa, of which 32% are from Nigeria, 8% from South Africa, 6% from Botswana and 5% from each of the following countries: Cameroon, Niger, Malawi, Tanzania and Congo. In addition to IITA cowpea collections, which represent the world's largest collection, major world collections of cowpea are also maintained at USDA (Griffin, Georgia) and UCR (Riverside, California) with 7146 accessions from 50 countries and 4876 accessions from 45 countries, respectively. About 200 wild species are also available in these gene banks. There are a considerable number of duplicates in all these major cowpea world collections. About 10,323 (65%), 1393 (20%) and 1639 (34%) accessions are estimated to be unique in IITA, USDA and UCR, respectively (Ehlers, personal communication). To ensure safe duplication of cowpea accessions, IITA has also sent about 11,761 and 10,921 accessions of cowpea to Svalbard (Norway) and Saskatoon (Canada), respectively, for long-term conservation. In addition, 1517 accessions of wild *Vigna* were sent to Svalbard and 1564 to Saskatoon for the same purpose.

6.5 Germplasm Evaluation and Maintenance

The genetic resource centre (GRC) has been characterizing and evaluating the cowpea germplasm maintained in the gene bank for its agro-morphological traits, including resistance to major biotic and abiotic stresses. About 52 and 56 descriptors have been developed for cultivated cowpea and wild *Vigna*, respectively. In collaboration with breeders, entomologists and pathologists of the institute, germplasm accessions were evaluated for insect pest and disease resistance. From 1984 to 1988 more than 8500 accessions of cowpea were evaluated for resistance to Maruca pod borer and pod-sucking bugs, more than 4000 accessions for resistance to flowering thrips and bruchid and several hundred accessions for resistance to virus diseases. Many traits have been used in genetic studies and have identified over 200 genes (Fery, 1985; Fery & Singh, 1997; Singh & Matsui, 2002) that control important characters including plant pigmentation; plant type; plant height; leaf type; growth habit; photosensitivity and maturity; nitrogen fixation; fodder quality; heat and drought tolerances; root architecture; resistance to major bacterial, fungal and viral diseases; resistance to root-knot nematode; resistance to aphid, bruchid and thrips; resistance to parasitic weeds such as *S. gesnerioides* and *A. vogelii*; pod traits; seed traits and grain quality. To characterize the cowpea germplasm well, a core collection of about 2062 accessions was defined based on geographical, agronomical and botanical descriptors (Mahalakshmi, Ng, Lawson, & Ortiz, 2007). A mini-core set of 374 accessions was further defined that are being used intensively in several cowpea breeding programmes. Currently, GRC is characterizing about 270 additional accessions of wild cowpea relatives, using both agro-morphological descriptors and molecular tools. The main objectives of GRC are to evaluate the entire cowpea germplasm for priority traits and complete the agro-morphological description of wild *Vigna* accessions. Primary production constraints, which will be targeted, include drought and heat stresses, insects (flower thrips, pod-sucking bugs, cowpea aphid), diseases (viral, fungal, bacterial and nematode) and *Alectra* and *Striga* parasitic weeds. With the recent advances in high-throughput single nucleotide polymorphism (SNP) genotyping, germplasm diversity characterization and collection management (elimination of duplicates, identification of core sets) will be conducted to take advantage of the opportunity to enhance cowpea production and productivity through molecular advances.

6.6 Use of Germplasm in Crop Improvement

6.6.1 Resistance to Bacterial Blight

Bacterial blight, caused by *Xanthomonas campestris* pv. *vignicola* [Burkholder] is a serious disease that causes appreciable yield loss in cowpea. It is the most widespread disease and has been reported from the different countries where cowpea is grown (Emechebe & Florini, 1997). The best way to control this disease would be developing varieties that are resistant and beneficial to the farmers. One of the traditional farmers' varieties, Danila, has been found to be resistant to bacterial blight. It has been used as the parent in crosses with other lines for transferring resistance to the improved varieties and breeding lines. Bacterial blight resistant lines have been selected from the advanced segregating populations resulting from such crosses.

6.6.2 Resistance to Virus Diseases

Virus diseases caused significant yield reduction in the susceptible cowpea cultivars. Cowpea is susceptible to over 140 viruses, about 20 different virus species are known to naturally infect cowpea around the world and be capable of economic damage (Hampton, Thottappilly & Rossel, 1997; Taiwo & Shoyinka, 1998). At least 15 of these viruses are transmitted through cowpea seeds. The most economically important virus species infecting cowpea in SSA include Blackeye cowpea mosaic virus (genus, *Potyvirus*), Cucumber mosaic virus (genus, *Cucumovirus*), Cowpea aphid-borne mosaic virus (genus, *Potyvirus*), Cowpea mottle virus (genus, *Carmovirus*), Cowpea mosaic virus (genus, *Comovirus*), Southern bean mosaic virus (genus, *Sobemovirus*) and Cowpea golden mosaic virus (genus, *Begomovirus*). They cause mosaic, mottling, necrosis and stunting, ultimately affecting seed production. Mixed infection with more than one virus is frequent in cowpea. Infection with multiple viruses results in much more severe symptoms and dramatic reduction in yield (Taiwo, Kareem, Nsa, & Hughes, 2007). Virus diseases are best controlled through the use of resistant varieties. Resistance to two potyviruses was found in germplasm accessions, namely TVu401, TVu1453 and TVu1948, and in advanced breeding lines, IT82D-885, IT28D-889 and IT82E-60 (Gumedzoe, Rossel, Thottappily, Asselin, & Huguenot, 1998). In addition, recent studies identified multiple virus resistance and tolerance to three virus species in breeding lines IT98K-1092-1 and IT97K-1042-3 (Ogunsola, Fatokun, Boukar, Ilori, & Kumar, 2010). Cowpea varieties with resistance to multiple virus infection are yet to be found. At IITA, research work continues to identify durable resistant cowpea varieties and also determine the genetic determinants of virus resistance in cowpea germplasm.

6.6.3 Tolerance to Flower Bud Thrips

Flower thrips (*Megalurothrips sjostedti*) cause considerable grain yield loss in cowpea, if it is not controlled by spraying with appropriate insecticides. The insects suck young flower buds which then abort prematurely. Browning of stipules and shortening of peduncles are the symptoms of damage. However entomologists have screened some of the available germplasm lines for resistance to flower thrips. A line TVu 1509 was found to exhibit a fairly good level of tolerance to flower thrips. The improved breeding line TVx 3236, which showed tolerance to flower thrips, was derived from a cross between TVu 1509 and Ife Brown. Another local line from Ghana called 'sanzi' has been found to be resistant to flower bud thrips.

6.6.4 Tolerance to Drought and Heat

Cowpea is comparatively tolerant to drought. Despite this characteristic, however, drought can still cause considerable yield loss. Efforts have been made to screen cowpea germplasm to identify lines with better drought tolerance than the currently available varieties. According to Watanabe, Hakoyama, Terao, and Singh (1997), some germplasm lines, especially TVu 11979 and TVu 14914, were consistently highly drought tolerant under real field conditions. Research has been intensified in recent times to develop cowpea varieties with enhanced level of drought tolerance. This led to the evaluation of over 1280 germplasm lines under drought stress condition in the field and screen-house. Following evaluation, some additional lines have been reported as potential parents in the development of new improved breeding lines with drought tolerance (Fatokun, Boukar, & Muranaka, 2012). Drought can occur early in the season, mid-season or at the podding stage of crop development. Studies have shown that cowpea plants can show drought tolerance at the vegetative stage (Singh & Matsui, 2002) and reproductive stage (Hall et al., 2003). Some cowpea lines exhibit stay-green characteristic, also referred to as delayed leaf senescence (DLS), which can help plants to tolerate mid-season and terminal drought (Hall et al., 2003).

6.6.5 Seed Coat Colour and Texture

Most of the traditional cowpea varieties in SSA have white or light brown seed coats. In different communities of consumers preference can be for brown, red or white seed coat colour. Cowpea consumers in southwestern parts of Nigeria prefer brown-seeded grains, while in Ghana some consumers choose red grains when consuming cowpea and rice cooked together. Cowpea cultivars with black seed coat are not preferred in Africa, whereas in Cuba and some other Latin American communities, black-seeded cowpea is most preferred. Cowpea grains with rough coat texture are preferred by many consumers, because they soak up water and cook faster. It is also easier to remove the rough textured seed coat when processing cowpea grains into some food products.

6.6.6 Resistance to Aphids, Maruca and Other Insect Pests

Several of the cultivated cowpea germplasm lines in the gene bank have been evaluated for resistance to insect pests, but none was found to have the desired level of resistance to maruca pod borer and pod-sucking bugs. Few accessions of the wild *Vigna* species have also been screened for resistance to insect pests of cowpea. Many accessions of *V. vexillata* were found to show high levels of resistance to pod-sucking bugs, storage weevil and moderate resistance to maruca pod borer (Singh, Jackai, Thottappilly, Cardwell, & Myers, 1992). The dense hairs found on the different parts of *V. vexillata* have been associated with resistance to pod-sucking bugs and pod borer (Oghiakhe, Jackai, Makanjuola, & Hodgson, 1992). In addition, *Striga* does not attack the plants of *V. vexillata*, while their edible tuberous roots also enhance drought tolerance. *V. vexillata* should therefore be a good source of desirable genes that could be beneficial to the cultivated cowpea. Some efforts have been made to cross cowpea and *V. vexillata*, but these efforts have not been successful because of a strong barrier to compatibility between them (Fatokun, 2002). This has constituted a major limitation to the transfer of desirable genes present in the wild *V. vexillata* to cultivated background. The strong barrier to cross compatibility between *V. vexillata* and cowpea necessitated the development of transgenic plants.

6.7 Limitations in Germplasm Use

The genetic resource of over 15,000 accessions of cultivated cowpea in the global gene bank of IITA is conserved for the international community. Besides cultivated cowpea germplasm, there are also some accessions of wild relatives conserved in the gene bank at IITA, which could be used for widening the genetic base of cultivated varieties through pre-breeding. Progress in the development of improved crop varieties that are better in performance depends on the availability of germplasm with desired traits. Some of these cowpea wild relatives have been evaluated for their potentials in terms of genes that may be desirable in cowpea improvement. However, the basic need for exploiting the wild relatives is its cross compatibility with cultivated cowpea. It is possible that some of the available wild cowpea lines belong to the same or different gene pools. The subspecies or varieties that constitute the primary and secondary gene pools for cowpea are not yet well defined. Cross compatibility studies have shown that lines which can hybridize successfully with cultivated species are found only among members of the subspecies *unguiculata*, i.e. those belonging to section *Catiang* in the genus *Vigna*.

6.8 Germplasm Enhancement Through Wide Crosses

Cowpea has an intrinsically narrow genetic base and that situation limits breeders' progress today (Hall, Singh, & Ehlers, 1997). The low level of genetic diversity was also revealed when RFLP markers were used to differentiate between a cowpea line

(IT-84S-2246-2) and a wild relative *V. unguiculata* ssp. *dekindtiana* (TVNu 1963), in which only about 22% of 400 genomic clones hybridized were polymorphic between them (Fatokun, Danesh, Young, et al., 1993). Some wild relatives have been screened for certain agro-morphological traits that are desired in cultivated varieties for widening the genetic base. The genes sought from these wild lines include those that confer resistance to insect pests, especially maruca pod borer, pod-sucking bugs, bruchids and aphids, among others. Hanchinal, Goud, Habib, and Bhumannavar (1976) evaluated three wild *Vigna* species, namely *V. vexillata*, *V. unguiculata* var. *cylindrica* and *V. parviflora*, for resistance to the pod borer *Cydia ptychora* Meyr. There are not many reports in the literature on the use of wild cowpea relatives for the genetic improvement of cultivated varieties. The relatively low level of utilization of wild cowpea relatives in the development of improved cowpea varieties may be due to some factors such as linkage drag, in which some undesirable genes may be closely linked to the desirable ones. Such linkages may be difficult to break and this may prolong the time needed for the development and release of the improved variety.

6.9 Cowpea Genomic Resources

6.9.1 Genetic Diversity

With the development of biochemical-based analytical techniques and molecular markers, several studies were undertaken to characterize genetic variation in domesticated cowpea and its wild ancestors, as well as their relationships, in order to complement early analysis using morphological and physiological traits (Ehlers & Hall, 1996; Fery, 1985; Perrino, Laghetti, Spagnoletti, & Monti, 1993). All types of molecular markers were used to characterize DNA variation patterns within cultivated cowpea and closely related wild species. These include allozymes (Panella & Gepts, 1992; Pasquet, 1993, 1999, 2000; Vaillancourt, Weeden, & Barnard, 1993), seed storage proteins (Fotso, Azanza, Pasquet, & Raymond, 1994), chloroplast DNA polymorphism (Vaillancourt & Weeden, 1992), RFLP (Fatokun, Danesh, Young, et al., 1993), amplified fragment length polymorphisms (AFLP) (Fang, Chao, Roberts, & Ehlers, 2007; Fatokun, Young, & Myers, 1997; Tosti & Negri, 2002), DNA amplification fingerprinting (DAF) (Simon, Benko-Iseppon, Resende, Winter, & Kahl, 2007; Spencer et al., 2000), random amplified polymorphic DNA (RAPD) (Ba, Pasquet, & Gepts, 2004; Diouf & Hilu, 2005; Fall, Diouf, Fall-Ndiaye, Badiane, & Gueye, 2003; Mignouna, Ng, Ikea, & Thottapilly, 1998; Nkongolo, 2003; Xavier, Martins, Rumjanek, & Filho, 2005; Zannou et al., 2008), simple sequence repeats (SSRs) (Ogunkanmi, Ogundipe, Ng, & Fatokun, 2008; Uma, Hittalamani, Murthy, & Viswanatha, 2009; Wang, Barkley, Gillaspie, & Pederson, 2008; Xu et al., 2010), cross species SSRs from Medicago (Sawadogo, Ouédraogo, Gowda, & Timko, 2010), inter-simple sequence repeats (Ghalmi et al., 2010) and sequence tagged microsatellite sites (STMS) (Abe, Xu, Suzuki, Kanazawa, & Shimamoto, 2003; Choumane, Winter, Weigand, & Kahl, 2000; He, Poysa, & Yu, 2003; Li, Fatokun, Ubi, Singh, & Scoles, 2001). All these studies have contributed greatly to the understanding of cowpea genome organization and its evolution. In general, molecular taxonomic procedures confirmed early classifications based on classical taxonomic criteria, such as morphological and reproductive traits (Fatokun, Danesh, Young, et al., 1993; Kaga, Tomooka, Egawa, Hosaka, & Kamijima, 1996; Vaillancourt & Weeden, 1992; Vaillancourt et al., 1993). In addition to the taxonomic classification, these studies led to the use of genetic variation to identify duplicates or genetic contamination in gene bank or breeding programmes. Through the manipulation of the molecular marker technologies, several authors have detected low levels of polymorphism in cowpea (Badiane et al., 2004; Diouf & Hilu, 2005; Li et al., 2001; Tosti & Negri, 2002). They attributed this finding to the result of a genetic bottleneck induced by a single domestication event in cowpea, in addition to the inherent nature of the self-pollination mechanism (Badiane et al., 2012). The total genetic diversity in cultivated cowpea reported from these studies was lower than that reported in many other crops (Doebley, 1989).

6.9.2 Genetic Linkage Mapping

The development of molecular markers has also provided an opportunity to construct linkage maps in cowpea. Fatokun, Danesh, Menancio-Hautea, and Young (1993) have developed the first comprehensive linkage map for cowpea using a mapping population of 58 F_2 plants, derived from a cross between an improved cultivar IT84S-2246-4 and a wild relative TVu 1963 (*V. unguiculata* ssp. *dekindtiana*). This first map was based on 87 random genomic and 5 cDNA RFLPs, 5 RAPDs and some morphological traits representing 10 linkage groups (LGs) spanning 680 cM, although cowpea has a chromosome number of $n=11$. The resolution of the map was approximately 7.0 cM between loci. This map has also been used to locate two quantitative trait loci (QTLs) accounting for 52% of the variation in seed weight (Fatokun, Menancio-Hautea, Danesh, & Young, 1992). The markers flanking these QTLs in cowpea were the same as those identified for seed weight QTLs in mung bean (*V. radiata*). This map also comprised two markers associated with aphid resistance genes in cowpea (Myers, Fatokun, & Young, 1996). The second genetic linkage map of cowpea was constructed using 94 F_8 recombinant inbred lines (RILs) derived from a cross between two cultivated genotypes IT84S-2049 and 524B (Menéndez, Hall, & Gepts, 1997). This map consisted of 181 loci, comprising 133 RAPDs, 19 RFLPs, 25 AFLPs and 3 each of morphological and biochemical markers. These markers are assigned to 12 LGs spanning 972 cM with an average distance of 6.4 cM between markers. Two traits, earliness and seed weight, were mapped to LGs 2 and 5, respectively. Seed weight is significantly associated with a RAPD marker. Ouédraogo, Gowda, Jean, Close, and Ehlers (2002) improved this map based on segregation of various molecular markers (AFLP, RFLP, RAPD) and resistance traits (resistance to *S. gesnerioides* race 1 and 3, resistance to CPMV, CPSMV, BICMV, SBMV, *Fusarium* wilt, and root-knot nematode). Using 27 selective primer combinations, an additional 242 new markers were used in this mapping population and mapped in different LGs of an improved map. The resulting map consisted of 11 LGs spanning a total of 2670 cM, with an average distance of 6.43 cM between markers. A large portion of LG1 was discovered, mainly composed of 54 AFLP markers. In this new genetic map, the previously recognized LGs were simply expanded in size by the addition of new markers. A third genetic map of cowpea was reported using 94 F_8 RILs derived from the inter-subspecific cross between IT84S-2246-4, an improved cowpea line and TVu 110-3A, a wild relative (*Vigna unguiculata* spp. *dekindtiana* var. *pubescens*) (Ubi, Mignouna, & Thottappilly, 2000). This map spans 669.8 cM of the genome and comprises 80 mapped loci (77 RAPD and 3 morphological loci), making 12 LGs with an average distance of 9.9 cM between marker loci. QTLs for several agronomical and morphological traits, including days to flowering, days to maturity, pod length, seeds/pod, leaf length, leaf width, primary leaf length, primary leaf width, and the derived traits such as leaf area and primary leaf area were mapped in this genetic linkage map.

Through the Tropical Legumes I project of the Generation Challenge Program at the University of California, Riverside, cowpea genomics activities are being conducted and the tools developed will be used in cowpea breeding programme. A high-throughput SNP genotyping platform based on Illumina 1536 GoldenGate Assay was developed and has resulted in 1375 SNPs with 89.55% success rate. These SNPs were applied to develop a high-density SNP consensus map based on the genotyping of 741 members of six RILs populations derived from the following crosses: 524B×IT84S-2049, CB27×24-125B-1, CB46×IT93K-503-1, Dan Ila×TVu-7778, TVu-14676×IT84S-2246-4 and Yacine×58-77. The resulting consensus map contained 928 SNP markers on 619 unique map positions distributed over 11 LGs, covering a total genetic distance of 680 cM (Muchero, Ehlers, et al., 2009; Muchero, Diop, et al., 2009). The resolution of this map is an average marker distance of 0.73 cM, or 1 SNP per 668 kbp considering the cowpea genome to be 620 Mbp.

More recently, a 1536 SNP assay was applied to 13 breeding populations consisting of 11 RILs (from UCR–US, IITA–Nigeria, ISRA–Senegal, ZAAS–China) and 2 F_4 populations (from UCR) to generate a high-quality consensus genetic map (Lucas et al., 2011). The 11 RILs were derived from the following crosses, namely CB27×UCR 779, CB27×IT97K-566-6, 524B×IT84S-2049, Yacine×58-77, CB27×IT82E-18, Sanzi×Vita 7, CB46×IT93K-503-1, TVu14676×IT84S-2246-4, CB27×24-125B-1, Dan Ila×TVu-7778 and LB30#1×LB1162 #7. The two F_4 populations are obtained from the crosses of IT84S-2246×Mouride and IT84S-2246×IT93K-503. A total of 1293 individuals from 13 breeding populations were used to construct this consensus genetic map, which possesses 1107 EST-derived SNP markers (856 bins). This new map has 33% more bins, 19% more markers and an improved order compared to the consensus genetic map constructed using 6 RILs and 741 individuals (Muchero, Ehlers, et al., 2009).

6.9.3 Molecular Breeding

The application of DNA marker technologies in cowpea improvement has been very slow, when compared to many other crops. Most of the available reports in the literature on the use of molecular markers in cowpea are for taxonomic relationships and genetic linkage mapping, as described in the above sections. In these genetic maps, several QTLs and markers associated with genes of interest have been identified. Marker-assisted selection (MAS) could be used to accelerate the selection procedure and increase the selection efficiency in cowpea cultivar development. RFLPs with only a limited number of markers could not be used in QTL identification (Fatokun et al., 1992; Fatokun, Danesh, Menancio-Hautea, et al., 1993; Fatokun, Danesh, Young, et al., 1993; Menendez et al., 1997; Myers et al., 1996). Although RAPD markers were used in several genetic diversity studies (Ba et al., 2004; Badiane et al., 2004; Diouf & Hilu, 2005; Menendez et al., 1997; Mignouna et al., 1998), their use in MAS is limited by its poor level of reproducibility. AFLPs were found to be the most attractive and useful, and were used successfully in many studies (Boukar, Kong, Singh, Murdock, & Ohm, 2004; Coulibaly, Pasquet, Papa, & Gepts, 2002; Fatokun et al., 1997; Ouédraogo, Gowda, et al., 2002; Ouédraogo, Maheshwari, Berner, St-Pierre, & Belzile, 2001; Ouédraogo, Tignegre, Timko, & Belzile, 2002; Tosti & Negri, 2002). Unfortunately, their use required more skill and they could not be used in a breeding programme. Two sequence characterized amplified region (SCAR) markers, SEACT/MCTM83/84 (Boukar et al., 2004) and 61R (E-ACT/M-CAA) (Timko, Gowda, Ouédraogo, & Ousmane, 2007), derived from AFLP markers associated to *Striga* resistance offered an opportunity for MAS in cowpea. The latter SCAR was further improved into a SCAR marker called Mahse2 (Timko, personal communication), recently identified as 61R-M2 (Ouédraogo, Ouédraogo, Gowda, & Timko, 2012).

With the current generation of consensus genetic linkage maps, a genomic framework is established for QTLs identification, map-based cloning, and assessment of genetic diversity, association mapping and applied breeding in MAS schemes. These new developments in cowpea research build a strong basis for molecular breeding in cowpea. Areas of potential application include comparative genomics, quantitative trait characterization, and map-based cloning. Establishing synteny with crops like soybean will help in the exploitation of considerable progress made in basic gene discovery and gene regulation in these crops. Initial studies related to QTL and trait-linked markers (drought tolerance, foliar thrips, stem blight, bacterial blight, root-knot nematode, etc.) are being reported (Agbicodo et al., 2010; Muchero, Diop, et al., 2009; Muchero, Ehlers, et al., 2009). Modern breeding of cowpea is ready to use tools such as whole genome assembly, MAS and association mapping to complement and strengthen the progress achieved by conventional breeding. A MAGIC population is also under development that will be an invaluable community resource for trait discovery and breeding as well.

6.9.4 Genetic Transformation

As discussed in the previous section on use of germplasm in cowpea improvement programmes, high levels of resistance to several insects and diseases exist in wild species, but cross incompatibility with cultivated species is the biggest bottleneck preventing the transfer of genes into cultivated cowpea. Genetic transformation was suggested as one of the most important approaches to overcome these limitations. Several procedures for plant transformation in cowpea were attempted. The transfer of genes from one species to another using genetic engineering techniques requires (a) the setting up of effective bioassays for discovering resistant genes for specific pests, (b) the use of those bioassays to search through the plant, fungal, animal, and microbial kingdoms for suitable genes and (c) the understanding of insects' physiological and biochemical systems that are vulnerable to resistant genes (Monti, Murdock, & Thottappilly, 1997). However, transformation by *Agrobacterium tumefaciens* (Garcia, Hille, & Goldbach, 1986a,b) or embryo imbibition with or without subsequent electroporation (Akella & Lurquin, 1993; Penza, Akella, & Lurquin, 1992) has contributed to the development of transgenic cowpea calli or chimeric plantlets from leaf discs, auxiliary buds, or embryos. However, attempts to produce mature transgenic plants failed in all these cases (Kononowicz et al., 1997). Authors have reported the development of transformation systems using either microprojectile bombardment or *Agrobacterium* cocultivation that seem to have given some promising results. The coculturing of de-embryonated cotyledons with *A. tumefaciens* resulted in selection of four plants on hygromycin. Muthukumar, Mariamma, Veluthambi, and Gnanam (1996) avoided the callus regeneration route. Stable transformation was confirmed by Southern analysis in only one of the transgenic plants, whose seeds unfortunately failed to germinate. Similarly, Sahoo, Sushma, Sugla, Singh, and Jaiwal (2000) succeeded in producing transgenic shoots but could not show evidence of stable integration. Using microprojectile bombardment (biolistics), several researchers achieved the introduction of foreign DNA into cowpea leaf tissues and embryos and obtained high levels of transient expression of the ß-glucuronidase (gus) transgene, but were unable to regenerate plantlets from the transformed cells (Kononowicz et al., 1997). Identical results were obtained when using electroporation of embryos with plasmid DNA (Akella & Lurquin, 1993). Ikea, Ingelbrecht, Uwaifo, and Thottappilly (2003) used the biolistic approach and observed transformation in cowpea, but no evidence of stable transformation with transmission of transgenes to progeny was provided. Popelka, Gollasch, Moore, Molvig, and Higgins (2006) reported the first genetic transformation of cowpea and stable transmission of the transgenes to progeny. Their system used cotyledonary nodes from developing or mature seeds as explants and a tissue culture medium lacking auxins in the early stages, but including the cytokinin BAP at low levels during shoot initiation and elongation. Other parameters used included the addition of thiol compounds during infection and coculture with *Agrobacterium* and the use of bar gene for selection with phosphinothricin. These authors have now reported the development of cowpea with the *Bt* gene being field-tested during the last 3 years in Nigeria, Burkina Faso and Ghana. Chaudhury et al. (2007) have reported a transformation efficiency of 0.76%, better than the 0.05–0.15% obtained by Popelka et al. (2006). These researchers also used cotyledonary nodal explants as Popelka's group did, but they wounded the nodal cells by stabbing them with a sterile needle prior to *Agrobacterium* infection. In addition, they introduced a second selection regime at the rooting stage, which was described as a very important procedure in the transformation of *V. mungo* L. (Saini, Sonia, & Jaiwal, 2003). Recently Ivo, Nascimento, Vieira, Campos, and Aragão (2008), using biolistic methods, reported the first work on the use of this method of gene transfer leading to the development of transgenic plants. The transformation efficiency obtained by this group is 0.9%. Obembe (2009) cited much higher transformation efficiencies of 1.64–1.67% which have been obtained recently by Sahoo's group in India.

6.10 Conclusions

A large amount of cowpea germplasm of both cultivated and wild species has been collected and is being preserved in the global gene bank at IITA Nigeria. However, the genetic materials conserved at different gene banks need to be maintained nicely and evaluated for their use in breeding programmes. The useful variability detected for key biotic stresses should be used to develop suitable cultivars with multiple resistances to attain stable yield. In recent years tremendous progress has been made, including completion of whole genome sequencing of cowpea (Timko et al., 2008), which in combination with genomic information from model legumes and bioinformatics tools should make it possible to dissect genes that govern agronomically important traits. Advances have also been made in the area of genetic transformation, which could be used to understand the gene regulations and also to develop transgenic products. In addition, numerous genomic resources such as EST and transcriptome sequence data sets are available, which in combination with advances in next-generation sequencing technology could be applied to develop novel strategies to identify key genes of targeted traits for further marker-assisted breeding.

References

Abate, T. (2012). *Four seasons of learning and engaging smallholder farmers: Progress of Phase 1*. Nairobi, Kenya: International Crops Research Institute for the Semi-Arid Tropics. p. 255.

Abe, J., Xu, D., Suzuki, Y., Kanazawa, A., & Shimamoto, Y. (2003). Soybean germplasm pools in Asia revealed by nuclear SSRs. *Theoretical and Applied Genetics*, *106*, 445–453.

Agbicodo, E. M., Fatokun, C. A., Bandyopadhyay, R., Wydra, K., Diop, N. N., Muchero, W., et al. (2010). Identification of markers associated with bacterial blight resistance loci in cowpea [*Vigna unguiculata* (L.) Walp.]. *Euphytica*, *175*, 215–226.

Akella, V., & Lurquin, P. F. (1993). Expression in cowpea seedlings of chimeric transgenes after electroporation into seed-derived embryos. *Plant Cell Report*, *12*, 110–117.

Allen, D. J., Thottappilly, G., Emechebe, A. M., & Singh, B. B. (1998). Diseases of cowpea. In D. J. Allen & J. M. Lenne (Eds.), *The Pathology of Food and Pasture Legumes* (pp. 267–308). Wallingford, Oxon, UK: CAB International.

Amaza, P. (2011). *Early cowpea adoption of improved varieties in northern Nigeria* (p. 60). Ibadan, Nigeria: International Institute of Tropical Agriculture (IITA).

Ba, F. S., Pasquet, R. S., & Gepts, P. (2004). Genetic diversity in cowpea [*Vigna unguiculata* (L.) Walp.] as revealed by RAPD markers. *Genetic Resources and Crop Evolution*, *51*, 539–550.

Badiane, F. A., Diouf, D., Sané, D., Diouf, O., Goudiaby, V., & Diallo, N. (2004). Screening cowpea [*Vigna unguiculata* (L.) Walp.] varieties by inducing water deficit and RAPD analyses. *African Journal of Biotechnology*, *3*, 174–178.

Badiane, F. A., Gowda, B. S., Cissé, N., Diouf, D., Sadio, O., & Timko, M. P. (2012). Genetic relationship of cowpea (*Vigna unguiculata*) varieties from Senegal based on SSR markers. *Genetics of Molecular Research, 11*(1), 292–304.

Baudoin, J.P. & Marechal, R. (1988). Taxonomy and evolution of the genus *Vigna*. In *Mungbean: Proceedings of the second international symposium Asian vegetable research development center*, Shanhua, Taiwan, pp. 2–12.

Boukar, O., Kong, L., Singh, B. B., Murdock, L., & Ohm, H. W. (2004). AFLP and AFLP-derived SCAR markers associated with *Striga gesnerioides* resistance in cowpea. *Crop Science, 44*, 1259–1264.

Boukar, O., Massawe, F., Muranaka, S., Franco, J., Maziya-Dixon, B., Singh, B. B., et al. (2011). Evaluation of cowpea germplasm lines for protein and mineral concentrations in grains. *Plant Genetic Resources: Characterization and Utilization, 9*(4), 515–522.

Chaudhury, D., Madanpotra, S., Jawail, R., Saini, R., Kumar, P. A., & Jaiwal, P. K. (2007). *Agrobacterium tumefaciens*-mediated high frequency genetic transformation of an Indian cowpea (*Vigna unguiculata* L.) cultivar and transmission of transgenes into progeny. *Plant Science, 172*, 692–700.

Choumane, W., Winter, P., Weigand, F., & Kahl, G. (2000). Conservation and variability of sequence tagged microsatellites sites (STMS) from chickpea (*Cicer aerietinum* L.) within the genus *Cicer*. *Theoretical and Applied Genetics, 101*, 269–278.

Coulibaly, S., Pasquet, R. S., Papa, R., & Gepts, P. (2002). AFLP analysis of the phenetic organization and genetic diversity of cowpea [*Vigna unguiculata* (L.) Walp.] reveals extensive gene flow between wild and domesticated types. *Theoretical and Applied Genetics, 104*, 258–266.

Diouf, D., & Hilu, K. W. (2005). Microsatellites and RAPD markers to study genetic relationship among cowpea breeding lines and local varieties in Senegal. *Genetic Resources and Crop Evolution, 52*, 1057–1067.

Doebley, J. (1989). Isozymic evidence and the evolution of crop plants. In D. E. Soltis & P. S. Soltis (Eds.), *Isozymes in plant biology* (pp. 165–191). Portland, Oregon: Dioscor- ides Press.

Ehlers, J. D., & Hall, A. E. (1996). Genotypic classification of cowpea based on responses to heat and photoperiod. *Crop Sciences, 36*, 673–679.

Emechebe, A. M., & Florini, D. A. (1997). Shoot and pod diseases of cowpea induced by fungi and bacteria. In B. B. Singh, D. R. Mohan Raj, K. E. Dashiell, & L. E. N. Jackai (Eds.), *Advances in cowpea research* (pp. 176–192). Ibadan, Nigeria: IITA. Copublication of International Institute of Tropical Agriculture (IITA) and Japan International Research Center for Agricultural Sciences (JIRCAS).

Emechebe, A.M. & Lagoke, S.T.O. (2002). Recent advances in research of cowpea disease. In C. A. Fatokun, S. A. Tarawali, B. B. Singh, P. M. Komawa, & M. Tamò, (Eds.), *Challenges and opportunities for enhancing sustainable cowpea production*, pp. 94–123.

Fall, L., Diouf, D., Fall-Ndiaye, M. A., Badiane, F. A., & Gueye, M. (2003). Genetic diversity in cowpea [*Vigna unguiculata* (L.) Walp.] varieties determined by ARA and RAPD techniques. *African Journal of Biotechnology, 2*, 48–50.

Fang, J., Chao, C. C. T., Roberts, P. A., & Ehlers, J. D. (2007). Genetic diversity of cowpea [*Vigna unguiculata* (L.) Walp.] in four West African and USA breeding programmes as determined by AFLP analysis. *Genetic Resources and Crop Evolution, 54*, 1197–1209.

Fatokun, C. A. (2002). Breeding cowpea for insect pests: Attempted crosses between cowpea and *Vigna vexillata*. In C. A. Fatokun, S. A. Tarawali, B. B. Singh, P. M. Kormawa, & M. Tamo (Eds.), *Challenges and opportunities for enhancing sustainable cowpea production. Proceedings of the world cowpea conference, 4–8 September 2000* (pp. 52–61). Ibadan, Nigeria: International Institute of Tropical Agriculture IITA.

Fatokun, C. A., Boukar, O., & Muranaka, S. (2012). Evaluation of cowpea (*Vigna unguiculata* (L.) Walp.) germplasm lines for tolerance to drought. *Plant Genetic Resources: Characterization and Utilization, 10*, 171–176.

Fatokun, C. A., Danesh, D., Menancio-Hautea, D. I., & Young, N. D. (1993). A linkage map for cowpea (*Vigna unguiculata* [L.] Walp.) based on DNA markers. In S. J. O'Brien. (Ed.), *A compilation of linkage and restriction maps of genetically studied organisms. Genetic maps 1992* (pp. 6256–6258). Cold Spring Harbor, NY, USA: Cold Spring Harbor Laboratory Press.

Fatokun, C. A., Danesh, D., Young, N. D., & Stewart, E. L. (1993). Molecular taxonomic relationships in the genus *Vigna* based on RFLP analysis. *Theoretical and Applied Genetics, 86*, 97–104.

Fatokun, C. A., Menancio-Hautea, D. I., Danesh, D., & Young, N. D. (1992). Evidence for orthologous seed weight genes in cowpea and mung bean based on RFLP mapping. *Genetics, 132*, 841–846.

Fatokun, C. A., Young, N. D., & Myers, G. O. (1997). Molecular markers and genome mapping in cowpea. In B. B. Singh, D. R. Mohan Raj, K. E. Dashiell, & L. E. N. Jackai (Eds.), *Advances in cowpea research* (pp. 352–360). Ibadan, Nigeria: IITA. Copublication of International Institute of Tropical Agriculture (IITA) and Japan International Research Center for Agricultural Sciences (JIRCAS).

Fery, R. L. (1985). The genetics of cowpeas: a review of the world literature. In S. R. Singh & K. O. Rachie (Eds.), *Cowpea research, production and utilization* (pp. 25–62). Chichester, UK: John Wiley and Sons.

Fery, R. L., & Singh, B. B. (1997). Cowpea genetics: A review of recent literature. In B. B. Singh, D. R. Mohan Rah, K. E. Dashiell, & L. E. N. Jackai (Eds.), *Advances in cowpea research* (pp. 13–29). Ibadan, Nigeria: IITA-JIRCAS.

Flight, C. (1976). The Kimtampo culture and its place in the economic prehistory of West Africa. In J. R. Harlan, J. M. J. de Wet, & A. B. L. Stemler (Eds.), *Origin of African plant domestication* (pp. 212–221). Hague, the Netherlands: Mouton.

Food and Agriculture Organization, (2012). The second report on the state of the world's plant genetic resources for food and agriculture.

Fotso, M., Azanza, J. L., Pasquet, R., & Raymond, J. (1994). Molecular heterogeneity of Cowpea (*Vigna unguiculata* Fabaceae) seed storage proteins. *Plant Systematics Evolution, 191*, 39–56.

Garcia, J. A., Hille, J., & Goldbach, R. (1986a). Transformation of cowpea (*Vigna unguiculata*) cells with an antibiotic-resistance gene using a Ti-plasmid-derived vector. *Plant Sciences, 44*, 37–46.

Garcia, J. A., Hillie, J., & Goldbach, R. (1986b). Transformation of cowpea (*Vigna unguiculata*) cells with a full length DNA copy of cowpea mosaic virus mRNA. *Plant Sciences, 44*, 89–98.

Ghalmi, N., Malice, M., Jacquemin, J. M., Ounane, S. M., Mekliche, L., & Baudoin, J. P. (2010). Morphological and molecular diversity within Algerian cowpea (*Vigna unguiculata* (L.) Walp.) landraces. *Genetic Resources and Crop Evolution, 57*(3), 371–386.

Gumedzoe, M. Y. D., Rossel, H. W., Thottappily, G., Asselin, A., & Huguenot, C. (1998). Reaction of cowpea (*Vigna unguiculata* L. Walp.) to six isolates of blackeye cowpea mosaic virus (BlCMV) and cowpea aphid-borne mosaic virus (CAMV), two potyviruses infecting cowpea in Nigeria. *International Journal of Pest Management, 44*, 11–16.

Hall, A. E., Cisse, N., Thiaw, S., Elawad, H. O. A., Ehlers, J. D., Ismail, A. M., et al. (2003). Development of cowpea cultivars and germplasm by the bean/cowpea CRSP. *Field Crops Research, 82*, 103–134.

Hall, A. E., Singh, B. B., & Ehlers, J. D. (1997). Cowpea breeding. *Plant Breeding Reviews, 15*, 215–274.

Hampton, R. O., Thottappilly, G., & Rossel, H. W. (1997). Viral disease of cowpea and their control by resistance-conferring genes. In B. B. Singh, D. R. Mohan Raj, K. E. Dashiell, & L. E. N. Jackai (Eds.), *Advances in cowpea research* (pp. 159–175). Ibadan, Nigeria/Ibaraki, Japan: International Institute of Tropical Agriculture (IITA)/International Research Center for Agricultural Sciences (JIRCAS).

Hanchinal, R. R., Goud, J. V., Habib, A. F., & Bhumannavar, B. S. (1976). Interspecific and intraspecific response of *Vigna* to pod borer *Cydia ptychora* Meyr. (Totricidae: Lepidoptera). *Current Research, 5*(9), 154–156.

He, C., Poysa, V., & Yu, K. (2003). Development and characterization of simple sequence repeat (SSR) markers and their use in determining relationships among *Lycopersicon esculentum* cultivars. *Theoretical and Applied Genetics, 106*, 363–373.

Ikea, J., Ingelbrecht, I., Uwaifo, A., & Thottappilly, G. (2003). Stable gene transformation in cowpea (*Vigna unguiculata* L.Walp.) using particle gun method. *African Journal of Biotechnology, 2*, 211–218.

Ivo, N. L., Nascimento, C. P., Vieira, L. S., Campos, F. A. P., & Aragão, F. J. L. (2008). Biolistic-mediated genetic transformation of cowpea (*Vigna unguiculata*) and stable Mendelian inheritance of transgenes. *Plant Cell Reports, 27*, 1475–1483.

Kaga, A., Tomooka, N., Egawa, Y., Hosaka, K., & Kamijima, O. (1996). Species relationships in the subgenus *Ceratotropis* (genus *Vigna)* as revealed by RAPD analysis. *Euphytica, 88*, 17–24.

Kononowicz, A. K., Cheah, K. T., Narasimhan, M. L., Murdock, L. L., Shade, R. E., Chrispeels, R. J., et al. (1997). Developing a transformation system for cowpea (*Vigna unguiculata* [L.] Walp.). In B. B. Singh (Ed.), *Advances in cowpea research* (pp. 361–371). Ibadan, Nigeria: IITA. Copublication of IITA and JIRCAS.

Li, C. D., Fatokun, C. A., Ubi, B., Singh, B. B., & Scoles, G. J. (2001). Determining genetic similarities and relationships among cowpea breeding lines and cultivars by microsatellite primers. *Crop Science, 41*, 189–197.

Lucas, M. R., Diop, N. N., Wanamaker, S., Ehlers, J. D., Roberts, P. A., & Close, T. J. (2011). Cowpea–soybean synteny clarified through an improved genetic map. *Plant Genome Journal, 4*, 218–225.

Mahalakshmi, V., Ng, Q., Lawson, M., & Ortiz, R. (2007). Cowpea [*Vigna unguiculata* (L.) Walp.] core collection defined by geographical, agronomical and botanical descriptors. *Plant Genetic Resources: Characterization and Utilization, 5*, 113–119.

Marechal, R., Mascherpa, J. M., & Stainier, F. (1978). Etude taxonomique d'un groupe d'especes des genres *Phaseolus* et *Vigna* (*Papilionaceae*) sur la base donnees morphologiques, traits pour l'analyse informatique. *Boissiera, 28*, 1–273.

Maynard, D. N. (2008). Underutilized and underexploited horticultural crops (*Vigna unguiculata* (L.) Walp.). *Horticulture Science, 43*, 279.

Menéndez, C. M., Hall, A. E., & Gepts, P. (1997). A genetic linkage map of cowpea (*Vigna unguiculata*) developed from a cross between two inbred domesticated lines. *Theoretical and Applied Genetics, 95*, 1210–1217.

Mignouna, H. D., Ng, Q., Ikea, J., & Thottapilly, G. (1998). Genetic diversity in cowpea as revealed by random amplified polymorphic DNA. *Journal of Genetics & Breeding, 52*, 151–159.

Monti, L. M., Murdock, L. L., & Thottappilly, G. (1997). Opportunities for biotechnology in cowpea. In B. B. Singh (Ed.), *Advances in cowpea research* (pp. 341–351). Ibadan, Nigeria: IITA. Copublication of IITA and JIRCAS.

Muchero, M., Diop, N. N., Bhat, P. R., Fenton, R. D., Wanamaker, S., Pottorff, M., et al. (2009). A consensus genetic map of cowpea [*Vigna unguiculata* (L.) Walp.] and synteny based on EST-derived SNPs. *PNAS*, *106*, 18159–18164.

Muchero, M., Ehlers, J. D., Close, T. J., & Roberts, P. A. (2009). Mapping QTL for drought stress-induced premature senescence and maturity in cowpea [*Vigna unguiculata* (L.) Walp.]. *Theoretical and Applied Genetics*, *118*, 849–863.

Muthukumar, B., Mariamma, M., Veluthambi, K., & Gnanam, A. (1996). Genetic transformation of cotyledon explants of cowpea (*Vigna unguiculata* L. Walp.) using *Agrobacterium tumefaciens*. *Plant Cell Reports*, *15*, 980–985.

Myers, G. O., Fatokun, C. A., & Young, N. D. (1996). RFLP mapping of an aphid resistance gene in cowpea (*Vigna unguiculata* [L.] Walp.). *Euphytica*, *91*, 181–187.

Ng, N. Q., & Marechal, R. (1985). Cowpea taxonomy, origin and germplasm. In S. R. Singh & K. O. Rachie (Eds.), *Cowpea research, production and utilization* (pp. 11–21). Chichester, UK: John Wiley and Sons.

Nkongolo, K. K. (2003). Genetic characterization of Malawian cowpea [*Vigna uncuguilata* (L.) Walp.] landraces: diversity and gene flow among accessions. *Euphytica*, *129*, 219–228.

Obembe, O. O. (2009). Exciting times for cowpea genetic transformation research. *Australian Journal of Basic and Applied Sciences*, *3*(2), 1083–1086.

Oghiakhe, S., Jackai, L. E. N., Makanjuola, W. A., & Hodgson, G. J. (1992). Morphology, distribution and the role of trichomes in cowpea (*Vigna vexillata*) resistance to the legume borer, *Maruca testulalis* (Lepidoptera: Pyralidae). *Bulletin Environmental Research*, *82*, 499–505.

Ogunkanmi, L. A., Ogundipe, O. T., Ng, N. Q., & Fatokun, C. A. (2008). Genetic diversity in wild relatives of cowpea (*Vigna unguiculata*) as revealed by simple sequence repeats (SSR) markers. *Journal of Food Agriculture and Environment*, *6*, 253–268.

Ogunsola, K. E., Fatokun, C. A., Boukar, O., Ilori, C. O., & Kumar, P. L. (2010). Characterizing genetics of resistance to multiple virus infections in cowpea (*Vigna unguiculata* L. Walp.): *Abstracts: fifth world cowpea conference: improving livelihoods in the cowpea value chain through advancement of science, September 27–October 1, Saly, Senegal.* (pp. 26–27). Nigeria: IITA.

Ouédraogo, J. T., Gowda, B. S., Jean, M., Close, T. J., & Ehlers, J. D. (2002). An improved genetic linkage map for cowpea (*Vigna unguiculata* L.) combining AFLP, RFLP, RAPD, biochemical markers and biological resistance traits. *Genome*, *45*, 175–188.

Ouédraogo, J. T., Maheshwari, V., Berner, D., St-Pierre, C. A., & Belzile, F. (2001). Identification of AFLP markers linked to resistance of cowpea (*Vigna unguiculata* L.) to parasitism by *Striga gesnerioides*. *Theoretical and Applied Genetics*, *102*, 1029–1036.

Ouédraogo, J. T., Ouédraogo, M., Gowda, B. S., & Timko, M. P. (2012). Development of sequence characterized amplified region (SCAR) markers linked to race-specific resistance to *Striga gesnerioides* in cowpea (*Vigna unguiculata* L.). *African Journal of Biotechnology*, *11*(62), 12555–12562.

Ouédraogo, J. T., Tignegre, J. B., Timko, M. P., & Belzile, F. J. (2002). AFLP markers linked to resistance against *Striga gesnerioides* race 1 in cowpea (*Vigna unguiculata*). *Genome*, *45*, 787–793.

Padulosi, S., & Ng, N. Q. (1997). Origin, taxonomy and morphology of *Vigna unguiculata* (L.) Walp. In B. B. Singh, D. R. Mohan Raj, K. E. Dashiell, & L. E. N. Jackai (Eds.), *Advances in cowpea research* (pp. 1–12). Ibadan, Nigeria: IITA. Copublication of International Institute of Tropical Agriculture (IITA) and Japan International Research Center for Agricultural Sciences (JIRCAS).

Panella, L., & Gepts, P. (1992). Genetic relationships within [*Vigna unguiculata* (L.) Walp.] based on isoenzyme analyses. *Genetic Resources and Crop Evolution*, *39*, 71–88.

Pasquet, R. S. (1993). Variation at isoenzyme loci in wild [*Vigna unguiculata* (L.) Walp.] (Fa baceae, Phaseoleae). *Plant Systematics Evolution*, *186*, 157–173.

Pasquet, R. S. (1999). Genetic relationships among subspecies of *Vigna unguiculata* (L.) Walp. based on allozyme variation. *Theoretical and Applied Genetics*, *98*, 1104–1119.

Pasquet, R. S. (2000). Allozyme diversity of cultivated cowpea *Vigna unguiculata* (L) Walp. *Theoretical and Applied Genetics*, *1*, 211–219.

Penza, R., Akella, V., & Lurquin, P. F. (1992). Transient expression and histological localization of a gus chimeric gene after direct transfer to mature cowpea embryos. *Biotechniques*, *13*, 576–580.

Perrino, P., Laghetti, G., Spagnoletti, Z. P. L., & Monti, L. M. (1993). Diversification of cowpea in the Mediterranean and other centers of cultivation. *Genetic Resources and Crop Evolution*, *40*, 121–132.

Popelka, J. C., Gollasch, S., Moore, A., Molvig, L., & Higgins, T. J. V. (2006). Genetic transformation of cowpea (*Vigna unguiculata* L.) and stable transmission of transgenes to progeny. *Plant Cell Reports*, *25*, 304–312.

Sahoo, L., Sushma, , Sugla, T., Singh, N. D., & Jaiwal, P. K. (2000). *In vitro* plant regeneration and recovery of cowpea (*Vigna unguiculata*) transformants via Agrobacterium-mediated transformation. *Plant Cell Biotechnology and Molecular Biology*, *1*, 47–51.

Saini, R., Sonia, , & Jaiwal, P. K. (2003). Stable genetic transformation of *Vigna mungo* L. Hepper via *Agrobacterium tumefaciens*. *Plant Cell Reports*, *21*, 851–859.

Sawadogo, M., Ouédraogo, J. T., Gowda, B. S., & Timko, M. P. (2010). Genetic diversity of cowpea (*Vigna unguiculata* L. Walp.) cultivars in Burkina Faso resistant to *Striga gesnerioides*. *African Journal of Biotechnology*, *9*(48), 8146–8153.

Simon, M. V., Benko-Iseppon, A. M., Resende, L. V., Winter, P., & Kahl, G. (2007). Genetic diversity and phylogenetic relationships in Vigna Savi germplasm revealed by DNA amplification fingerprinting. *Genome*, *50*(6), 538–547.

Singh, B. B., & Matsui, T. (2002). Cowpea varieties for drought tolerance. In C. A. Fatokun, S. A. Tarawali, B. B. Singh, P. M. Kormawa, & M. Tamo (Eds.), *Challenges and opportunities for enhancing sustainable cowpea production. Proceedings of the world cowpea conference III, 4–8 September 2000* (pp. 287–300). Ibadan, Nigeria: International Institute of Tropical Agriculture (IITA).

Singh, S. R., Jackai, L. E. N., Thottappilly, G., Cardwell, K. F., & Myers, G. O. (1992). Status of research on constraints to cowpea production. In G. Thottappilly, L. M. Monti, D. T. Mohan Raj, & A. W. Moore (Eds.), *Biotechnology: Enhancing research on tropical crops in Africa* (pp. 21–26). Ibadan, Nigeria: IITA. CTA/IITA copublication.

Spencer, M. M., Ndiaye, M. A., Gueye, M., Diouf, D., Ndiaye, M., & Gresshoff, P. M. (2000). DNA-based relatedness of cowpea (*Vigna unguiculata* (L.) Walp.) genotypes using DNA amplification fingerprinting. *Physiology Molecular Biological Plants*, *6*, 81–88.

Taiwo, M. A., Kareem, K. T., Nsa, I. Y., & Hughes, D. A. (2007). Cowpea viruses: effect of single and mixed infections on symptomatology and virus concentration. *Virology Journal*, *4*, 95. doi:10.1186/1743-422X-4-95.

Taiwo, M. A., & Shoyinka, S. A. (1998). Viruses infecting cowpeas in Africa with special emphasison the potyviruses. In A. O. Williams, A. L. Mbiele, & N. Nkouka (Eds.), *Virus diseases of plants in Africa* (pp. 93–115). Lagos, Nigeria: OAU/STRC Scientific Publication.

Timko, M. P., Gowda, B. S., Ouédraogo, J., & Ousmane, B. (2007). Molecular markers for analysis of resistance to *Striga gesnerioides* in cowpea. In G. Ejeta & J. Gressell (Eds.),

Integrating new technologies for striga control: Towards ending the witch-hunt (pp. 115–128). Singapore: World Scientific Publishing Co. Pte Ltd.

Timko, M. P., Rushton, P. J., Laudeman, T. W., Bokowiec, M. T., Chipumuro, E., Cheung, F., et al. (2008). Sequencing and analysis of the gene-rich space of cowpea. *BMC Genomics, 103*, 9.

Tosti, N., & Negri, V. (2002). Efficiency of three PCR-based markers in assessing genetic variation among cowpea (*Vigna unguiculata* subsp. *unguiculata*) landraces. *Genome, 45*, 268–275.

Ubi, B. E., Mignouna, H., & Thottappilly, G. (2000). Construction of a genetic linkage map and QTL analysis using a recombinant inbred population derived from an intersubspecific cross of a cowpea (Vina unguiculata (L.) Walp). *Breeding Science, 50*, 161–173.

Uma, M. S., Hittalamani, S., Murthy, B. C. K., & Viswanatha, K. P. (2009). Microsatellite DNA marker aided diversity analysis in cowpea (*Vigna unguiculata* (L.) Walp.). *Indian Journal of Genetics & Plant Breeding, 69*, 33–35.

Vaillancourt, R. E., & Weeden, N. F. (1992). Chloroplast DNA polymorphism suggests Nigerian center of domestication for the cowpea, *Vigna unguiculata* (Leguminosae). *American Journal of Botany, 79*, 1194–1199.

Vaillancourt, R. E., Weeden, N. F., & Barnard, J. D. (1993). Isozyme diversity in the cowpea species complex. *Crop Science, 33*, 606–613.

van de Wouw, M., van Hintum, T., Kik, C., van Treuren, R., & Visser, B. (2010). Genetic erosion in crops: concept, research results and challenges. *Plant Genetic Resources: Characterization and Utilization, 8*, 1–15.

Verdcourt, B. (1970). Studies in the Leguminosae-Papilionoideae for the flora of tropical East *Africa. IV. Kew Bulletin, 24*, 507–569.

Wang, M. L., Barkley, N. A., Gillaspie, G. A., & Pederson, G. A. (2008). Phylogenetic relationships and genetic diversity of the USDA *Vigna* germplasm collection revealed by gene-derived markers and sequencing. *Genetic Resources, 90*, 467–480.

Watanabe, I., Hakoyama, S., Terao, T., & Singh, B. B. (1997). Evaluation methods for drought tolerance of cowpea. In B. B. Singh, D. R. Mohan Raj, K. E. Dashiell, & L. E. N. Jackai (Eds.), *Advances in cowpea research* (pp. 141–146). Ibadan, Nigeria: IITA. Copublication of International Institute of Tropical Agriculture (IITA) and Japan International Research Center for Agricultural Sciences (JIRCAS).

Xavier, G. R., Martins, L. M. V., Rumjanek, N. G., & Filho, F. R. F. (2005). Variabilidade genética em acessos de caupi analisada por meio de marcadores RAPD. *Pesquisa Agropecuária Brasileira, 40*, 353–359.

Xu, P., Wu, X., Wang, B., Liu, Y., Quin, D., Ehlers, J. D., et al. (2010). Development and polymorphism of *Vigna unguiculata* ssp. *unguiculata* microsatellite markers used for phylogenetic analysis in asparagus bean (*Vigna unguiculata* ssp. *sesquipedialis* (L.) Verdc. *Molecular Breeding, 25*, 675–684.

Zannou, A., Kossou, D. K., Ahanchédé, A., Zoundjihékpon, J., Agbicodo, E., Struik, P. C., et al. (2008). Genetic variability of cultivated cowpea in Benin assessed by random amplified polymorphic DNA. *African Journal of Biotechnology, 7*, 4407–4414.

7 Lentil

Clarice Coyne[1] *and Rebecca McGee*[2]

[1]United States Department of Agriculture, Agricultural Research Service, Plant Introduction and Testing Unit, Washington State University, Pullman, WA, [2]United States Department of Agriculture, Agricultural Research Service, Grain Legume Breeding and Physiology Unit, Washington State University, Pullman, WA

7.1 Introduction

Lentil ranks among the oldest and most appreciated grain legumes of the Old World (Smartt, 1990). Worldwide, production has increased over the last few decades (FAO, 2010); however, direct and indirect human activities have posed imminent threats to the integrity of the genetic diversity of indigenous germplasm in many areas of the world, including the Mediterranean region, Western Asia, Ethiopia and the Indian subcontinent. Approximately 37,000 accessions have been collected and are conserved *ex situ* by national and international gene banks. The genus *Lens* Miller is part of the family Fabaceae (Leguminosae). It is placed variously in either subfamily Faboideae, tribe Fabeae (Soltis et al., 2011), or in subfamily Papilionaceae, tribe Vicieae (Sonnante, Hammer, & Pignone, 2009). Lentil is an annual, self-pollinating, diploid ($2n = 2x = 14$) species with an estimated genome size of 4063 Mbp/C (Arumuganathan & Earle, 1991). In this chapter, the genetic and genomic resources of lentil are reviewed. We discuss the origin, distribution, diversity and taxonomy. We also address the conservation, evaluation and maintenance of germplasm and its uses and limitations in crop improvement.

7.2 Origin, Distribution, Diversity and Taxonomy

Lentil is one of the eight founder grain crops that started agriculture in Southwest Asia (the Levant) during the Pre-Pottery Neolithic period, some 11,000–10,000 years ago (Weiss & Zohary, 2011). The Levant includes most of modern Lebanon, Syria, Jordan, Israel, Palestinian Authority, Cyprus, Turkey's Hatay Province and some regions of Iraq or the Sinai Peninsula areas that are now confirmed by the archaeobotanical record. Described as the 'richest sites', these sites include *c.* 10,200–9550 BP Tell Aswad, Syria; *c.* 10,200–8700 BP Tell Abu Hureyra, Syria; *c.* 10,250–9500 BP Jericho, Palestine; *c.* 10,600–9900 BP Çayönü, Turkey; *c.* 9600–8800 BP Ali Kosh, Iran; *c.* 10,400–9450 BP Yiftah'el, northern Israel; *c.* 9450–9300 BP Jarmo, Iraq;

Genetic and Genomic Resources of Grain Legume Improvement. DOI: http://dx.doi.org/10.1016/B978-0-12-397935-3.00007-4

c. 9250–9000 cal BP Tell Ramad, Syria; *c.* 8200–7800 cal BP Hacilar, Turkey and *c.* 8350–7750 BP Tepe Sabz, Deh Luran Valley, Iran (Weiss & Zohary, 2011). However, these archaeological remains do not provide direct diagnostic traits (such as indehiscent pod) to determine the origin of lentil domestication, though lentil seed size suggests selection.

Lentil derivation has not always been clear, as Zohary noted in his seminal publication (1972). The accepted dogma until 1973 proposed by Barulina (1930) put the origin of lentil cultivation between the Hindu Kush and the Himalaya. Further, Kislev and Bar-Yosef (1988) suggested lentils as the earliest domesticated plants in the Near East based on the presence of pulses among the charred plant remains retrieved from archaeological sites, but cautioned that there was not sufficient evidence to support this intriguing claim. This may not be the case, as wheat, but not lentil, was found at Nevali Çori in southeastern Turkey, a 10,500-year-old archaeological site (Balter, 2007). However, two large samples of lentil were found about 11,000 BP in Jerf el Ahmar, Syria, and Netiv Hagdud, near Jericho (Weiss, Kislev, & Hartmann, 2006). Morphological change can no longer be held as the first indication of domestication; rather, a long period of increasingly intensive human management typically precedes the manifestation of archaeologically detectable morphological change in managed crops (Zeder, 2011). Further, agriculture in the Near East arose in the context of broad-based systematic human efforts of cultivating plant resources (Zeder, 2011). There is a current controversy over slow or fast rate (duration) of the process of domestication (Allaby, Fuller, & Brown, 2008; Balter, 2007; Heun, Abbo, Lev-Yadun, & Gopher, 2012). Domestication took place across the entire Fertile Crescent during a period of dramatic post-Pleistocene climate and environmental change, with a range of resources being manipulated by humans (Zeder, 2011). Ladizinsky (1987) suggested the 'pulse domestication before cultivation' model for lentil based on the identification of free germinating genotypes among wild legume populations that must have predated any cultivation experiments. A fast rate of plant domestication is supported by biological evidence of Near Eastern wild and domesticated lentil (Abbo, Lev-Yadun, & Gopher, 2011). Initial domestication of lentils occurred in southeastern Turkey or northern Syria based on genetic and archaeological evidence (Ladizinsky, 1979b; 1993; 1999). Lentil as a crop spread quickly from here into the southern Levant; however, a separate southern Levantine domestication cannot be ruled out (Weiss et al., 2006). Zohary (1999) hypothesized a monophyletic origin and tethered this theory to lentils being 'very likely taken into cultivation only once or – at most – a very few times', but did not consider published allozyme data (Ladizinsky, Cohen, & Muehlbauer, 1985; Pinkas, Zamir, & Ladizinsky, 1985).

Allozymes were the first biomarker in support of polyphyly in crops such as lentils (Allaby et al., 2008). Recent studies based on data of eight founder species suggest that domestication happened in a small region of the southern Levant (Sonnante et al., 2009). Further, botanical, genetic and archaeological evidence points to a small core area of domestication in present-day southeastern Turkey and northern Syria, near the Tigris and Euphrates rivers (Sonnante et al., 2009). Alo, Furman, Akhunov, Dvorak, and Gepts (2011) concluded that the study of wild and cultivated lentil further supports the hypothesis of a polycentric origin of domestication. Abbo et al. (2012) cautioned that 'only detailed phylogenetic studies of representative collections of wild and domesticated forms can determine the place of origin and their phylogeny'.

Wild *Lens* taxa are widely distributed in the Mediterranean basin; it was thought that only in Aegean and southwestern Turkey do the distributions of wild taxa overlap (Ferguson, Acikgoz, Ismail, & Cinsoy, 1996) (Figure 7.1). Maxted, Hargreaves et al. (2010) performed an *in situ* and *ex situ* gap analysis using taxonomic, ecological, geographic and conservation information for 672 wild *Lens* collated from ICARDA (International Center for Agricultural Research in the Dry Areas) and GBIF (Global Biodiversity Information Facility) data sets as well as data sets collected by the authors over 25 years. Gap analysis, a process by which the distribution of taxon and vegetation types are compared, assists in identifying biodiversity to protect either *in situ* or *ex situ* (Scott et al., 1993). Maxted's gap analysis refined the regions of highest *Lens* species richness (three to four species) to the Crimea Peninsula and along southeastern Turkey through the eastern Mediterranean countries of Syria, Jordan, Israel and Palestinian Authority (Figure 7.1). Regions with two species include Mediterranean Spain, Mediterranean Balkans, Albania, Greece and western Turkey (Maxted, Kell et al., 2010).

Lens culinaris ssp. *orientalis* (Boiss.) Ponert has an eastern distribution from Turkey, Cyprus and Palestine across to Uzbekistan. *Lens culinaris* subsp. *odemensis* (Ladiz.) M.E. has a more restricted distribution in the east, extending from Turkey southwards to Syria and Palestine (Ferguson, Maxted, Slageren, & Robertson, 2000). A single population of *Lens culinaris* subsp. *tomentosus* (ladiz.) M.E. has been found in Libya. *Lens ervoides* (Brign.) Grande has a broad distribution from Spain to Ukraine and south to Jordan. Outlier populations have also been found in Ethiopia and Uganda. *Lens nigricans* (M. Bieb.) Godr. grows in diffuse small colonies on stony hillsides and shallow rocky soils in pine forest clearings (Zohary, 1972). *L. nigricans* has a western distribution from Spain to Turkey and south to Morocco (Ferguson et al., 1996) and east to Crimea and the eastern shore of the Mediterranean Sea (Zohary, 1972). *Lens lamottei* (Czefr.) grows in Morocco (van Oss, Aron, & Ladizinsky, 1997). It is only in Aegean and southwestern portions of Turkey that the distributions of all wild taxa overlap. Unfortunately, Turkey, like other Mediterranean countries, is suffering the rapid loss of many of its valuable genetic resources. These resources, which have the potential to provide useful genetic material for plant breeding efforts, are being eroded primarily by habitat destruction (Solh & Erskine, 1981). Ferguson et al. (1996) noted 'the poor competitive ability and palatability of *Lens* species, together with the fact that they occur in small disjunct populations, intensifies this threat'.

Molecular diversity evaluations of *ex situ* germplasm collections include studies completed with DNA-based markers, such as random amplified polymorphic DNA (RAPDs), inter-simple sequence repeats (ISSRs), amplified fragment length polymorphisms (AFLPs) and simple sequence repeats (SSRs). Studies of national collections tend to be smaller in terms of genotypes and number of accessions sampled. Two studies of Ethiopian lentil accessions, one using ISSR markers alone and the other using nine morphological and four ISSR markers of 10 accessions including

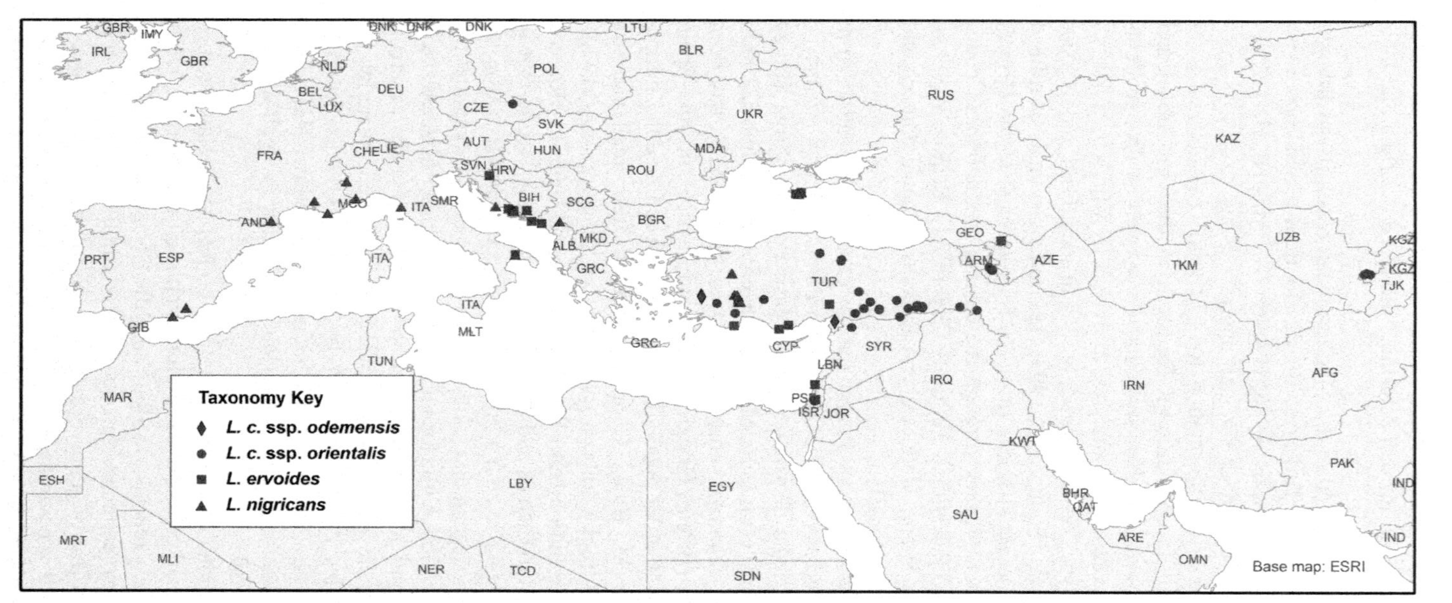

Figure 7.1 Map location of the wild *Lens* species and subspecies held by the USDA ARS, Pullman, WA, USA. Maxted's gap analysis will be helpful to fill out the lentil CWR in this national collection (Maxted, Kell et al., 2010).

L. culinaris varieties revealed useful variations, where the average gene diversity was 0.2734 (Fikiru, Tesfaye, & Bekele, 2010, 2011). A similarly sized study was conducted on six Bangladeshi lines using 10 RAPD markers, where the average gene diversity was 0.0552 (Hoque & Hasan, 2012). Larger studies have been published for Bangladeshi, Italian and Indian lentil germplasm collections. Ten RAPDs were used on 19 lines and seed protein profiles of 144 accessions were used to characterize and collected from Bangladesh (Sultana & Ghafoor, 2008). However, 14 Italian lines were studied using 31 traits measured over environments, including 9 agromorphological, 5 post-harvest seed traits, seed protein profiles and 16 SSR markers, which uncovered exploitable diversity (Zaccardelli et al., 2012). A second Italian study looked at 19 agronomic traits of 28 Italian landraces over environments and studied the genetic diversity using AFLPs (Torricelli et al., 2012). Datta et al. (2011) examined 30 Indian lines with 39 SSRs using cross-genera legume markers.

Further, international germplasm collection molecular diversity is presented in three published studies. Central Asia and Caucasian cultivated lentil germplasm were genotyped with five SSR markers and clustered into six groups (Babayeva et al., 2009). Fourteen newly reported SSR markers were used to examine the molecular diversity of 109 accessions including both cultivated lines and wild *Lens* species (Hamwieh, Udupa, Sarker, Jung, & Baum, 2009). They found that the wild accessions were rich in alleles (151 alleles) compared to cultigens (114 alleles). These lines also clustered into two groups, one cultivated and the other wild germplasm. The largest study published to date consisted of 133 domesticated lentil and 175 wild lentil accessions conducted by ICARDA using 22 cross-genera SSR markers (Alo et al., 2011). Structure analysis revealed eight haplotype groups (K=8) (Pritchard, Stephens, & Donnelly, 2000). All groups consisted of one taxon except one, which had all taxa except domesticated *Lens* (Alo et al., 2011). Linkage disequilibrium (LD) was calculated and varied across the individual groups, with the higher LD in the cultivated lines as found in other domesticated crop species.

However, other biochemical genetic diversity research has been conducted on lentil genetic resources. Examples include a study on the diversity of lentil seed starch and complex carbohydrates, where the diversity discovered invites researchers, especially breeders, to exploit the variability uncovered (Chibbar, Ambigaipalan, & Hoover, 2010). Two studies have looked at the seed protein profiles of 144 accessions, mainly landraces of Pakistan (Sultana, Ghafoor, & Ashraf, 2006) and 14 cultivars of Turkey (Yüzbaşioğlu, Açik, & Özcan, 2008). Both studies have identified useful diversity. The lentil seed proteome was determined for two Italian landraces; multivariate analysis of 145 differentially expressed protein spots demonstrated that 52 proteins are required to discriminate (Ialicicco et al., 2012). Taxonomically, lentil holds an intermediate position between *Vicia* and *Lathyrus*. Zohary (1972) reported five species of cultivated *L. culinaris* Medikus. [*L. esculenta* Moench] and four wild species: *Lens monbretti* (Fisch. & May) Davis and Plitm. [*L. kotschyanum* (Boiss.) Nab.; *L. kotschyaya* (Boiss.) Alef.]; *Lens nigricans* (Bieb.) Godr. [*Ervum nigricans* Bieb.]; *Lens ervoides* (Brign.) Grande [*L. lenticula* (Schreb,) Alef.] and *Lens orientalis* (Boiss.) Hand.-Mazz. During that time, all the morphological evidence indicated five lentil species. *L. monbretti* (Fisch. & Mey.) is morphologically and cytologically different from the other *Lens* species and was moved back to the genus *Vicia* (Ladizinsky & Sarker, 1982). Pinkas et al. (1985) proposed five *Lens* species, namely *L. culinaris*, *L. orientalis*, *L. odemensis*, *L. ervoides* and *L. nigricans*, based on allozyme divergence. Hoffman, Soltis, Muehlbauer, and Ladizinsky (1986) proposed two species and five taxa, namely *L. culinaris* with three subspecies *Lens culinaris* subsp. *culinaris*, *Lens culinaris* subsp. *orientalis* and *Lens culinaris* subsp. *odemensis*; *L. nigricans* with two subspecies *L. nigricans* subsp. *nigricans* and *L. nigricans* subsp. *ervoides*. Ladizinsky updated the *Lens* taxa (1997) and defined four species by reducing *L. orientalis* to a subspecies and creating two new *Lens* species, namely *L. lamottei* Czefr. and *L. tomentosus* Ladiz. Chloroplast DNA marker variation briefly concluded there are six species in the *Lens* taxa (van Oss et al., 1997).

Further molecular phylogeny analysis both clarifies and confuses *Lens* taxonomy regarding species and subspecies. From the period between 1979 and 2005, numerous studies focussed on the phylogeny of *Lens* using the molecular tools of various marker classes, including isozymes (Ferguson, Newbury, Maxted, Ford-Lloyd, & Robertson, 1998; Hoffman et al., 1986; Ladizinsky, 1979a), restriction fragment length polymorphism (RFLPs) (Havey & Muehlbauer, 1989), RAPDs (Abo-Elwafa, Murai, & Shimada, 1995; Ahmad, Fautrier, Burritt, & McNeil, 1997; Ahmad & McNeil, 1996; Sharma, Dawson, & Waugh, 1995), AFLPs (Sharma, Knox, & Ellis, 1996). Fortunately, all the studies indicate that *Lens culinaris* spp. *orientalis* is the closest progenitor of cultivated lentil.

Ferguson et al. (2000) proposed the taxonomy of four species, reducing *L. odemensis* and *L. orientalis* to subspecies of *L. culinaris* based on morphological, isozyme and RAPD marker data combined (Table 7.1). The contemporary literature is fraught with differing interpretations of the exact number of taxa and splits (e.g. Tahir, Båga, Vandenberg, & Chibbar, 2012). The taxonomy is understandably difficult given the close relationships between the *Lens* taxa (Ferguson et al., 2000). This taxonomic description for *Lens* is accepted by the USDA for use in GRIN. This study is given heavy weight by the taxonomic community as it combines the molecular characterization with botanical descriptors of the species and subspecies for the classification of the herbarium samples.

Table 7.1 The Latest Taxonomy of *Lens* Comprising Seven Taxa Split into Four Species (Ferguson et al., 2000)

GRIN Taxonomy[a]	Gene Pool[b]
Lens culinaris Medik.	Primary
Lens culinaris subsp. *culinaris*	Primary
Lens culinaris subsp. *odemensis* (Ladiz.)	Primary
Lens culinaris subsp. *orientalis* (Boiss.) Ponert	Primary
Lens culinaris subsp. *tomentosus* (Ladiz.) M.E.	Primary
Lens ervoides (Brign.) Grande	Secondary/tertiary
Lens lamottei Czefr.	Secondary/tertiary
Lens nigricans (M. Bieb.) Godr.	Secondary/tertiary

[a]Germplasm Resources Information Network: http://www.ars-grin.gov/cgi-bin/npgs/html/tax_search.pl.
[b]Tullu, Bett et al. (2011) and Tullu, Diederichsen et al. (2011).

Gene-based phylogenic studies of the *Lens* taxa were conducted from 1994 to 2012 using genes favoured by the botanic taxonomists for studying plant evolution across the plant kingdom.

Muench, Slinkard, and Scoles (1991) and Mayer and Soltis (1994) both examined chloroplast RFLPs, while the 1994 study looked at far more accessions. Both told the same story as mentioned in Table 7.1 except that subspecies *tomentosus* was not represented. Similarly, studies using RFLPs of ITS region of ribosomal DNA (Mayer & Bagga, 2002; Sonnante, Galasso, & Pignone, 2003) resulted in some differences at the time of divergence, but not grouping. Recent sequencing data will continue to shed light on the species and taxa status of *Lens* (Schaefer et al., 2012). Finally, using maximum likelihood and Bayesian phylogeny analysis based on six chloroplast gene sequences (rbcL, matK, trnL/trnL-trnF, trnS-trnG, psbA-trnH) and one nuclear gene sequence (ribosomal ITS) of the legume tribe Fabae finds *Lens* nested in the middle of the *Vicia* clade. *Lens* diverged from its nearest *Vicia* ancestors 14.9–12.6 million years ago. The sequence data of these seven genes also confirmed the monophyly origin of *Lens* and that *Lens culinaris* spp. *orientalis* is the closest progenitor of cultivated lentil. The authors suggested that based on sequence analysis lentil may be placed within the *Vicia* genera (Schaefer et al., 2012).

7.3 Biosystematics

Of course the most interesting question is which species or subspecies is the progenitor of cultivated *L. culinaris*. Zohary and Hopf (1973) ruled out *L. monbretti* based on taxonomy. Using Zohary and Hopf (1973) species classification also ruled out *L. ervoides* and *L. nigricans* based on species distribution and suggested *Lens culinaris* subsp. *orientalis*, as it manifested the closest morphological similarity to cultivated lentil. Cubero et al. (2009) suggested that 'some populations of *orientalis* were unconsciously subjected to automatic selection' in the region of southern Turkey to northern Syria and gave rise to *L. culinaris*. The strongest evidence to date is the data provided by the phylogenetic study based on sequencing seven genes, which supports the morphological data of *Lens culinaris* subsp. *orientalis* as the progenitor of cultivated lentil (Schaefer et al., 2012).

7.4 Status of Germplasm Resources Conservation

7.4.1 *Ex Situ Conservation*

The world collection is held by ICARDA; most of the other national collections hold some portion of subsets of this collection and vice versa (Table 7.2). ICARDA also holds the largest collection of the wild *Lens* accessions from 46 countries (Furman, Coyne, Redden, Sharma, & Vishnyakova, 2009; Table 7.3). It is difficult to determine exactly the overlap, duplication or redundancy due to the lack of consistent access to

Table 7.2 The World *Ex Situ Lens* Collection Held by the ICARDA with Significant *Lens* Germplasm with Other National Gene Bank Collections of 2000+ Accessions

Institution	Accessions	Website
ICARDA[a]	10,822	http://www.icarda.org/
ECPGR[b]	4598	http://www.ecpgr.cgiar.org/germplasm_databases.html
India[c]	7712	http://www.nbpgr.ernet.in/
ATFCC[d]	5254	http://www.dpi.vic.gov.au/
USDA ARS[e]	3187	http://www.ars-grin.gov/npgs/
Iran[f]	3000	http://en.spii.ir/seSPII/
Russian Federation[g]	2556	http://www.vir.nw.ru/

(modified from Tullu, Bett et al., 2011; Tullu, Diederichsen et al., 2011).
[a]International Center for Agricultural Research in the Dry Areas, Aleppo, Syria.
[b]European Cooperative Program for Plant Genetic Resources includes Russian Federation.
[c]National Bureau of Plant Genetic Resources (NBPGR), New Delhi, India.
[d]Australian Temperate Field Crops Collection, Horsham, will be consolidated into the new Australian Grains Gene Bank, Horsham, Victoria, Australia.
[e]United States Department of Agriculture, Agricultural Research Service, Pullman, WA, USA.
[f]Seed and Plant Improvement Institute (SPII), Karaj, Iran.
[g]N.I. Vavilov All-Russian Scientific Research Institute of Plant Industry (VIR), St. Petersburg, Russia.

Table 7.3 Wild *Lens* Conserved *Ex Situ* with the World Collection Held by ICARDA and One National Gene Bank of USDA ARS NPGS

Taxon	USDA	ICARDA
Lens culinaris ssp. *orientalis*	92	268
Lens culinaris ssp. *odemensis*	8	65
Lens culinaris ssp. *tomentosus*	0	11
Lens ervoides	61	166
Lens lamottei	0	10
Lens nigricans	37	63
Total	198	583

databases, lack of cross-reference to other gene bank accession identification within databases (i.e. accession names/numbers) and lack of data *per se* (Potan, 2009; Tullu, Diederichsen, Suvorova, & Vandenberg, 2011). The Australian Temperate Field Crops Collection (ATFCC) database has made the most progress in cross-referencing by name/number identification across national gene banks including the world lentil collection at ICARDA and is available by request (Redden, personal communications at ATFCC). Fortunately, the world crop genetic resources community is addressing the database issue directly through efforts within the Consultative Group on International Agricultural Research (CGIAR) system, through Bioversity International, through conferences, particularly the conference series International Symposium on Genomics of Plant Genetic Resources and white papers under development by the Global Crop Diversity Trust (http://www.croptrust.org/). One white paper developed was the

'Global Strategy for the Ex Situ Conservation of Lentil (*Lens* Miller) (2008)' which includes a goal to assemble passport data on major pulses, including lentil, from collections worldwide into a single database linked with geographical information system (GIS) data. While not the largest lentil collection by far, the USDA ARS stands out in the accessibility of its database and seed samples and will be used as an example of a national database in comparison with the world collection (Table 7.3). Recent collections include two plant explorations in Crimea and Ukraine. Diederichsen, Rozhkov, Korzhenevsky, and Boguslavsky (2012) collected genetic resources of crop wild relatives (CWR) including eight wild *Lens* species and Bockelman (1999) collected one each of *L. ervoides* and *L. nigricans* accessions.

7.4.2 In Situ Conservation

The number of accessions preserved *ex situ* from the regions of origin and diversity has been increasing. Seed has been collected from each taxon and used in further study to determine within-population diversity. This will help to establish the potential of *in situ* conservation for wild *Lens* species (Ferguson & Robertson, 1996). Unfortunately, many areas of greatest interest for *in situ* conservation (e.g. Turkey and other Mediterranean countries) are suffering from rapid loss of invaluable genetic resources due to habitat destruction (Solh & Erskine, 1981). The relatively poor competitive ability and high palatability of *Lens* species, together with the fact that they occur in small disjunct populations, intensifies this threat (Ferguson et al., 1996). Important areas to target for *in situ* conservation include west Turkey for *L. nigricans*, southeast Turkey, northwest Syria, south Syria and Jordan for *L. culinaris* ssp. *orientalis*, south Syria for *L. culinaris* ssp. *odemensis* and the coastal border region between Turkey and Syria stretching along the Syrian coast for *L. ervoides* (Ferguson, Ford-Lloyd, Robertson, Maxted, & Newbury, 1998).

7.5 Germplasm Evaluation and Maintenance

Cultivated lentil experienced a genetic bottleneck with low amounts of molecular variation in the lentil germplasm collections (Alo et al., 2011; Alvarez, García, & Pérez de la Vega, 1997; Ferguson et al., 2000; Ford, Pang, & Taylor, 1997; Mayer & Soltis, 1994; Muench et al., 1991). Erskine, Sarker, and Ashraf (2011) used traits of flowering time and yield to reconstruct the genetic bottleneck of lentil into south Asia. Nonetheless, useful variation in cultivated lentil has led to significant breeding advances. Future genetic gains will be dependent on introgressing useful alleles from landraces and other wild *Lens* relatives for widening the genetic base of cultivated species. Lentil evaluation descriptors were published in 1985 by the International Board for Plant Genetic Resources (now Bioversity International) and ICARDA (IBPGR, 1985). Abiotic and biotic stress resistance screening are summarized in Table 7.4. Several studies have been conducted and published on multilocational trials of landrace accessions for agronomic (descriptor) traits. Lentil core and composite collections allow for the sampling of diverse lines and provide an efficient method

Table 7.4 Sources of Foreign Genes from the Landraces and Wild Relatives for Introgression into Lentil

Useful Trait(s)	Wild Relative	References
Anthracnose resistance	*L. ervoides*, *L. lamottei*, *L. nigricans*	Tullu et al. (2006), Tullu, Banniza, Taran, Warkentin, and Vandenberg (2010), Fiala, Tullu, Banniza, Séguin-Swartz, and Vandenberg (2009), Vail and Vandenberg (2011) and Vail, Strelioff, Tullu, and Vandenberg, (2012)
Ascochyta blight resistance	*L. ervoides*, *L. culinaris* ssp. *orientalis*, *L. odemensis*, *L. nigricans*, *L. lamottei*	Bayaa et al. (1994), Nguyen, Taylor, Brouwer, Pang, and Ford (2001) and Tullu et al. (2006, 2010)
Colletotrichum truncatum resistance	*L. culinaris*	Buchwaldt, Anderson, Morrall, Gossen, and Bernier (2004) and Shaikh et al. (2012)
Stemphylium blight	*L. ervoides*, *L. culinaris* ssp. *orientalis*, *L. tomentosus*, *L. nigricans*, *L. odemensis*, *L. lamottei*	Podder, Banniza, and Vandenberg (2012)
Fusarium wilt resistance	*L. culinaris* ssp. *orientalis*, *L. ervoides*	Bayaa et al. (1995), Gupta and Sharma (2006) and Mohammadi, Puralibaba, Goltapeh, Ahari, and Sardrood (2012)
Powdery mildew resistance	*L. culinaris* ssp. *orientalis*, *L. nigricans*	Gupta and Sharma (2006)
Rust resistance	*L. culinaris* ssp. *orientalis*, *L. ervoides*, *L. nigricans*, *L. odemensis*	Gupta and Sharma (2006)
Drought tolerance	*L. odemensis*, *L. ervoides*, *L. nigricans*	Hamdi and Erskine (1996)
Cold tolerance	*L. culinaris* ssp. *orientalis*	Hamdi, Küsmenoĝlu, and Erskine (1996)
Heat tolerance	*L. culinaris*	Roy, Tarafdar, Das, and Kundagrami (2012)
Yield attributes	*L. culinaris* ssp. *orientalis*	Gupta and Sharma (2006)
Resistance to *Orobanche*	*L. culinaris*, *L. ervoides*, *L. odemensis*, *L. orientalis*	Fernández-Aparicio, Sillero, Pérez-De-Luque, and Rubiales (2008) and Fernández-Aparicio, Sillero, and Rubiales (2009)
Resistance to sitona weevils	*L. odemensis*, *L ervoides*, *L. nigricans*, *L. culinaris* ssp. *orientalis*	El-Bouhssini, Sarker, Erskine, and Joubi (2008)
Resistance to bruchid weevils	*L. culinaris* ssp. *orientalis*, *L. nigricans*, *L. lamottei*	Laserna-Ruiz, De-Los-Mozos-Pascual, Santana-Méridas, Sánchez-Vioque, and Rodríguez-Conde (2012)

Source: Adapted from Kumar et al. (2011). Taxonomic designations are those of the authors.

for finding sources of new traits (Furman, 2006; Simon & Hannan, 1995). The USDA lentil core collection of 287 *L. culinaris* accessions was characterized for phenology, morphology, biomass and seed yields over two seasons (Tullu, Kusmenoglu, McPhee, & Muehlbauer, 2001). Thirty landraces of Pakistan were evaluated for flowering and yield components also over two seasons to determine diversity for breeding strategies (Tyagi & Khan, 2011). Morphological and phenological variation was also assessed in 310 accessions of the wild relatives of lentil (Ferguson & Robertson, 1999). ICARDA has established a composite collection of 1000 accessions to represent genetic diversity and the agro-climatological range of lentil and this will be used for intensive phenotyping and genotyping purposes (Furman, 2006).

Lentil is a naturally self-pollinated species due to its cleistogamous flowers (Wilson, 1972) and usually has <0.8% natural cross pollination (Wilson & Law, 1972). Outcrossing in lentil depends on cultivar, location and year, and varies within cultivars (Horneburg, 2006). For regeneration and backup storage, bioversity recommends a base collection of accessions in long-term storage used for regeneration, an active collection in less stringent conditions accessible for distribution and a security backup collection at a different location (Engels & Visser, 2003). Similarly, a guide is published for regeneration guidelines of lentil (Sackville Hamilton & Chorlton, 1997). Lentil seed can be stored for relatively long periods of time at −18°C (Walters, Wheeler, & Grotenhuis, 2005). Seed handling conditions from harvest to storage temperature and relative humidity are critical components affecting seed longevity (Walters, Wheeler, & Stanwood, 2004). Long-term storage temperatures are an important (neglected) factor given conventional seed bank temperatures (Li & Pritchard, 2009).

7.6 Use of Germplasm in Crop Improvement

The wild relatives of lentil are a dynamic resource of unique genes/alleles that are not present in cultivated lines. Many economically important traits, such as resistance to biotic and abiotic stresses, are not currently represented in *L. culinaris* ssp. *culinaris*, but are found in the wild relatives. Introgression of these useful genes will greatly enhance the genetic base of cultivated lentil. Deploying these genes from the secondary and tertiary gene pools frequently requires techniques of embryo rescue and tissue culture. Initial development of lentil varieties was via single plant selection within landraces. Landraces are defined by their historical origin, recognizable identity, lack of formal genetic improvements, high genetic diversity, local adaptation and association with traditional farming systems (Villa, Maxted, Scholten, & Ford-Lyod, 2006). Lentil landraces have existed since domestication and over time have genetically responded to selection pressures of biotic and abiotic stresses. Cultivars developed from pure-line selection within landraces include Uthfala (Sarker, Rahman, Rahman, & Zaman, 1992), Laird (Slinkard & Bhatty, 1979), Eston (Slinkard, 1981), ILL 5582 (Idlib 1; Jordan 3; El Safsaf 3 and Baraka) (Erskine,

Saxena, & Malhotra, 1996), Bichette (Sakr et al., 2004), Crimson (Muehlbauer, 1991) and Ozbek (Aydoğan et al., 2008).

Germplasm lines derived from pure-line selection of landraces have also played a prominent role as a source of novel alleles in traditional breeding programs. For example, Uthfala (Barimasur-1=ILL 5888) was used as a parent in Bangladesh to develop varieties with improved resistance to *Fusarium* wilt and *Ascochyta* blight (Sarker, Erskine, Hassan, Afzal, & Murshed, 1999). Nonelite germplasm has been used extensively as parents in mapping populations developed to identify sources of resistance to *Stemphylium* blight (Saha, Sarker, Chen, Vandemark, & Muehlbauer, 2010a), *Fusarium* vascular wilt (Eujayl, Erskine, Bayaa, Baum, & Pehu, 1998; Hamwieh et al., 2005), Anthracnose (Tullu, Buchwaldt, Warkentin, Taran, & Vandenberg, 2003), *Aschochyta* blight (Ford, Pang, & Taylor, 1999; Taylor, Ades, & Ford, 2006; Tullu et al., 2006) and lentil rust (Kant, Sharma, Sharma, & Basandrai, 2004; Saha, Sarker, Chen, Vandemark, & Muehlbauer, 2010b). They have also been used to study the earliness and plant height (Tullu, Tar'an, Warkentin, & Vandenberg, 2008) and cold tolerance (Eujayl, Erskine, Baum, & Pehu, 1999; Kahraman et al., 2004). Genetic resources of lentil's wild relatives have become recognized as the source of many economically useful genes (Table 7.4, modified from Kumar, Imtiaz, Gupta, & Pratap, 2011) and will contribute to the success of breeding new cultivars adapted to major biotic and abiotic stresses.

7.7 Limitations in Germplasm Use

Issues to be addressed in terms of limitations of lentil germplasm use are access, precise phenotypic data, breeding efficiencies and available diversity preserved *ex situ* and *in situ*.

The first issue is access. Lentil is covered by the Convention on Biological Diversity (CBD, 1994) and the International Treaty for Plant Genetic Resources of Food and Agriculture (IT-PGRFA or IT, 2004). These treaties are part of the evolving standards that regulate access to genetic resources and define benefit sharing (Ghijsen, 2009). In 2006 a standard material transfer agreement (SMTA) was agreed for the IT, in which the requirements for access to the genetic resources of the 64 food, feed and forage crops, including lentil (annex I of IT) was established and the ways of benefit sharing are enumerated (Ghijsen, 2009). Lentil germplasm is freely available from the world collection held by ICARDA under the SMTA set in place by the 2006 IT treaty and is now used by some national gene banks. For example, requesters of germplasm from CGIAR centres such as ICARDA accessioned with USDA in GRIN after 2006 must agree to the SMTA stipulations. However, lentil germplasm donated or collected and directly accessioned in GRIN is not covered by SMTA, nor is CGIAR material received prior to 2006. Breeders must have efficient methods and gene-based methods to introduce positive new alleles locked in nonelite, unadapted or wild germplasm held *ex situ* or *in situ* or not yet collected. New methodologies such as genomic selection and genome-wide association studies have created opportunities for breeders to mine lentil germplasm for needed genes/alleles. Currently, lentil suffers from one of the poorest genomic resources of the top six grain legumes in production. Not unexpectedly, this is currently changing at an exponential rate (Varshney, Close, Singh, Hoisington, & Cook, 2009). Additionally, recent meeting reports of transcriptomes of diversity panels, single nucleotide polymorphisms (SNPs) discovered, dense gene-based maps (Sharpe et al., 2013), single nucleotide polymorphism (SNP) panels, a 10X lentil bacterial artificial chromosome (BAC) library and high-throughput genomic sequencing (Bett, personal communication) will soon put lentil in the realm of published genomic resources for crop improvement.

Tremendous lentil genetic diversity is currently unavailable either in *ex situ* or under *in situ* conservation. This treasure of lentil germplasm is held in populations poorly or incompletely sampled or even completely unsampled wild lentil taxa. This gap not only limits the use, it renders precious genetic diversity inaccessible and vulnerable to erosion or extinction. Fortunately, this is well recognized, and international efforts led by the Bioversity organization are in progress to conduct gap analyses (Scott et al., 1993) on CWR including lentil and develop comprehensive strategies for wild relative germplasm conservation (Maxted, Kell, Ford-Lloyd, Dulloo, & Toledo 2012). Grain legume gap analysis (Maxted et al., 2012) illustrates how existing georeferenced passport data associated with accessions of *Lens* species from ICARDA and GBIF (http://www.gbif.org/) can be used to identify gaps in current *ex situ* conservation and develop a more systematic *in situ* conservation strategy. It might be expected that all of the species closely related to crops have already been well sampled, but some that are the closest CWR of the crops, such as *Lathyrus amphicarpos*, *La. belinensis*, *La. chrysanthus*, *La. hirticarpus*, *Medicago hybrida*, *Lens culinaris* subsp. *tomentosus* (Maxted et al., 2012), have fewer than 10 samples conserved *ex situ*. It is evident that wild *Lens* species provide an invaluable gene source for the improvement of lentil cultivars (Maxted & Bennett, 2001).

7.8 Germplasm Enhancement Through Wide Crosses

The domesticated lentil, *Lens culinaris* subsp. *culinaris*, is readily crossable with the wild *Lens culinaris* subsp. *orientalis* (Fratini, Ruiz, & Pérez de la Vega, 2004; Gupta & Sharma, 2007; Muehlbauer, Weeden, & Hoffman, 1989; Vaillancourt & Slinkard, 1993; Vandenberg & Slinkard, 1989; Singh et al., 2013) and the wild *Lens culinaris* subsp. *odemensis* (Abbo & Ladizinsky, 1991; Singh et al., 2013). The resulting hybrids are fertile or partially fertile, as a result of chromosome rearrangements (Abbo & Ladizinsky, 1991). The same holds true for crosses between *L. ervoides* and *L. nigricans* (Abbo & Ladizinsky, 1991). However, almost all hybrids abort within 2 weeks in crosses between *L. ervoides* and *L. nigricans* and all *L. culinaris* subspecies (Ladizinsky, Braun, Goshen, & Muehlbauer, 1984). Also reported were rare hybrid seeds, which were albino and died shortly after germination (Abbo & Ladizinsky, 1991).

Interspecific crosses within the genus *Lens* generally abort and embryo rescue techniques are necessary to recover hybrids (Tullu, Bett, Saha, Vail, & Vandenberg, 2011). The first lentil embryo rescue protocol (Cohen, Ladizinsky, Ziv, & Muehlbauer, 1984) allowed the recovery of interspecific hybrids between the cultivated lentil and *L ervoides* and *L. nigricans*. Later, using the same embryo culture technique, Ladizinsky et al. (1985) again obtained hybrids of the cultivated lentil with *L. ervoides*. Fratini and Ruiz (2006, 2011) successfully recovered interspecific hybrids between the cultivated lentil and *L. odemensis*, *L. ervoides* and *L. nigricans* using embryo rescue techniques. 'The *in vitro* culture procedure to rescue interspecific hybrid embryos consists of at least four different stages: (i) *in ovule* embryo culture, (ii) embryo culture, (iii) plantlet development and finally (iv) the gradual habituation to *ex vitro* conditions of the recovered interspecific hybrid plantlets' (Fratini & Ruiz, 2011). Viable interspecific hybrids were also obtained between the cultivated lentil and *L. odemensis*, *L. ervoides* and *L. nigricans* without the use of embryo rescue by applying gibberellic acid after pollination (Ahmad, Fautrier, McNeil, Burritt, & Hill, 1995).

7.9 Lentil Genomic Resources

Unlike major crops, genomic resources for lentil have lagged behind (Varshney et al., 2009), effectively preventing the application of genomics to characterize lentil germplasm and mine the cultivated and wild accessions for novel new alleles. Leveraging genomics model species such as *Medicago truncatula* has assisted lentil (Alo et al., 2011; Gepts, 2012; Gupta et al., 2012; Choi, Luckow, Doyle, & Cook, 2006; Choi et al., 2004; Phan et al., 2007; Zhu, Choi, Cook, & Shoemaker, 2005). However recent reductions in the costs of developing the much more effective lentil-specific genomic resources will result in better gene-specific characterization of lentil germplasm. Several transcriptomes have been developed and the sequences available through gene banks, first by researchers in Australia (Kaur et al., 2011) and now also in Canada (Bett, 2012). Kaur et al. (2011) used their transcriptome to identify gene-specific microsatellites (expressed sequence tag (EST)-SSRs) and Bett (2012) used their transcriptomes from eight lentil lines to identify SNPs. Bett (2012) have developed 8533 SNP assays (Illumina) and KASPar SNP assays (KBiosystems) for characterizing lentil germplasm, while Tanyolac (2013) reported the further development of 1095 high-quality Illumina SNP assays for lentil.

Using high-throughput gene-based assays will now allow for association mapping and eventually genome-wide association studies using lentil germplasm collections (Rafalski, 2010). Conditions for this to move forward include the completion of the structure of underlying relationships in germplasm collections and uncovering the LD found in cultivated lentil and in the lentil wild relatives. Several curated databases are under development to improve the access to useful information regarding genomic data of gene banks (Table 7.5). Finally, the question put forth last century

Table 7.5 Web-Based Databases Containing Lentil Genetic and Genomic Data

Databases	Website	Tools
LIS[a]	http://lencu.comparative-legumes.org/	GBrowse sequenced legumes and other legumes
KnowPulse[b]	http://knowpulse2.usask.ca/portal/	GBrowse with lentil track
CSFL genome[c]	http://coolseasonfoodlegume.org/	GBrowse with lentil track
IBP[d]	http://www.integratedbreeding.net/	Lentil crop and genomic information (under construction)

[a]Legume Information System, National Center for Genome Resources, Santa Fe, NM, USA.
[b]KnowPulse, hosted by University of Saskatchewan Pulse Crop Research Group.
[c]Cool Season Food Legume Genome Database, hosted by Washington State University.
[d]Integrated Breeding Platform (Varshney et al., 2012).

by Tanksley and McCouch (1997) has now been answered: there are now genome-wide association studies to effectively mine and deploy positive alleles from germplasm collections for efficient lentil crop improvement.

7.10 Conclusions

The opportunities for lentil improvement through the use of collected germplasm appear to be quite good. Future improvements and discoveries of useful variation speak to the need for continuing to collect for *ex situ* preservation in addition to *in situ* reserves, so that natural selection can continue, given the environmental challenges predicted during climate change (Yadav, Redden, Hatfield, Lotze-Campen, & Hall, 2011). Lentil CWR have been proven to provide for needed genetic diversity for crop improvement and to counteract biotic and abiotic stresses besides agronomic performance, and their conservation *ex situ* and *in situ* is paramount (Maxted et al., 2012). Kilian and Graner (2012) reviewed the deployment of next-generation sequencing technologies for the analysis of plant genetic resources, in order to identify patterns of genetic diversity, map quantitative traits and mine novel alleles from the vast amount of genetic resources maintained in gene banks worldwide. In the near future, lentil will be completely sequenced, providing the necessary reference sequence upon which massive resequencing of diverse lines and wild germplasm can commence, similar to the efforts in rice and other crops. Resequencing 50–100 germplasm lines allows for the precise movement of positive wild alleles to cultivated phenotypes (Xu et al. 2011) and genomic selection (Jannick, Lorenz, & Iwata, 2010). Genomic selection combined with high-throughput phenotyping will also create efficiencies in moving new positive alleles to advanced breeding populations and lines (Cabrera-Bosquet, Crossa, von Zitzewitz, Dolors Serret, & Araus, 2012).

References

Abbo, S., & Ladizinsky, G. (1991). Anatomical aspects of hybrid embryo abortion in the genus *Lens* L.. *Botanical Gazette*, *152*(3), 316–320.

Abbo, S., Lev-Yadun, S., & Gopher, A. (2011). Origin of Near Eastern plant domestication: Homage to Claude Levi-Strauss and 'La Pensée Sauvage'. *Genetic Resources and Crop Evolution*, *58*(2), 175–179.

Abbo, S., Lev-Yadun, S., & Gopher, A. (2012). Plant domestication and crop evolution in the Near East: On events and processes. *Critical Reviews in Plant Sciences*, *31*(3), 241–257.

Abo-Elwafa, A., Murai, K., & Shimada, T. (1995). Intra- and inter-specific variations in *Lens* revealed by RAPD markers. *Theoretical and Applied Genetics*, *90*(3), 335–340.

Ahmad, M., Fautrier, A. G., Burritt, D. J., & McNeil, D. L. (1997). Genetic diversity and relationships in *Lens* species and their F1 interspecific hybrids as determined by SDS-PAGE. *New Zealand Journal of Crop and Horticultural Science*, *25*(2), 99–108.

Ahmad, M., Fautrier, A. G., McNeil, D. L., Burritt, D. J., & Hill, G. D. (1995). Attempts to overcome postfertilization barrier in interspecific crosses of the genus *Lens*. *Plant Breeding*, *114*(6), 558–560.

Ahmad, M., & McNeil, D. L. (1996). Comparison of crossability, RAPD, SDS-PAGE and morphological markers for revealing genetic relationships within and among *Lens* species. *Theoretical and Applied Genetics*, *93*(5), 788–793.

Allaby, R. G., Fuller, D. Q., & Brown, T. A. (2008). The genetic expectations of a protracted model for the origins of domesticated crops. *Proceedings of the National Academy of Sciences*, *105*(37), 13982–13986.

Alo, F., Furman, B. J., Akhunov, E., Dvorak, J., & Gepts, P. (2011). Leveraging genomic resources of model species for the assessment of diversity and phylogeny in wild and domesticated lentil. *Journal of Heredity*, *102*(3), 315–329.

Alvarez, M. T., García, P., & Pérez de la Vega, M. (1997). RAPD polymorphism in Spanish lentil landraces and cultivars. *Journal of Genetics and Breeding*, *51*(2), 91–96.

Arumuganathan, K., & Earle, E. D. (1991). Nuclear DNA content of some important plant species. *Plant Molecular Biology Reporter*, *9*(3), 208–218.

Aydo⊠an, A., Sarker, A., Aydin, N., Küsmenoğlu, I., Karagöz, A., & Erskine, W. (2008). Registration of 'Ozbek' lentil. *Journal of Plant Registrations*, *2*(1), 16.

Babayeva, S., Akparov, Z., Abbasov, M., Mammadov, A., Zaifizadeh, M., & Street, K. (2009). Diversity analysis of Central Asia and Caucasian lentil (*Lens culinaris* Medik.) germplasm using SSR fingerprinting. *Genetic Resources and Crop Evolution*, *56*(3), 293–298.

Balter, M. (2007). Seeking agriculture's ancient roots. *Science*, *316*(5833), 1830–1835.

Barulina, H. (1930). Lentils of the USSR and other countries. *Bulletin of Applied Genetics and Plant Breeding (Leningrad)*, *40*(Supplement), 1–319. in Russian.

Bayaa, B., Erskine, W., & Hamdi, A. (1994). Response of wild lentil to *Ascochyta fabae* f. sp. *lentis* from Syria. *Genetic Resources and Crop Evolution*, *41*(2), 61–65.

Bayaa, B., Erskine, W., & Hamdi, A. (1995). Evaluation of a wild lentil collection for resistance to vascular wilt. *Genetic Resources and Crop Evolution*, *42*(3), 231–235.

Bett, K. (2012). SNP-based genotyping in lentil: Linking sequence information with phenotypes. In: *Plant and animal genome XX conference*, 14–18 January 2012.

Bockelman, H. (1999). In: *Plant collection in the Crimea, Ukraine, 25 July–6 August 1999*. Unpublished report to USDA.

Buchwaldt, L., Anderson, K. L., Morrall, R. A. A., Gossen, B. D., & Bernier, C. C. (2004). Identification of lentil germplasm resistant to *Colletotrichum truncatum* and characterization of two pathogen races. *Phytopathology*, *94*(3), 236–243.

Cabrera-Bosquet, L., Crossa, J., von Zitzewitz, J., Dolors Serret, M., & Araus, J. L. (2012). High-throughput phenotyping and genomic selection: The frontiers of crop breeding converge. *Journal of Integrative Plant Biology*, *54*(5), 312–320.

Chibbar, R. N., Ambigaipalan, P., & Hoover, R. (2010). Molecular diversity in pulse seed starch and complex carbohydrates and its role in human nutrition and health. *Cereal Chemistry*, *87*(4), 342–352.

Choi, H., Luckow, M. A., Doyle, J., & Cook, D. R. (2006). Development of nuclear gene-derived molecular markers linked to legume genetic maps. *Molecular Genetics and Genomics*, *276*(1), 56–70.

Choi, H., Mun, J., Kim, D., Zhu, H., Baek, J., Mudge, J., et al. (2004). Estimating genome conservation between crop and model legume species. *Proceedings of the National Academy of Sciences*, *101*(43), 15289–15294.

Cohen, D., Ladizinsky, G., Ziv, M., & Muehlbauer, F. J. (1984). Rescue of interspecific *Lens* hybrids by means of embryo culture. *Plant Cell Tissue and Organ Culture*, *3*(121), 343–347.

Convention on Biological Diversity. (1994). <http://www.cbd.int/convention/convention.shtml/> Accessed 27.11.12.

Cubero, J. I., Pérez de la Vega, M., Fratini, R., Erskine, W., Muehlbauer, F. J., Sarker, A., et al. (2009). Origin, phylogeny, domestication and spread. *The Lentil: Botany, Production and Uses*, 13–33.

Datta, S., Tiwari, S., Kaashyap, M., Gupta, P. P., Choudhury, P. R., Kumari, J., et al. (2011). Genetic similarity analysis in lentil using cross-genera legume sequence tagged microsatellite site markers. *Crop Science*, *51*(6), 2412–2422.

Diederichsen, A., Rozhkov, R. V., Korzhenevsky, V. V., & Boguslavsky, R. L. (2012). Collecting genetic resources of crop wild relatives in Crimea, Ukraine, in 2009. *Crop Wild Relative*, *8*, 34–38.

El-Bouhssini, M., Sarker, A., Erskine, W., & Joubi, A. (2008). First sources of resistance to Sitona weevil (*Sitona crinitus* Herbst) in wild *Lens* species. *Genetic Resources and Crop Evolution*, *55*(1), 1–4.

Engels, J., & Visser, L. (Eds.). (2003). *A guide to effective management of germplasm collections*. Bioversity International. (No. 6).

Erskine, W., Sarker, A., & Ashraf, M. (2011). Reconstructing an ancient bottleneck of the movement of the lentil (*Lens culinaris* ssp. *culinaris*) into South Asia. *Genetic Resources and Crop Evolution*, *58*(3), 373–381.

Erskine, W., Saxena, M. C., & Malhotra, R. S. (1996). Registration of ILL 5582 lentil germplasm. *Crop Science*, *36*(4), 1079–1080.

Eujayl, I., Erskine, W., Baum, M., & Pehu, E. (1999). Inheritance and linkage analysis of frost injury in lentil. *Crop Science*, *39*(3), 639–642.

Eujayl, I., Erskine, W., Bayaa, B., Baum, M., & Pehu, E. (1998). Fusarium vascular wilt in lentil: Inheritance and identification of DNA markers for resistance. *Plant Breeding*, *117*(5), 497–499.

FAO, (2010). *FAOSTAT database agricultural production*. Rome: Food and Agriculture Organization of the United Nations. <http://faostat.fao.org/> Accessed 28.11.12.

Ferguson, M., Acikgoz, N., Ismail, A., & Cinsoy, A. (1996). An ecogeographic survey of wild *Lens* species in Aegean and south west Turkey. *Anadolu*, *6*(2), 159–166.

Ferguson, M. E., Ford-Lloyd, B. V., Robertson, L. D., Maxted, N., & Newbury, H. J. (1998). Mapping the geographical distribution of genetic variation in the genus *Lens* for the enhanced conservation of plant genetic diversity. *Molecular Ecology*, *7*(12), 1743–1755.

Ferguson, M. E., Maxted, N., Slageren, M. V., & Robertson, L. D. (2000). A re-assessment of the taxonomy of *Lens* Mill. (Leguminosae, Papilionoideae, Vicieae). *Botanical Journal of the Linnean Society*, *133*(1), 41–59.

Ferguson, M. E., Newbury, H. J., Maxted, N., Ford-Lloyd, B. V., & Robertson, L. D. (1998). Population genetic structure in *Lens* taxa revealed by isozyme and RAPD analysis. *Genetic Resources and Crop Evolution*, *45*(6), 549–559.

Ferguson, M. E., & Robertson, L. D. (1996). Genetic diversity and taxonomic relationships within the genus *Lens* as revealed by allozyme polymorphism. *Euphytica*, *91*(2), 163–172.

Ferguson, M. E., & Robertson, L. D. (1999). Morphological and phenological variation in the wild relatives of lentil. *Genetic Resources and Crop Evolution*, *46*(1), 3–12.

Fernández-Aparicio, M., Sillero, J. C., Pérez-De-Luque, A., & Rubiales, D. (2008). Identification of sources of resistance to crenate broomrape (*Orobanche crenata*) in Spanish lentil (*Lens culinaris*) germplasm. *Weed Research*, *48*(1), 85–94.

Fernández-Aparicio, M., Sillero, J. C., & Rubiales, D. (2009). Resistance to broomrape in wild lentils (*Lens* spp.). *Plant Breeding*, *128*(3), 266–270.

Fiala, J. V., Tullu, A., Banniza, S., Séguin-Swartz, G., & Vandenberg, A. (2009).). Interspecies transfer of resistance to anthracnose in lentil (*Lens culinaris* Medic.). *Crop Science*, *49*(3), 825–830.

Fikiru, E., Tesfaye, K., & Bekele, E. (2010). Genetic diversity and population structure of Ethiopian lentil (*Lens culinaris* Medikus) landraces as revealed by ISSR marker. *African Journal of Biotechnology*, *6*(12), 1460–1468.

Fikiru, E., Tesfaye, K., & Bekele, E. (2011). Morphological and molecular variation in Ethiopian lentil (*Lens culinaris* Medikus) varieties. *African Journal of Biotechnology*, *3*(4), 60–67.

Ford, R., Pang, E. C. K., & Taylor, P. W. J. (1997). Diversity analysis and species identification in *Lens* using PCR generated markers. *Euphytica*, *96*(2), 247–255.

Ford, R., Pang, E. C. K., & Taylor, P. W. J. (1999). Genetics of resistance to ascochyta blight (*Ascochyta lentis*) of lentil and the identification of closely linked RAPD markers. *Theoretical and Applied Genetics*, *98*(1), 93–98.

Fratini, R., & Ruiz, M. L. (2006). Interspecific hybridization in the genus *Lens* applying *in vitro* embryo rescue. *Euphytica*, *150*(1–2), 271–280.

Fratini, R., & Ruiz, M. L. (2011). Wide crossing in lentil through embryo rescue. *Methods in Molecular Biology*, *710*, 131–139.

Fratini, R., Ruiz, M. L., & Pérez de la Vega, M. (2004). Intra specific and inter-sub-specific crossing in lentil (*Lens culinaris* Medik.). *Canadian Journal of Plant Science*, *84*(4), 981–986.

Furman, B. J. (2006). Methodology to establish a composite collection: Case study in lentil. *Plant Genetic Resources: Characterization and Utilization*, *4*(1), 2–12.

Furman, B. J., Coyne, C., Redden, B., Sharma, S. K., & Vishnyakova, M. (2009). Genetic resources: Collection, characterization, conservation and documentation: *The lentil: Botany, production and uses*. (pp. 64–75). Wallingford, UK: *CABI*.

Gepts, P. (2012). Leveraging genomic resources of model species for the assessment of phylogeny in wild and domesticated lentil. In: *Plant and animal genome XX conference*, 14–18 January 2012.

Ghijsen, H. (2009). Intellectual property rights and access rules for germplasm: Benefit or straitjacket. *Euphytica*, *170*(1), 229–234.

Global Strategy for the Ex Situ Conservation of Lentil (Lens Miller) (2008). <http://www.croptrust.org/documents/cropstrategies/lentil.pdf/> Accessed 30.11.12.

Gupta, D., & Sharma, S. K. (2006). Evaluation of wild *Lens* taxa for agro-morphological traits, fungal diseases and moisture stress in North Western Indian Hills. *Genetic Resources and Crop Evolution*, *53*(6), 1233–1241.

Gupta, D., & Sharma, S. K. (2007). Widening the gene pool of cultivated lentils through introgression of alien chromatin from wild *Lens* subspecies. *Plant Breeding*, *126*(1), 58–61.

Gupta, D., Taylor, P. W., Inder, P., Phan, H. T., Ellwood, S. R., Mathur, P. N., et al. (2012). Integration of EST-SSR markers of *Medicago truncatula* into intraspecific linkage map of lentil and identification of QTL conferring resistance to ascochyta blight at seedling and pod stages. *Molecular Breeding*, *30*(1), 429–439.

Hamdi, A., & Erskine, W. (1996). Reaction of wild species of the genus *Lens* to drought. *Euphytica*, *91*(2), 173–179.

Hamdi, A., Küsmenoĝlu, I., & Erskine, W. (1996). Sources of winter hardiness in wild lentil. *Genetic Resources and Crop Evolution*, *43*(1), 63–67.

Hamwieh, A., Udupa, S. M., Choumane, W., Sarker, A., Dreyer, F., Jung, C., et al. (2005). A genetic linkage map of *Lens* sp. based on microsatellite and AFLP markers and the localization of fusarium vascular wilt resistance. *Theoretical and Applied Genetics*, *110*(4), 669–677.

Hamwieh, A., Udupa, S. M., Sarker, A., Jung, C., & Baum, M. (2009). Development of new microsatellite markers and their application in the analysis of genetic diversity in lentils. *Breeding Science*, *59*(1), 77–86.

Havey, M. J., & Muehlbauer, F. J. (1989). Variability for restriction fragment lengths and phylogenies in lentil. *Theoretical and Applied Genetics*, *77*(6), 839–843.

Heun, M., Abbo, S., Lev-Yadun, S., & Gopher, A. (2012). A critical review of the protracted domestication model for Near-Eastern founder crops: Linear regression, long-distance gene flow, archaeological, and archaeobotanical evidence. *Journal of Experimental Botany*, *63*(12), 4333–4341.

Hoffman, D. L., Soltis, D. E., Muehlbauer, F. J., & Ladizinsky, G. (1986). Isozyme polymorphism in *Lens* (Leguminosae). *Systematic Botany*, *11*(3), 392–402.

Hoque, M. E., & Hasan, M. M. (2012). Molecular diversity analysis of lentil (*Lens culinaris* Medik.) through RAPD markers. *Plant Tissue Culture and Biotechnology*, *22*(1), 51–58.

Horneburg, B. (2006). Outcrossing in lentil (*Lens culinaris*) depends on cultivar, location and year, and varies within cultivars. *Plant Breeding*, *125*(6), 638–640.

Ialicicco, M., Viscosi, V., Arena, S., Scaloni, A., Trupiano, D., Rocco, M., et al. (2012). *Lens culinaris* Medik. seed proteome: Analysis to identify landrace markers. *Plant Science*, *197*, 1–9.

International Board for Plant Genetic Resources (IPGRI) and International Center for Agricultural Research in the Dry Areas (ICARDA), (1985). *Lentil descriptors*. Rome, Italy: IPGRI Secretariat. (p. 15).

International Treaty on Plant Genetic Resources for Food and Agriculture. (2004). <http://www.cbd.int/doc/treaties/agro-pgr-fao-en.pdf/> Accessed 27.11.12.

Jannick, J. L., Lorenz, A. J., & Iwata, H. (2010). Genomic selection in plant breeding: from theory to practice. *Briefings in Functional Genomics*, *9*(2), 166–177.

Kahraman, A., Kusmenoglu, I., Aydin, N., Aydogan, A., Erskine, W., & Muehlbauer, F. J. (2004). QTL mapping of winter hardiness genes in lentil. *Crop Science*, *44*, 13–22.

Kant, A., Sharma, S. K., Sharma, R., & Basandrai, D. (2004). Identification of RAPD and AFLP markers linked with rust resistance gene in lentil. *Crop Improvement India*, *31*(1), 1–10.

Kaur, S., Cogan, N. O., Pembleton, L. W., Shinozuka, M., Savin, K. W., Materne, M., et al. (2011). Transcriptome sequencing of lentil based on second-generation technology permits large-scale unigene assembly and SSR marker discovery. *BMC Genomics*, *12*(1), 265.

Kilian, B., & Graner, A. (2012). NGS technologies for analyzing germplasm diversity in gene banks. *Briefings in Functional Genomics*, *11*(1), 38–50.

Kislev, M. E., & Bar-Yosef, O. (1988). The Legumes: The earliest domesticated plants in the Near East. *Current Anthropology*, *29*(1), 175–179.

Kumar, S., Imtiaz, M., Gupta, S., & Pratap, A. (2011). Distant hybridization and alien gene. *Biology and Breeding of Food Legumes*, 81–110.

Ladizinsky, G. (1979a). Species relationships in the genus *Lens* as indicated by seed-protein electrophoresis. *Botanical Gazette*, *140*(4), 449–451.

Ladizinsky, G. (1979b). The origin of lentil and its wild genepool. *Euphytica*, *28*(1), 179–187.

Ladizinsky, G. (1987). Pulse domestication before cultivation. *Economic Botany*, *41*(1), 60–65.

Ladizinsky, G. (1993). Lentil domestication: On the quality of evidence and arguments. *Economic Botany*, *47*(1), 60–64.

Ladizinsky, G. (1997). A new species of *Lens* from south⊠east Turkey. *Botanical Journal of the Linnean Society*, *123*(3), 257–260.

Ladizinsky, G. (1999). Identification of the lentil's wild genetic stock. *Genetic Resources and Crop Evolution*, *46*(2), 115–118.

Ladizinsky, G., Braun, D., Goshen, D., & Muehlbauer, F. J. (1984). The biological species of the genus *Lens* L. (*Lens nigricans*). *Botanical Gazette*, *145*(2), 253–261.

Ladizinsky, G., Cohen, D., & Muehlbauer, F. J. (1985). Hybridization in the genus *Lens* by means of embryo culture. *Theoretical and Applied Genetics*, *70*(1), 97–101.

Ladizinsky, G., & Sarker, D. (1982). Morphological and cytogenetical characterization of *Vicia montbretii* Fisch. & Mey. (Synonym: *Lens montbretii* (Fisch. & Mey.) Davis & Plitmann. *Botanical Journal of the Linnean Society*, *85*(3), 209–212.

Laserna-Ruiz, I., De-Los-Mozos-Pascual, M., Santana-Méridas, O., Sánchez-Vioque, R., & Rodríguez-Conde, M. F. (2012). Screening and selection of lentil (*Lens* Miller) germplasm resistant to seed bruchids (*Bruchus* spp.). *Euphytica*, *73*, 1–10.

Li, D. Z., & Pritchard, H. W. (2009). The science and economics of *ex situ* plant conservation. *Trends in Plant Science*, *14*(11), 614–621.

Mayer, M. S., & Bagga, S. K. (2002). The phylogeny of *Lens* (Leguminosae): New insight from ITS sequence analysis. *Plant Systematics and Evolution*, *232*(3), 145–154.

Mayer, M. S., & Soltis, P. S. (1994). Chloroplast DNA phylogeny of *Lens* (Leguminosae): Origin and diversity of the cultivated lentil. *Theoretical and Applied Genetics*, *87*(7), 773–781.

Maxted, N., & Bennett, S. J. (Eds.), (2001). *Plant genetic resources of legumes in the Mediterranean* (Vol. 39). Dordrecht, The Netherlands: Kluwer Academic Publishers.

Maxted, N., Hargreaves, S., Kell, S. P., Amri, A., Street, K., Shehadeh, A., et al. (2010). Temperate forage and pulse legume genetic gap analysis. *XIII OPTIMA Meeting in Antalya, Turkey*, 22–26.

Maxted, N., Kell, S., Ford-Lloyd, B., Dulloo, E., & Toledo, Á. (2012). Toward the systematic conservation of global crop wild relative diversity. *Crop Science*, *52*(2), 774–785.

Maxted, N., Kell, S., Toledo, Á., Dulloo, E., Heywood, V., Hodgkin, T., et al. (2010). A global approach to crop wild relative conservation: Securing the gene pool for food and agriculture. *Kew Bulletin*, *65*(4), 561–576.

Mohammadi, N., Puralibaba, H., Goltapeh, E. M., Ahari, A. B., & Sardrood, B. P. (2012). Advanced lentil lines screened for resistance to *Fusarium oxysporum* f. sp. *lentis* under greenhouse and field conditions. *Phytoparasitica*, *40*(1), 69–76.

Muehlbauer, F. J. (1991). Registration of 'Crimson' lentil. *Crop Science*, *31*(4), 1094–1095.

Muehlbauer, F. J., Weeden, N. F., & Hoffman, D. L. (1989). Inheritance and linkage relationships of morphological and isozyme loci in lentil (*Lens* Miller). *Journal of Heredity*, *80*(4), 298–303.

Muench, D. G., Slinkard, A. E., & Scoles, G. J. (1991). Determination of genetic variation and taxonomy in lentil (*Lens* Miller) species by chloroplast DNA polymorphism. *Euphytica*, *56*(3), 213–218.

Nguyen, T. T., Taylor, P. W., Brouwer, J. B., Pang, E. C., & Ford, R. (2001). A novel source of resistance in lentil (*Lens culinaris* ssp. *culinaris*) to ascochyta blight caused by *Ascochyta lentis*. *Australasian Plant Pathology*, *30*(3), 211–215.

Phan, H. T. T., Ellwood, S. R., Hane, J. K., Ford, R., Materne, M., & Oliver, R. P. (2007). Extensive macrosynteny between *Medicago truncatula* and *Lens culinaris* ssp. *culinaris*. *Theoretical and Applied Genetics*, *114*(3), 549–558.

Pinkas, R., Zamir, D., & Ladizinsky, G. (1985). Allozyme divergence and evolution in the genus *Lens*. *Plant Systematics and Evolution*, *151*(1), 131–140.

Podder, R., Banniza, S., & Vandenberg, A. (2012). Screening of wild and cultivated lentil germplasm for resistance to stemphylium blight. *Plant Genetic Resources: Characterization and Utilization*, *1*, 1–10.

Potan, A. (2009). Biodiversity informatics and the plant conservation baseline. *Trends in Plant Science*, *14*(11), 629–637.

Pritchard, J., Stephens, M., & Donnelly, P. (2000). Inference of population structure using multilocus genotype data. *Genetics*, *155*, 945–959.

Rafalski, J. A. (2010). Association genetics in crop improvement. *Current Opinion in Plant Biology*, *13*(2), 174–180.

Roy, C. D., Tarafdar, S., Das, M., & Kundagrami, S. (2012). Screening lentil (*Lens culinaris* Medik.) germplasms for heat tolerance. *Trends in Biosciences*, *5*(2), 143–146.

Sackville Hamilton, N. S., & Chorlton, K. H. (1997). *Regeneration of accessions in seed collections: A decision guide*. Bioversity International. No. 5.

Saha, G. C., Sarker, A., Chen, W., Vandemark, G. J., & Muehlbauer, F. J. (2010a). Identification of markers associated with genes for rust resistance in *Lens culinaris* Medik. *Euphytica*, *175*(2), 261–265.

Saha, G. C., Sarker, A., Chen, W., Vandemark, G. J., & Muehlbauer, F. J. (2010b). Inheritance and linkage map positions of genes conferring resistance to stemphylium blight in lentil. *Crop Science*, *50*(5), 1831–1839.

Sakr, B., Sarker, A., El Hassan, H., Kadah, N., Karim, B. A., & Erskine, W. (2004). Registration of 'Bichette' lentil. *Crop Science*, *44*(2), 686–687.

Sarker, A., Erskine, W., Hassan, M. S., Afzal, M. A., & Murshed, A. N. M. M. (1999). Registration of 'Barimasur-4' lentil. *Crop Science*, *39*(3), 876.

Sarker, A., Rahman, M., Rahman, A., & Zaman, W. (1992). Uthfala: A lentil variety for Bangladesh. *LENS*, *19*(1), 14–15.

Schaefer, H., Hechenleitner, P., Santos-Guerra, A., de Sequeira, M. M., Pennington, R. T., Kenicer, G., et al. (2012). Systematics, biogeography, and character evolution of the legume tribe Fabeae with special focus on the middle-Atlantic island lineages. *BMC Evolutionary Biology*, *12*(1), 250.

Scott, J. M., Davis, F., Csuti, B., Noss, R., Butterfield, B., Groves, C., et al. (1993). Gap analysis: A geographic approach to protection of biological diversity. *Wildlife Monographs*, *123*, 3–41.

Shaikh, R., Diederichsen, A., Harrington, M., Adam, J., Conner, R. L., & Buchwaldt, L. (2012). New sources of resistance to *Colletotrichum truncatum* race Ct0 and Ct1 in

Lens culinaris Medikus subsp. *culinaris* obtained by single plant selection in germplasm accessions. *Genetic Resources and Crop Evolution*, 1–9.

Sharma, S. K., Dawson, I. K., & Waugh, R. (1995). Relationships among cultivated and wild lentils revealed by RAPD analysis. *Theoretical and Applied Genetics*, *91*(4), 647–654.

Sharma, S. K., Knox, M. R., & Ellis, T. H. (1996). AFLP analysis of the diversity and phylogeny of *Lens* and its comparison with RAPD analysis. *Theoretical and Applied Genetics*, *93*(5), 751–758.

Sharpe, A. G., Ramsay, L., Sanderson, L. -A., Fedoruk, M. J., Clarke, W. E., Li, R., et al. (2013). Ancient orphan crop joins modern era: gene-based SNP discovery and mapping in lentil. *BMC Genomics*, *14*(1), 192.

Simon, C. J., & Hannan, R. M. (1995). Development and use of core subsets of cool-season food legume germplasm collections. *Horticulture Science*, *30*(4), 907.

Singh, M., Rana, M. K., Kumar, K., Bisht, I. S., Dutta, M., Gautam, N. K., et al. (2013). Broadening the genetic base of lentil cultivars through inter-sub-specific and interspecific crosses of Lens taxa. *Plant Breeding* 10.1111/pbr.12089.

Slinkard, A. E. (1981). Eston lentil. *Canadian Journal of Plant Science*, *61*(3), 733–734.

Slinkard, A. E., & Bhatty, R. S. (1979). Laird lentil. *Canadian Journal of Plant Science*, *59*(2), 503–504.

Smartt, J. (1990). *Grain legumes: Evolution and genetic resources*. Cambridge: Cambridge University Press.

Solh, M., & Erskine, W. (1981).. In C. Webb & G. C. Hawtin (Eds.), *Genetic resources* (pp. 54–67). Slough, UK.: Commonwealth Agricultural Bureaux.

Soltis, D. E., Smith, S. A., Cellinese, N., Wurdack, K. J., Tank, D. C., Brockington, S. F., et al. (2011). Angiosperm phylogeny: 17 genes, 640 taxa. *American Journal of Botany*, *98*(4), 704–730.

Sonnante, G., Galasso, I., & Pignone, D. (2003). ITS sequence analysis and phylogenetic inference in the genus *Lens* mill. *Annals of Botany*, *91*(1), 49–54.

Sonnante, G., Hammer, K., & Pignone, D. (2009). From the cradle of agriculture a handful of lentils: History of domestication. *Rendiconti Lincei*, *20*(1), 21–37.

Sultana, T., & Ghafoor, A. (2008). Genetic diversity in *ex⊠situ* conserved *Lens culinaris* for botanical descriptors, biochemical and molecular markers and identification of landraces from indigenous genetic resources of Pakistan. *Journal of Integrative Plant Biology*, *50*(4), 484–490.

Sultana, T., Ghafoor, A., & Ashraf, M. (2006). Geographic patterns of diversity of cultivated lentil germplasm collected from Pakistan, as assessed by seed protein assays. *Acta Biologica Cracoviensia, Series Botanica, Poland*, *48*(1), 77–84.

Tahir, M., Båga, M., Vandenberg, A., & Chibbar, R. N. (2012). An assessment of Raffinose family oligosaccharides and sucrose concentration in genus *Lens*. *Crop Science*, *52*(4), 1713–1720.

Tanksley, S. D., & McCouch, S. R. (1997). Seed banks and molecular maps: Unlocking genetic potential from the wild. *Science*, *277*(5329), 1063–1066.

Tanyolac, M.B. (2013). Single nucleotide polymorphism (SNP) discovery in lentil using illumina platform. In: *Plant and animal genome XXI conference*.

Taylor, P. W. J., Ades, P. K., & Ford, R. (2006). QTL mapping of resistance in lentil (*Lens culinaris* ssp. *culinaris*) to ascochyta blight (*Ascochyta lentis*). *Plant Breeding*, *125*(5), 506–512.

Torricelli, R., Silveri, D. D., Ferradini, N., Venora, G., Veronesi, F., & Russi, L. (2012). Characterization of the lentil landrace Santo Stefano di Sessanio from Abruzzo, Italy. *Genetic Resources and Crop Evolution*, *59*(2), 261–276.

Tullu, A., Banniza, S., Tar'an, B., Warkentin, T., & Vandenberg, A. (2010). Sources of resistance to ascochyta blight in wild species of lentil (*Lens culinaris* Medik.). *Genetic Resources and Crop Evolution*, *57*(7), 1053–1063.

Tullu, A., Bett, K., Saha, S., Vail, S., & Vandenberg, A. (2011). Revisiting strategies in lentil breeding: Wild species update. *Pisum Genetics*, *43*, 49–50.

Tullu, A., Buchwaldt, L., Lulsdorf, M., Banniza, S., Barlow, B., Slinkard, A. E., et al. (2006). Sources of resistance to anthracnose (*Colletotrichum truncatum*) in wild *Lens* species. *Genetic Resources and Crop Evolution*, *53*(1), 111–119.

Tullu, A., Buchwaldt, L., Warkentin, T., Taran, B., & Vandenberg, A. (2003). Genetics of resistance to anthracnose and identification of AFLP and RAPD markers linked to the resistance gene in PI 320937 germplasm of lentil (*Lens culinaris* Medikus). *Theoretical and Applied Genetics*, *106*(3), 428–434.

Tullu, A., Diederichsen, A., Suvorova, G., & Vandenberg, A. (2011). Genetic and genomic resources of lentil: Status, use and prospects. *Plant Genetic Resources*, *9*(1), 19–29.

Tullu, A., Kusmenoglu, I., McPhee, K. E., & Muehlbauer, F. J. (2001). Characterization of core collection of lentil germplasm for phenology, morphology, seed and straw yields. *Genetic Resources and Crop Evolution*, *48*(2), 143–152.

Tullu, A., Tar'an, B., Warkentin, T., & Vandenberg, A. (2008). Construction of an intraspecific linkage map and QTL analysis for earliness and plant height in lentil. *Crop Science*, *48*(6), 2254–2264.

Tyagi, S. D., & Khan, M. H. (2011). Correlation, path-coefficient and genetic diversity in lentil (*Lens culinaris* Medik) under rainfed conditions. *International Research Journal of Plant Science*, *2*(27), 191–200.

Vail, S., Strelioff, J. V., Tullu, A., & Vandenberg, A. (2012). Field evaluation of resistance to *Colletotrichum truncatum* in *Lens culinaris*, *Lens ervoides*, and *Lens ervoides* x *Lens culinaris* derivatives. *Field Crops Research*, *126*, 145–151.

Vail, S., & Vandenberg, A. (2011). Genetic control of interspecific-derived and juvenile resistance in lentil to *Colletotrichum truncatum*. *Crop Science*, *51*(4), 1481–1490.

Vaillancourt, R. E., & Slinkard, A. E. (1993). Linkage of morphological and isozyme loci in lentil, *Lens culinaris* L.. *Canadian Journal of Plant Science*, *73*(4), 917–926.

van Oss, H., Aron, Y., & Ladizinsky, G. (1997). Chloroplast DNA variation and evolution in the genus *Lens* Mill. *Theoretical and Applied Genetics*, *94*(3), 452–457.

Vandenberg, A., & Slinkard, A. E. (1989). Inheritance of four new qualitative genes in lentil. *Journal of Heredity*, *80*(4), 320–322.

Varshney, R. K., Close, T. J., Singh, N. K., Hoisington, D. A., & Cook, D. R. (2009). Orphan legume crops enter the genomics era!. *Current Opinion in Plant Biology*, *12*(2), 202–210.

Varshney, R. K., Ribaut, J. M., Buckler, E. S., Tuberosa, R., Rafalski, J. A., & Langridge, P. (2012). Can genomics boost productivity of orphan crops. *Nature Biotechnology*, *30*(12), 1172–1176.

Villa, T. C. C., Maxted, N., Scholten, M., & Ford-Lyod, B. (2006). Defining and identifying crop landraces. *Plant Genetic Resources*, *3*(3), 373–384.

Walters, C., Wheeler, L. M., & Grotenhuis, J. M. (2005). Longevity of seeds stored in a genebank: Species characteristics. *Seed Science Research*, *15*(1), 1–20.

Walters, C., Wheeler, L., & Stanwood, P. C. (2004). Longevity of cryogenically stored seeds. *Cryobiology*, *48*(3), 229–244.

Weiss, E., Kislev, M. E., & Hartmann, A. (2006). Autonomous cultivation before domestication. *Science*, *132*(5780), 1608–1610.

Weiss, E., & Zohary, D. (2011). The Neolithic Southwest Asian founder crops: Their biology and archaeobotany. *Current Anthropology*, *52*(S4), S237–S254.

Wilson, V. E. (1972). Morphology and technique for crossing *Lens esculenta* Moench. *Crop Science, 12*, 231–232.

Wilson, V. E., & Law, A. G. (1972). Natural crossing in *Lens esculenta* Moench. *Journal American Society of Horticultural Science, 97*, 142–143.

Xu, X., Liu, X., Ge, S., Jensen, J. D., Hu, F., Li, X., et al. (2011). Resequencing 50 accessions of cultivated and wild rice yields markers for identifying agronomically important genes. *Nature Biotechnology, 30*(1), 105–114.

Yadav, S. S., Redden, R., Hatfield, J. L., Lotze-Campen, H., & Hall, A. (2011). *Crop adaptation to climate change*. Chichester, U.K: Wiley-Blackwell.

Yüzba⊠io⊠lu, E., Açik, L., & Özcan, S. (2008). Seed protein diversity among lentil cultivars. *Biologia Plantarum, 52*(1), 126–128.

Zaccardelli, M., Lupo, F., Piergiovanni, A. R., Laghetti, G., Sonnante, G., Daminati, M. G., et al. (2012). Characterization of Italian lentil (*Lens culinaris* Medik.) germplasm by agronomic traits, biochemical and molecular markers. *Genetic Resources and Crop Evolution, 59*(5), 727–738.

Zeder, M. A. (2011). The origins of agriculture in the Near East. *Current Anthropology, 52*(4), 221–235.

Zhu, H., Choi, H., Cook, D. R., & Shoemaker, R. C. (2005). Bridging model and crop legumes through comparative genomics. *Plant Physiology, 137*(4), 1189–1196.

Zohary, D. (1972). The wild progenitor and the place of origin of the cultivated lentil *Lens culinaris*. *Economic Botany, 26*(4), 326–332.

Zohary, D. (1999). Monophyletic vs. polyphyletic origin of the crops on which agriculture was founded in the Near East. *Genetic Resources and Crop Evolution, 46*(2), 133–142.

Zohary, D., & Hopf, M. (1973). Domestication of pulses in the old World. *Science, 182*, 887–894.

8 Pigeon Pea

Hari D. Upadhyaya, Shivali Sharma, K.N. Reddy, Rachit Saxena, Rajeev K. Varshney and C.L. Laxmipathi Gowda

International Crops Research Institute for the Semi-Arid Tropics (ICRISAT), Patancheru, Hyderabad, India

8.1 Introduction

Pigeon pea [*Cajanus cajan* (L.) Millspaugh] is a short-lived perennial shrub that is traditionally cultivated as an annual grain legume crop in tropical and subtropical regions of the world. It is known by various names, such as red gram and congo bean (English), tur and arhar (Hindi), guand (Portuguese), gandul (Spanish), poid d'Angole and poid de Congo (French) and ervilba de Congo in Angola, and is grown primarily as a food crop. Dry whole seed and dehulled and split seed (dhal) are used for cooking various dishes. Besides its use as a food crop, there are also forage, fodder, fuel and medicine uses. The crushed dry seeds are fed to animals, while the green leaves form a quality fodder. In rural areas, dry stems of pigeon pea are used for fuel, thatching, basket-making, etc. The plants are also used to culture lac insects. Pigeon pea has a deep root system which helps it to withstand drought, and is grown on mountain slopes to bind the soil and reduce soil erosion. Due to its deep root system, pigeon pea offers little competition to associated crops and is therefore extensively used in intercropping systems with cereals, such as millets, sorghum and maize; it also provides a good means to improve fertility in fallows. In a cropping season, the plants fix about 40 kg/ha atmospheric nitrogen and add valuable organic matter to the soil through fallen leaves (up to 3.1 t/ha of leaf dry matter) (Rupela, Gowda, Wani, & Ranga Rao, 2004). Its roots help in releasing soil-bound phosphorus to make it available for plant growth. Pigeon pea seed protein content (on average approximately 21%) compares well with that of other important grain legumes. Owing to several unique characteristics and benefits, pigeon pea has become an ideal crop for sustainable agricultural systems in rainfed areas. Because of the large temporal variation (90–300 days) for maturity, four major durations for pigeon pea varieties exist: extra short (mature in <100 days), short (100–120 days), medium (140–180 days) and long duration (>200 days). Each group is suited to a particular agro-ecosystem, which is defined by altitude, temperatures, latitude and day length. Invariably, the traditional pigeon pea cultivars and landraces are long duration types and grown as intercrops with other more early maturing cereals and legumes. Extra short and short varieties have the potential for inclusion as sole crop into rotation as an alternative to rice within the rice–wheat systems of the Indo-Gangetic Plain

Genetic and Genomic Resources of Grain Legume Improvement. DOI: http://dx.doi.org/10.1016/B978-0-12-397935-3.00008-6

in Asia, especially during periods of water shortage, price incentives and problems of soil fertility. Further, pigeon pea production is affected by several biotic and abiotic stresses. Among biotic factors, important diseases such as sterility mosaic, *Fusarium* wilt (FW), *Phytophthora* blight, root rot, stem canker and *Alternaria* blight in the Indian subcontinent; wilt and *Cercospora* leaf spot in eastern Africa and witches' broom in the Caribbean and Central America cause considerable yield losses. The distribution of these diseases is geographically restricted. For example, sterility mosaic disease (SMD), the most important disease of Indian subcontinent, is not found in eastern Africa. Similarly witches' broom is absent from the two major pigeon pea-growing regions, the Indian subcontinent and eastern Africa. Besides diseases, the seeds and other parts of the plant are fed upon by many insects, with over 200 species having been recorded in India alone. Some of these insects cause sufficient crop losses to be regarded as major pests, but the majority are seldom abundant enough to cause much damage, or are of sporadic or localized importance, and regarded as minor pests. The pod-damaging insects (pod borers and pod fly) cause significant yield losses in pigeon pea and therefore are the most important pests of this crop.

8.2 Origin, Distribution, Diversity and Taxonomy

The name pigeon pea was first reported from Barbados, where the seeds were used to feed pigeons (Plukenet, 1692). There are several theories about the true origin of pigeon pea (reviewed in Saxena, Kumar, Reddy, & Arora, 2003). However, based on the range of genetic diversity of the crop in India, Vavilov (1951) concluded that pigeon pea originated in India. Several authors considered eastern Africa to be the centre of origin of pigeon pea, as it occurs there in wild form. However, based on the large diversity among the crop varieties, the presence of several related wild species, including the progenitor species, linguistic evidence and wide usage in daily cuisine, most of the researchers have agreed on India as the original home of pigeon pea. India is now unequivocally accepted as the primary centre of origin and Africa as the secondary centre of origin of pigeon pea (De, 1974; Royes Vernon, 1976; van der Maesen, 1980). Most probably in the nineteenth century, immigrants from India introduced the crop into East Africa (Hillocks, Minja, Nahdy, & Subrahmanyam, 2000). Thereafter, pigeon pea moved into the Nile valley, then into West Africa and eventually to the Americas (Odeny, 2007). It is now widely grown in the Caribbean region. Further, Reddy (1973) and De (1974) also postulated that the genus *Cajanus* probably originated from an advanced *Atylosia* (now reclassified as *Cajanus*) species through single gene mutation. It is now well known that this advanced species is *C. cajanifolius*, the most probable progenitor of pigeon pea, found only in India. Besides *C. cajanifolius*, 16 species of *Cajanus*, including cultivated species *C. cajan*, occur in India.

At present, pigeon pea is cultivated in the tropical and subtropical areas between 30°N and 30°S latitude on 4.71 million hectares with an annual production of 3.69 million metric tons and productivity of 783 kg/ha (FAOSTAT, 2010). The pigeon

Table 8.1 Major Pigeon Pea-Growing Countries of the World

Continent	Country	Area (ha)	Productivity (kg/ha)	Production (tonnes)
Asia	Bangladesh	811	951	772
	India	3,530,000	696	2,460,000
	Myanmar	581,200	1246	724,200
	Nepal	21,296	875	18,647
	Pakistan	0		0
	Philippines	684	1244	851
Africa	Burundi	1900	1000	1900
	Comoros	540	592	320
	Democratic Republic of the Congo	10,139	582	5901
	Kenya	158,746	650	103,324
	Malawi	190,437	1013	193,005
	Uganda	98,200	947	93,000
	United Republic of Tanzania	75,000	733	55,000
America	Bahamas	230	565	130
	Dominican Republic	23,461	1068	25,070
	Grenada	640	765	490
	Haiti	7200	333	2400
	Jamaica	723	1036	749
	Panama	4400	447	1969
	Puerto Rico	344	755	260
	Trinidad and Tobago	1300	769	1000
	Venezuela (Bolivarian Republic of)	1900	789	1500
World		4,709,151	783	3,690,488

pea is widely grown in the Indian subcontinent, which accounts for about 88% of the global pigeon pea production. The major pigeon pea-growing countries in the region are India followed by Myanmar and Nepal. India alone represents about 75% of the area and about 67% of the global pigeon pea production. Africa, including major pigeon pea-growing countries, such as Malawi, Kenya and Uganda, accounts for about 11% of the global production. The Americas and the Caribbean produce about 1% of the total pigeon pea of the world (Table 8.1). Pigeon pea is often cross-pollinated, with an insect-aided natural out-crossing range from 20% to 70% (Saxena, Singh, & Gupta, 1990), with chromosome number $2n=2x=22$ and genome size 1C = 858 Mbp. It belongs to the family Leguminosae, subfamily Papilionoideae, tribe *Phaseoleae* and the subtribe *Cajaninae*. The tribe *Phaseoleae* comprises many edible bean species (*Phaseolus*, *Vigna*, *Cajanus*, *Lablab*, etc.) of which the members of subtribe *Cajaninae* are well distinguished by the presence of vesicular glands on the leaves, calyx and pods. Currently, 11 genera are grouped under the subtribe *Cajaninae*, including *Rhynchosia* Lour., *Eriosema* (DC.), G. Don, *Dunbaria*, W. & A. and *Flemingia* Roxb. ex Aiton, but the cultivated pigeon pea *C. cajan* is the only domesticated species in *Cajaninae*. The word '*Cajanus*' is derived from a Malay word 'katschang' or 'katjang' meaning pod or bean. The members of the earlier genus *Atylosia* closely resemble the genus *Cajanus* in vegetative and reproductive characters. However, they were relegated to two separate genera mainly on the basis of the presence or absence of seed strophiole. In 1980, van der Maesen revised the taxonomy of both the genera and merged the genus *Atylosia* into *Cajanus* following systematic analysis of morphological, cytological and chemotaxonomical data, which indicated the congenicity of the two genera (van der Maesen, 1980). The revised genus *Cajanus* currently comprises 18 species from Asia, 15 species from Australia and 1 species from West Africa. Of these, 13 are found only in Australia, 8 in the Indian subcontinent, and 1 in West Africa, with the remaining 14 species occurring in more than 1 country. Based on growth habit, leaf shape, hairiness, structure of corolla, pod size and presence of strophiole, van der Maesen (1980) grouped the genus *Cajan* into six sections. The 18 erect species were placed under three sections: seven species in section *Atylosia*, nine species in section *Fruticosa* and two species in section *Cajanus*, which consists of the cultivated pigeon pea along with its progenitor, *C. cajanifolius*. Eleven climbing and creeping species were arranged in two sections, section *Cantharospermum* (5) and section *Volubilis* (6); the remaining three trailing species were classified under section *Rhynchosoides*. Three *Cajanus* species have been further subdivided into botanical varieties: *C. scarabaeoides* var. *pedunculatus* and var. *scarabaeoides*; *C. reticulatus* var. *grandifolius*, var. *reticulatus*, and var. *maritimus*; and *C. volubilis* var. *burmanicus* and var. *volubilis*.

On the basis of success in hybridization between pigeon pea and its wild relatives, van der Maesen (1990) placed cultigens in the primary gene pool, all 10 cross-compatible species *C. acutifolius*, *C. albicans*, *C. cajanifolius*, *C. lanceolatus*, *C. latisepalus*, *C. lineatus*, *C. reticulatus*, *C. scarabaeoides*, *C. sericeus* and *C. trinervius* in the secondary gene pool, and the cross-incompatible species *C. goensis*, *C. heynei*, *C. kerstingii*, *C. mollis*, *C. platycarpus*, *C. rugosus*, *C. volubilis* and other *Cajaninae* such as *Rhynchosia* Lour., *Dunbaria* W. and A., *Eriosema* (DC.) Reichenb in the tertiary gene pool.

8.3 Erosion of Genetic Diversity from the Traditional Areas

The contribution of landraces as source material for crop improvement has been substantial. In the past, most released pigeon pea varieties have been developed through selection from landraces. To meet the challenges in crop improvement, efforts were made to widen the genetic base by collecting and conserving germplasm across the world before it is lost forever, which led to the assembly of large collections at the national and international gene banks. The gene bank at the International Crops Research Institute for the Semi-Arid Tropics (ICRISAT), serving as a world repository for genetic resources of its mandate crop including pigeon pea, holds 13,771 accessions from 74 countries. Landraces and wild relatives are the best sources of resistance to the biotic and abiotic stresses and contribute towards food security, poverty alleviation, environmental protection and sustainable development. Plant genetic resources (PGR) are finite and vulnerable to erosion due to the severe threats to world food security of replacement of landraces/traditional cultivars by modern varieties, natural catastrophes such as droughts, floods, fire hazards, urbanization and industrialization, and habitat loss due to irrigation projects, overgrazing, mining and climate change (Upadhyaya & Gowda, 2009). Therefore, there is an urgent need to assess the existing collection to identify geographical, trait-diversity and taxonomical gaps for planning future collection strategies for pigeon pea.

8.4 Status of Germplasm Resources Conservation

The CGIAR consortium represents the largest concerted effort towards collecting, preserving and utilizing global agricultural resources. CGIAR holds nearly 760,000 samples of the estimated 7.4 million accessions of different crops preserved globally (FAOSTAT, 2010). There are a number of gene banks conserving the pigeon pea germplasm worldwide. ICRISAT has the global responsibility of collecting, conserving and distributing the pigeon pea germplasm comprising of landraces, modern cultivars, genetic stocks, mutants and wild *Cajanus* species. It contains 13,216 accessions of cultivated pigeon pea and 555 accessions of wild species in the genus *Cajanus* from 60 countries. The collection includes 8315 landraces, 4830 breeding materials, 71 improved cultivars and 555 wild accessions. This is the single largest collection of pigeon pea germplasm assembled at any one place in the world. India is the major contributor with 9200 accessions. These accessions came from donations as well as from collecting missions launched in different countries. Other major gene banks holding pigeon pea germplasm are the National Bureau of Plant Genetic Resources (12,900 accessions), New Delhi, India; All India Coordinated Research Project on Pigeon pea (5195 accessions); NBPGR Regional Station Akola (2268 accessions), India; Indian Agricultural Research Institute (IARI; 1500 accessions), New Delhi and the National Gene Bank of Kenya, Crop Plant Genetic Resources Centre (1380 accessions), Muguga, Kenya (Table 8.2).

Table 8.2 Major Gene Banks Holding Pigeon Pea Germplasm

Country	Institute	Wild	Cultivated	Total
Australia	Australian Tropical Crops and Forages Genetic Resources Centre	352	406	758
Brazil	Embrapa Recursos Genéticos e Biotecnologia	3	279	282
Colombia	Centro Internacional de Agricultura Tropical	623	135	758
Ethiopia	International Livestock Research Institute	539	143	682
India	All India Coordinated Research Project on Pigeon pea		5195	5195
	Indian Agricultural Research Institute		1500	1500
	ICRISAT	555	13,216	13,771
	National Bureau of Plant Genetic Resources	41	12,859	12,900
	Regional Station Akola, NBPGR		2268	2268

(*Continued*)

Table 8.2 (Continued)

Country	Institute	Wild	Cultivated	Total
Indonesia	National Biological Institute		200	200
Kenya	National Genebank of Kenya, Crop Plant Genetic Resources Centre – Muguga	92	1288	1380
Nepal	Nepal Agricultural Research Council		228	228
Philippines	Institute of Plant Breeding, College of Agriculture, University of the Philippines, Los Baños		629	629
Thailand	Thailand Institute of Scientific and Technological Research		201	201
Uganda	Serere Agriculture and Animal Production Research Institute		200	200

8.5 Germplasm Characterization and Evaluation

Germplasm collection is of little value unless it is characterized, evaluated and documented properly to enhance its utilization in crop improvement. A multidisciplinary approach is followed at ICRISAT gene bank; the data generated in various disciplines are fed to the pigeon pea germplasm characterization database. The characterization was done at the ICRISAT Research Farm in Patancheru on 18 qualitative characters (plant vigor, growth habit, plant pigmentation, stem thickness, flower base colour, streak colour, streak pattern, flowering pattern, pod colour, pod shape, pod hairiness, seed colour pattern, primary seed colour, secondary seed colour, seed eye colour, seed eye colour width, seed shape and seed hilum) and 16 quantitative characters were recorded following the 'Descriptors for Pigeon pea' (IBPGR & ICRISAT, 1993). Observations on all qualitative and six quantitative characters (days to 50% flowering, days to 75% maturity, 100-seed weight, harvest index, shelling percentage and plot seed yield) were recorded on a plot basis. Observations on the remaining 10 quantitative traits (leaf size, plant height, number of primary, secondary and tertiary branches, number of racemes, pod bearing length, pods per plant, pod length, seeds per pod) were recorded on three representative plants from each plot. To realize the true potential of the accessions and to facilitate the selection of genotypes by researchers, sets of selected pigeon pea germplasm, such as core and mini-core collections, were evaluated for important agronomic characters at different locations in India and several other countries in Africa during suitable seasons.

8.5.1 Diversity in the Collection

To study the geographical patterns of diversity in the collection, data of 14 qualitative and 12 quantitative traits of 11,402 accessions from 54 countries were analysed. The accessions were grouped based on geographical proximity and similarity of climate (Reddy, Upadhyaya, Gowda, & Singh, 2005; Upadhyaya, Pundir, Gowda, Reddy,

Table 8.3 Range of Variation for Important Agronomic Traits in the World Collection of Pigeon Pea at ICRISAT Gene Bank, Patancheru, India

Character	Mean	Minimum	Maximum
Days to 50% flowering	133.5	52	237
Days to 75% maturity	192.1	100	299
Plant height (cm)	177.9	39	310
Primary branches (no.)	13.5	1	107
Secondary branches (no.)	31.3	0	145.3
Tertiary branches (no.)	8.8	0	218.7
Racemes per plant (no.)	150.3	6	915
Pod length (cm)	5.7	2.5	13.1
Pods per plant (no.)	287.3	9.3	1819.3
Seeds per pod (no.)	3.7	1.6	7.2
100-seed weight (g)	9.3	2.7	25.8
Seed protein (%)	21.3	13	30.8

& Singh, 2005). Large variation was observed in the entire collection for important agronomic traits (Table 8.3). The range of variation for quantitative traits in respect to the different regions was maximum for group AS 4 (south India, Maldives and Sri Lanka) and minimum for germplasm accessions from Europe and Oceania. The region AS 4 encompasses the area of the primary centre of diversity of pigeon pea; therefore, the high variation in the germplasm from that region is not surprising (Upadhyaya et al., 2005). The accessions from Africa were of longer duration, tall and producing large seeds. Accessions from India had medium plant height, high pod number, medium duration and high seed yield. Accessions from Oceania were conspicuous in their short growth duration, short height, few branches, small seeds and low seed yield. Shannon–Weaver diversity index (H') (Shannon & Weaver, 1949) indicates that the accessions from AS 6 (Indonesia, Philippines and Thailand) had the highest pooled H' for qualitative traits (0.349 + 0.059) and accessions from Africa the highest for quantitative traits (0.613 + 0.006) (Upadhyaya et al., 2005). African accessions also had highest pooled H' (0.464 + 0.039) over all the traits. The accessions from Oceania had the lowest pooled H' (0.337 + 0.037). The H' values across the regions were highest for primary seed colour (0.657 + 0.050) followed by flower streak pattern, seed protein content and shelling percentage, whereas it was lowest for flowering pattern (0.087 + 0.026). A hierarchical cluster analysis conducted on the first three PC scores (92.28% variation) resulted in three clusters. Cluster 1 comprised accessions from Oceania (60 accessions), cluster 2 comprised accessions from AS 1–5 containing 9648 accessions and cluster 3 comprised accessions from Africa, America, Caribbean countries, Europe and AS 6 containing 1694 accessions (Figure 8.1) (Upadhyaya et al., 2005). Semi-spreading growth habit, green stem colour, indeterminate (NDT) flowering pattern and yellow flower were predominant among the qualitative traits. Primary seed colour had maximum variability; orange colour

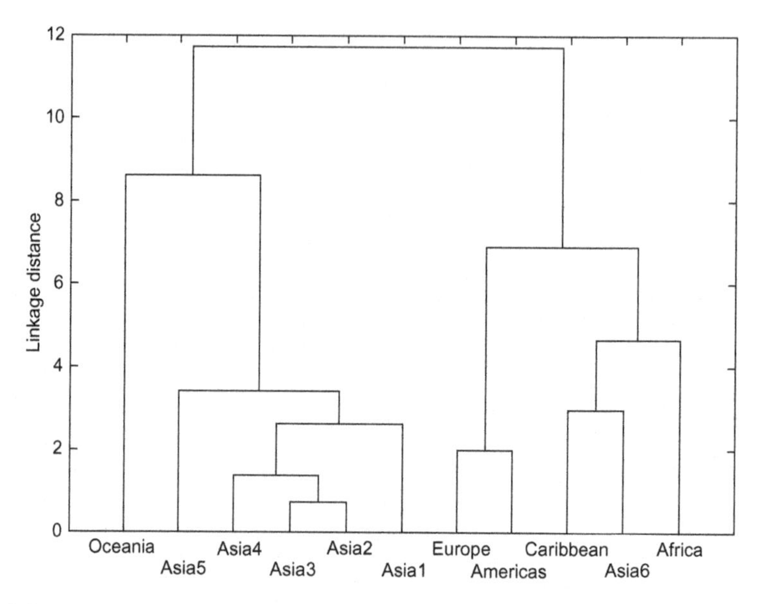

Figure 8.1 Dendrogram of 11 regions in the entire pigeon pea germplasm based on scores of the first three principal components (92.3% variation).

followed by cream were the two most frequent second colours in the collection. At ICRISAT a large number of pigeon pea accessions were tested for biotic and abiotic stresses and promising sources for resistance were identified.

8.6 Germplasm Maintenance

The ICRISAT gene bank ensures maintenance of germplasm at international standards and the continued availability of good-quality seeds of its mandate crops for research and development globally. Maintenance of germplasm includes maintenance of seed viability and seed quantity in the gene bank. Seed viability and quantity of germplasm accessions in medium-term store are monitored at regular intervals. Accessions are regenerated when the seed viability is below 85% and/or seed quantity <100g in medium-term store. Regeneration is the crucial process in gene bank management. Accessions with poor quality are given top priority. Objectives for regeneration include maximizing seed quality, optimizing seed quantity and maintaining the genetic integrity of accessions. Pigeon pea floral biology favors self-pollination. However, it is considered an often cross-pollinating species without crossing ranging from 20 to 70%, due to visits by bees (Saxena et al., 1990). Therefore, it is essential to preserve the accessions' integrity using effective pollination control methods. Controlling pollination is the most crucial part of the regeneration process. Methods to control pollination include: bagging individual plants, growing accessions in isolation, growing barrier crops, growing under

Figure 8.2 Field view of growing pigeon pea germplasm under insect-proof cages for regeneration.

insect-proof cages, 'polyhouses', etc. But the most common procedure is covering individual plants using muslin cloth bags and growing accessions under insect-proof cages (Figure 8.2). The pollination control method of growing accessions under insect-proof cages was three times cheaper than the traditional method of bagging individual plants. However, the regeneration cost depends largely on method of pollination control, availability and cost of materials in local markets, labour wages, quantity of seed required per accession in one cycle of regeneration, type of material to be regenerated, etc. Due to increased seed yield per plant, we can minimize the regeneration frequency (Reddy, Upadhyaya, Reddy, & Gowda, 2006). Minimizing the regeneration requirement of each accession can reduce maintenance costs of the total collection. Therefore, pigeon pea germplasm accessions are grown under insect-proof cages for regeneration at ICRISAT Research Farm, Patancheru, during the rainy season. In order to minimize the damage to the nylon net used for the cages by reducing the vegetative growth, particularly plant height, accessions are sown later during the crop season, during the first week of August, in Alfisol fields. Remanandan, Sastry, and Mengesha Melak (1988) reported that sowing pigeon pea in Alfisols close to the shortest day of the year results in reduced plant height. Each accession is grown on a single 9-m-long ridge, spaced 75 cm apart. Plant to plant spacing is 25 cm, accommodating about 72 plants in 36 hills. Adequate plant protection measures are taken inside the cage to reduce damage by pests and diseases.

At maturity, individual plants are harvested and an equal quantity of seeds from each plant is bulked to reconstitute the accession.

8.6.1 Regeneration of Wild Pigeon Pea Germplasm

Seeds of almost all species require scarification by making a small cut to the seed coat to improve water absorption and germination. Seeds are treated with Thiram or any other appropriate fungicide and initially sown in small cups or pots and transplanted to the field when they have three to four leaves. Climbers, such as *C. albicans*, *C. mollis* and *C. crassus*, are provided support using bamboo sticks or iron poles. At maturity, pods from individual plants are harvested and threshed, and seeds are cleaned. An equal quantity of seed from each plant is bulked to reconstitute an accession (Upadhyaya & Gowda, 2009).

8.6.2 Documentation

All information, such as method of viability test, initial viability, seed quantity, as well as the year of regeneration, pollination control method used, regeneration site, accession, field number, accession verification, number of plants harvested and seed quantity obtained are recorded and documented (Upadhyaya & Gowda, 2009).

8.7 Use of Germplasm in Crop Improvement

The small subsets, such as core and mini-core collections, are now international public goods and used by scientists globally. Many national programmes have shown interest in the mini-core collection and ICRISAT has supplied 19 sets of pigeon pea mini-core to National Agricultural Research Systems (NARS) in India (17), UAE (1) and USA (1). Using the mini-core collection, scientists at ICRISAT and NARS partners have identified several promising sources for agronomic, nutritional, biotic and abiotic traits (Upadhyaya, Dronavalli, Gowda, & Singh, 2012).

8.7.1 Biotic Stresses

8.7.1.1 Resistance to Diseases

Evaluation of a mini-core collection has resulted in the identification of six accessions (ICP 6739, ICP 8860, ICP 11015, ICP 13304, ICP 14638 and ICP 14819) resistant to FW (Sharma et al., 2012) and 24 accessions (ICP 3451, ICP 6739, ICP 6845, ICP 7869, ICP 8152, ICP 8860, ICP 9045, ICP 11015, ICP 11059, ICP 11230 and others) resistant to SMD (Sharma et al., 2012).

8.7.1.2 Resistance to Insects

Evaluation of a mini-core collection has resulted in the identification of 11 accessions (ICP 7, ICP 655, ICP 772, ICP 1071, ICP 3046, ICP 4575, ICP 6128, ICP

8860, ICP 12142, ICP 14471 and ICP 14701) reported moderately resistant to pod borer (damage rating 5.0 as compared to 9.0 in ICPL 87) under unprotected conditions, and also had no wilt incidence as compared to 38.2% wilt in ICP 8266 (ICRISAT Archival Report, 2010).

8.7.2 Abiotic Stresses

8.7.2.1 Waterlogging

Evaluation of a pigeon pea mini-core collection resulted in the identification of 23 accessions (ICP 1279, ICP 4575, ICP 5142, ICP 6370, ICP 6992, ICP 7057 and others) recorded tolerant to waterlogging conditions (Krishnamurthy, Upadhyaya, Saxena, & Vadez, 2011).

8.7.2.2 Salinity

Evaluation of a pigeon pea mini-core collection resulted in the identification of 16 accessions (ICP 2746, ICP 3046, ICP 6815, ICP 7260, ICP 7426, ICP 7803, ICP 8860 and others) selected for tolerance to salinity (Srivastava, Vadez, Upadhyaya, & Saxena, 2006).

8.7.3 Agronomic Traits

Evaluation of a pigeon pea mini-core collection resulted in the identification of eight accessions (ICP 1156, ICP 9336, ICP 14471, ICP 14832, ICP 14900, ICP 14903, ICP 15068 and ICP 16309) for early flowering (<85 days); three accessions (ICP 13139, ICP 13359 and ICP 14976) for large seed size (>15g/100 seed); one accession (ICP 8860) for more primary branches (>29) and three accessions (ICP 4167, ICP 8602 and ICP 11230) for high pod number per plant (>200 pods/plant) (Upadhyaya, Yadav, Dronavalli, Gowda, & Singh, 2010).

8.7.4 Nutritional Traits

Evaluation of a pigeon pea mini-core collection resulted in the identification of six accessions (ICP 4575, ICP 7426, ICP 8266, ICP 11823, ICP 12515 and ICP 12680) for high seed protein (>24%); eight accessions (ICP 4029, ICP 6929, ICP 6992, ICP 7076, ICP 10397, ICP 11690, ICP 12298 and ICP 12515) for high seed iron (>40 ppm) and four accessions (ICP 2698, ICP 11267, ICP 14444 and ICP 14976) for high seed zinc (>40 ppm).

8.8 Limitations in Germplasm Use

Very few germplasm accessions (<1%) have been used by plant breeders in crop improvement programmes (Upadhyaya, 2008). A large gap exists between availability and actual utilization of the germplasm. This was true both in the international programmes (CGIAR institutes) as well as in the national programmes. Extensive use of fewer and closely related parents in crop improvement could result in vulnerability of cultivars to pests and diseases. The main reason for low use of germplasm in crop improvement programmes is the lack of information on the large number of accessions, particularly for traits of economic importance, which display a great deal of genotype×environment interaction and require multilocation evaluation. To overcome the difficulties with large collections, ICRISAT scientists have developed a 'core collection' consisting of 1290 accessions (about 10% of entire collection), representing the genetic variability of the entire collection (Reddy et al., 2005).

When the entire collection is over 10,000 accessions, developing a core collection will not solve the problem of low use of germplasm, as even the size of the core collection would be unwieldy for meaningful evaluation and convenient exploitation. To overcome this, a seminal two-stage strategy was followed. The first stage involves developing a representative core collection (about 10%) from the entire collection using all the available information on origin, geographical distribution, and characterization and evaluation data of accessions. The second stage involves evaluation of the core collection for various morphological, agronomic and quality traits, and selecting a further subset of about 10% accessions from the core collection. Thus, the mini-core collection contains 10% of the core or approximately 1% of the entire collection and represents the diversity of the entire collection (Upadhyaya & Ortiz, 2001). In pigeon pea, a mini-core collection consisting of 146 accessions was constituted by evaluating a core collection of 1290 accessions for 34 morpho-agronomic traits (Upadhyaya, Reddy, Gowda, Reddy, & Singh, 2006). Due to their greatly reduced size, mini-core collections provide an easy access to the germplasm collections and scientists can evaluate the mini-core collection easily and economically for traits of economic importance.

8.9 Germplasm Enhancement Through Wide Crosses

Narrow genetic diversity in cultivated germplasm has hampered the effective utilization of conventional breeding as well as development and utilization of genomic tools, resulting in pigeon pea being often referred to as an 'orphan crop legume'. A number of wild *Cajanus* species, especially those from the secondary gene pool which are cross-compatible with cultivated pigeon pea, have been used for the genetic improvement of pigeon pea. The most significant achievement is the development of unique cytoplasmic nuclear male sterility systems (CMS). The CMS systems have been developed with cytoplasm derived from cultivated and wild *Cajanus* species. The A_1 cytoplasm is derived from *C. sericeus* (Ariyanayagam, Nageshwara, & Zaveri, 1995). The CMS lines derived from this source are temperature sensitive and the male sterile lines restore fertility under low temperature conditions (Saxena, 2005). The A_2 cytoplasm derived from *C. scarabaeoides* (Saxena & Kumar, 2003; Tikka, Parmar, & Chauhan, 1997) is a stable source of CMS but the fertility restoration (fr) is not consistent across environments, making it unsuitable for hybrid seed production. A_3 cytoplasm derived from *C. volubilis* (Wanjari, Patil, Manapure, Manjaya, & Manish, 2001) has a poor-quality fr system. The A_4 cytoplasm derived from *C. cajanifolius* (Saxena et al., 2005) is stable across environments with a good fr system and has been used to develop the world's first commercial pigeon pea hybrid, ICPH 2671 (Saxena et al., 2013). The A_5 cytoplasm derived from *C. cajan* (Mallikarjuna & Saxena, 2005) is still under development. The A_6 cytoplasm has been derived from *C. lineatus* and at present this CMS source is in BC_5F_1 generation with a perfect male sterility maintenance system available (Saxena, Sultana et al., 2010). The studies on A_7 CMS system derived from *C. platycarpus* are in progress. Recently, the A_8 CMS system derived from *C. reticulatus* has also been developed, but the detailed studies on this CMS system are in progress at ICRISAT.

Wild *Cajanus* species, especially, *C. scarabaeoides*, *C. acutifolius*, *C. platycarpus*, *C. reticulates*, *C. sericeus* and *C. albicans* have been reported to have resistance to pod borer, *Helicoverpa armigera* (Rao, Reddy, & Bramel, 2003; Sharma, Sujana, & Rao, 2009; Sujana, Sharma, & Rao, 2008). At ICRISAT, utilization of *C. acutifolius* as the pollen parent has resulted in the development of advanced generation population having resistance to pod borer (Mallikarjuna, Sharma, & Upadhyaya, 2007), variation in seed colour and high seed weight. Evaluation of wild *Cajanus* species has identified accessions having resistance to *Alternaria* blight (Sharma, Kannaiyan, & Saxena, 1987), *Phytophthora* blight (Rao et al., 2003), sterility mosaic virus (Kulkarni et al., 2003; Rao et al., 2003), pod fly (Rao et al., 2003; Saxena et al., 1990), pod fly and wasps (Sharma, Pampapathy, & Reddy, 2003), root-knot nematodes (Rao et al., 2003; Sharma, 1995; Sharma, Remanandan, & Jain, 1993; Sharma, Remanandan, & McDonald, 1993), and tolerance to salinity (Rao et al., 2003; Srivastava et al., 2006; Subbarao, 1988; Subbarao, Johansen, Jana, & Rao, 1991), drought (Rao et al., 2003), and photoperiod insensitivity (Rao et al., 2003).

Besides for CMS systems and as resistant/tolerant sources for biotic/abiotic stresses, utilization of wild *Cajanus* species has also contributed significantly towards the improvement of agronomic performance and nutritional quality of cultivated pigeon pea. Some wild *Cajanus* species, namely *C. scarabaeoides*, *C. sericeus*, *C. albicans*, *C. crassus*, *C. platycarpus* and *C. cajanifolius*, have higher seed protein content (average 28.3%) compared to pigeon pea cultivars (24.6%) (Singh & Jambunathan, 1981). A high protein line, ICPL 87162, was developed from the cross *C. cajan*×*C. scarabaeoides* (Reddy et al., 1997). This line contains 30–34% protein content compared to the control cultivar (23% protein). Breeding lines with high protein content have also been developed from *C. sericeus*, *C. albicans* and *C. scarabaeoides*. Utilization of wild *Cajanus* species has resulted in the development of several lines, such as HPL 2, HPL 7, HPL 40 and HPL 51, having high protein and high seed weight (Saxena, Faris, & Kumar, 1987). Recently, scientists at ICRISAT have generated segregants with high seed weight from the crosses between cultivated pigeon pea and *C. acutifolius*. Using wild *Cajanus* species, viable hybrids have been produced between pigeon pea and *C. platycarpus* (Mallikarjuna & Moss, 1995), *C. reticulatus* var. *grandifolius* (Reddy, Kameswara Rao, & Saxena, 2001), *C. acutifolius* (Mallikarjuna & Saxena, 2002) and *C. albicans* (Subbarao, Johansen, Kumar Rao, & Jana, 1990).

8.10 Pigeon Pea Genomic Resources

Pigeon pea breeders have developed varieties with several attributes with a major focus on productivity traits and as a result diversity has been lost in the elite gene pool; subsequently yield levels in pigeon pea have been stagnant during the last six decades. In order to meet future challenges and to enhance the yield levels, genomics interventions are required to identify the genes or quantitative trait loci (QTLs) responsible for resistance or tolerance to various economically important traits. A large amount of genomic and genetic resources have been developed by ICRISAT in collaboration with partners and have regularly been used in accelerating the genomics and breeding applications to increase the efficiency of pigeon pea improvement programmes. ICRISAT scientists have developed a number of marker systems and genetic linkage maps and identified marker-trait associations for a few important traits. Recently complete genome sequencing of pigeon pea has been accomplished (Varshney et al., 2012).

8.10.1 Mapping Populations

Genetic diversity among elite pigeon pea cultivars is very low (Saxena, Sultana et al., 2010) and hence selection of crossing parents is the most crucial step. In order to select a diverse set of parents, simple sequence repeats (SSRs) genotyping of elite cultivars was performed and a number of intraspecific biparental mapping populations, segregating for FW, SMD and fr have been developed (Saxena, Prathima et al., 2010; Saxena, Saxena, Kumar, Hoisington, & Varshney, 2010). One interspecific [ICP 28 (*C. cajan*)×ICPW 94 (*C. scarabaeoides*)] mapping population has also been developed (Saxena et al., 2012).

8.10.2 Molecular Markers

Recently several marker systems have been developed and used in pigeon pea (Table 8.4). Prior to PCR technologies, restriction fragment length polymorphisms (RFLPs) (Sivaramakrishnan, Seetha, & Reddy, 2002), protein isoforms and phenotypes were used. However, these markers present challenges for large-scale throughput because they are labour intensive, require large amounts of starting material (genomic DNA or protein) and are less informative as compared to the modern marker systems. The vast majority of markers now used for pigeon pea are PCR based, with the majority being microsatellite markers (SSR) (Bohra et al., 2011; Burns, Edwards, Newbury, Ford-Lloyd, & Baggott, 2001; Odeny et al., 2007; Saxena, Prathima et al., 2010; Saxena, Saxena, Kumar et al., 2010; Saxena, Saxena, & Varshney, 2010). Other potential marker systems, such as random amplified polymorphic DNA (RAPD) markers (Malviya & Yadav, 2010), single strand conformation polymorphisms (SSCPs) (Kudapa et al., 2012), amplified fragment length polymorphisms (AFLPs) (Panguluri, Janaiah, Govil, Kumar, & Sharma, 2006) and DArT (Yang et al., 2006, 2011) are also in use. By using an SSR-enriched library, several genomic DNA libraries enriched for di- and tri-nucleotide repeat motifs

Table 8.4 Available Genomic Resources in Pigeon Pea

Resource		References
Simple sequence repeats	29,000	Raju et al. (2010), Saxena, Sultana et al. (2010), Bohra et al. (2011), Dutta et al. (2011) and Varshney et al. (2012)
Single nucleotide polymorphisms (SNPs)	35,000	Saxena et al. (2012) and Varshney et al. (2012)
GoldenGate assays	768 SNPs	Unpublished
KASPar assays	1616 SNPs	Saxena et al. (2012)
Single feature polymorphisms (SFPs)	1131	Saxena et al. (2011)
Diversity arrays technology (DArT) markers	15,360	Yang et al. (2011)
Sanger ESTs	~20,000	Raju et al. (2010) and Dubey et al. (2011)
454/FLX reads	496,705	Dubey et al. (2011)
Tentative unique sequences (TUSs)	21,432	Dubey et al. (2011)
Illumina/454 reads (million reads)	>160	Dubey et al. (2011), Dutta et al. (2011) and Kudapa et al. (2012)

(CT, TG, AG, AAG, TCG, etc.) were also generated (Burns et al., 2001; Odeny et al., 2007; Saxena, Saxena, & Varshney, 2010). This approach involving SSR marker development has provided only 36 SSRs; however, subsequently SSRs were developed from bacterial artificial chromosome (BAC) end sequences (BESs) and found more effective. SSR development from BAC ends avoids the need for prior information about the repeat motifs within a species and offers genome-wide coverage. After examining 87,590 pigeon pea BESs, a total of 18,149 SSRs were identified in 14,001 BESs representing 6590 BAC clones. Excluding the mononucleotide repeats, a total of 3072 primer pairs were synthesized and tested (Bohra et al., 2011). The recent advent of affordable high-throughput technology for single nucleotide polymorphisms (SNPs), together with the reduction in sequencing costs, is resulting in a shift to SNP markers for trait mapping and association studies (Thudi, Li, Jackson, May, & Varshney, 2012). It is expected that within a couple of years the marker-based studies will be dominated by SNP markers. Three approaches were used for the identification of SNPs in pigeon pea. In the first approach, Illumina sequencing was carried out on parental genotypes of mapping populations of pigeon pea. RNA sequencing of 12 pigeon pea genotypes resulted in 128.9 million reads for pigeon pea (Kudapa et al., 2012). Alignment of these short reads onto transcriptome assembly (TA) has provided a large number of SNPs. The second approach, allele-specific sequencing of parental genotypes of the reference mapping population of pigeon pea using conserved orthologous sequence (COS) markers, has provided 768 SNPs for pigeon pea (Table 8.4). As a result, a large number of SNPs has become available for pigeon pea and cost-effective genotyping platforms have been developed.

8.10.3 Genotyping the Germplasm Collection

A composite collection of 1000 accessions was developed and profiled using 20 SSR markers. Analysis of molecular data for 952 accessions detected 197 alleles, of which 115 were rare and 82 common. Gene diversity varied from 0.002 to 0.726. There were 60 group-specific unique alleles in wild types and 64 in cultivated. Among the cultivated accessions, 37 unique alleles were found in NDT types. Geographically, 32 unique alleles were found in Asia 4 (southern Indian provinces, Maldives and Sri Lanka). Only two alleles differentiated Africa from other regions. Wild and cultivated types shared 73 alleles, DT (determinate) and NDT shared 10, DT and wild shared 4, and the NDT and wild shared 20 alleles. Wild types as a group were genetically more diverse than cultivated types. NDT types were more diverse than the other two groups based on flowering pattern (DT and SDT: semi-determinate). Reference sets consisting of the 300 most diverse accessions based on SSR markers, qualitative traits, quantitative traits and their combinations were formed and compared for allelic richness and diversity. A reference set based on SSR data captured 187 (95%) of the 197 alleles of the composite collection. Another reference set based on qualitative traits captured 87% of the alleles of the composite set. This demonstrates that both SSR markers and qualitative traits were equally efficient in capturing the allelic richness and diversity in the reference sets (Upadhyaya et al., 2008).

8.10.4 Linkage Maps and Trait Mapping

The first generation pigeon pea linkage map or reference map was developed using DArT markers for an interspecific mapping population (ICP 28×ICPW 94) of 79 F_2 individuals. The map is available in male and female forms, a total of 121 unique DArT maternal markers were placed on the maternal linkage map and 166 unique DArT paternal markers were placed on the paternal linkage map. The length of these two maps covered 437.3 cM and 648.8 cM, respectively (Yang et al., 2011). Another version of reference linkage map consisted of 239 SSR markers and spans 930.90 cM (Bohra et al., 2011). An interspecific mapping population (ICP 28×ICPW 94) relatively bigger in size (167 F_2s) was used for developing a comprehensive genetic map comprising 875 SNP loci (Saxena et al., 2012). The total length of this map was 967.03 cM with an average marker distance of 1.11 cM. This linkage map was a considerable improvement over the previous pigeon pea genetic linkage maps using SSR and DArT markers.

Construction of genetic maps for intraspecific mapping populations has also been performed and a total of six SSR-based intraspecific genetic maps were developed by using six F_2 mapping populations (Bohra et al., 2012; Gnanesh et al., 2011). Furthermore, all six intraspecific genetic maps were joined together into a single consensus genetic map providing map positions to a total of 339 SSR markers at map coverage of 1059 cM (Bohra et al., 2012). A few trait association efforts have been reported in pigeon pea for SMD and fr by using F_2 mapping populations. For instance, six QTLs explaining phenotypic variations in the range of 8.3–24.72%

(Gnanesh et al., 2011) for SMD and a total of four large effect QTLs explaining up to 24% of phenotypic variations for fr in pigeon pea (Bohra et al., 2012) were identified.

8.10.5 Transcriptomic Resources

To characterize the pigeon pea transcriptome, two NGS technologies, namely 454- and Illumina together with Sanger sequencing technology have been used. By using Sanger sequencing technology on FW and SMD, challenged cDNA libraries for pigeon pea 9888 expressed sequence tags (ESTs) were developed (Raju et al., 2010). To improve these transcriptomic resources further, 454/FLX sequencing was undertaken on normalized and pooled RNA samples collected from >20 tissues, generating 494,353 transcript reads for pigeon pea (Dubey et al., 2011). Cluster analysis of these transcript reads with Sanger ESTs generated at ICRISAT, as well as those available in the public domain, provided the first transcript assembly (TA) of pigeon pea (CcTA v1) with 127,754 transcriptional units (Dubey et al., 2011). 494,353 454/FLX transcript reads generated from Asha genotype and 128.9 million Illumina reads generated from 12 genotypes were analysed together with 18,353 Sanger ESTs and 1.696 million 454/FLX transcript reads (Dutta et al., 2011) with improved algorithms. As a result, an improved TA in pigeon pea referred to as CcTA v2, comprising 21,434 contigs, has been developed (Kudapa et al., 2012) (Table 8.4).

8.10.6 Genome Sequence

NGS (Illumina) was used to generate 237.2 Gbp of sequence that, along with Sanger-based BAC-end sequences and a genetic map, was assembled into scaffolds representing about 73% (605.78 Mb) of the 833 Mbp pigeon pea genome size. Genome analysis has resulted in the identification of 48,680 pigeon pea genes. High levels of synteny were observed between pigeon pea and soybean as well as between pigeon pea and *Medicago truncatula* and *Lotus japonicas*.

The genome sequence was also searched for the presence of tandem repeats and a total of 23,410 SSR primers were designed. Transcript reads from 12 different pigeon pea genotypes were aligned with the genome assembly for the identification of SNPs. As a result 28,104 novel SNPs were identified across 12 genotypes (Varshney et al., 2012). These developed resources will be used for germplasm characterization and to facilitate the identification of the genetic basis of important traits.

8.11 Conclusions

The narrow genetic base of pigeon pea, coupled with its susceptibility to a number of biotic and abiotic stresses, necessitates the use of diverse genetic resources for its improvement. Though a large number of germplasm accessions are conserved in different gene banks globally, only a small fraction (<1%) has been used in crop improvement programmes. The availability of trait-specific germplasm accessions will provide an opportunity for breeders to use new sources of variations in developing new cultivars with a broad genetic base. The utilization of wild *Cajanus* species has contributed significantly to the genetic enhancement of pigeon pea by providing resistance/tolerance to diseases, insect pests and drought, as well as good agronomic traits. The major contribution of wild relatives includes the development of diverse and unique CMS systems for pigeon pea improvement. The availability of rich genomic resources including genome sequence will further accelerate marker-assisted breeding for pigeon pea improvement.

References

Ariyanayagam, R. P., Nageshwara, A., & Zaveri, P. P. (1995). Cytoplasmic genic male sterility in interspecific matings of pigeonpea. *Crop Science*, *35*, 981–985.

Bohra, A., Dubey, A., Saxena, R. K., Penmetsa, R. V., Poornima, K. N., Kumar, N., et al. (2011). Analysis of BAC-end sequences (BESs) and development of BES-SSR markers for genetic mapping and hybrid purity assessment in pigeonpea (*Cajanus* spp.). *BMC Plant Biology*, *11*, 56.

Bohra, A., Saxena, R. K., Gnanesh, B. N., Saxena, K. B., Byregowda, M., Rathore, A., et al. (2012). An intra-specific consensus genetic map of pigeonpea [*Cajanus cajan* (L.) Millspaugh] derived from six mapping populations. *Theoretical and Applied Genetics*, *125*, 1325–1338.

Burns, M. J., Edwards, K. J., Newbury, H. J., Ford-Lloyd, B. V., & Baggott, C. D. (2001). Development of simple sequence repeat (SSR) markers for the assessment of gene flow and genetic diversity in pigeonpea (*Cajanus cajan*). *Molecular Ecology Notes*, *1*, 283–285.

De, D. N. (1974). Pigeonpea. In J. Hutchinson (Ed.), *Evolutionary studies in world crops: diversity and change in the Indian subcontinent* (pp. 79–87). London, UK: *Cambridge University*.

Dubey, A., Farmer, A., Schlueter, J., Cannon, S., Abernathy, B., Tuteja, R., et al. (2011). Defining the transcriptome assembly and its use for genome dynamics and transcriptome profiling studies in pigeonpea (*Cajanus cajan* L.). *DNA Research*, *18*, 153–164.

Dutta, S., Kumawat, G., Singh, B. P., Gupta, D. K., Singh, S., Dogra, V., et al. (2011). Development of genic-SSR markers by deep transcriptome sequencing in pigeonpea [*Cajanus Cajan* (L.) Millspaugh]. *BMC Plant Biology*, *11*, 17.

FAOSTAT, (2010). *The second report on the state of the world's plant genetic resources for food and agriculture*. Rome, Italy: Commission on Genetic Resources for Food and Agriculture (CGRFA), Food and Agriculture Organization of the United Nations (FAO). ISBN 978-92-5-106534-172.

Gnanesh, B. N., Bohra, A., Sharma, M., Byregowda, M., Pande, S., Wesley, V., et al. (2011). Genetic mapping and quantitative trait locus analysis of resistance to sterility mosaic disease in pigeonpea [*Cajanus cajan* (L.) Millsp.]. *Field Crops Research*, *123*, 53–61.

Hillocks, R. J., Minja, E., Nahdy, M. S., & Subrahmanyam, P. (2000). Diseases and pests of pigeonpea in eastern Africa. *International Journal of Pest Management*, *46*, 7–18.

IBPGR, (1993)., & ICRISAT, *Descriptors for pigeonpea [Cajanus cajan (L.) Millsp.]*. Patancheru, India/Rome, Italy: International Crops Research Institute for the Semi-Arid Tropics/International Board for Plant Genetic Resources.

ICRISAT Archival Report. (2010). <http://intranet/ddg/Admin%20Pages2009/Archival_Report_2010.aspx>

Krishnamurthy, L., Upadhyaya, H. D., Saxena, K. B., & Vadez, V. (2011). Variation for temporary water logging response within the mini core pigeonpea germplasm. *Journal of Agricultural Science*, *10*, 1–8.

Kudapa, H., Bharti, A. K., Cannon, S. B., Farmer, A. D., Mulaosmanovic, B., Kramer, R., et al. (2012). A comprehensive transcriptome assembly of pigeonpea (*Cajanus cajan*) using sanger and second-generation sequencing platforms. *Molecular Plant*, *5*, 1020–1028.

Kulkarni, N. K., Reddy, A. S., Lava Kumar, P., Vijaynarasimha, J., Rangaswamy, K. T., Muniyappa, V., et al. (2003). Broad-based resistance to pigeonpea sterility mosaic disease in accessions of *Cajanus scarabaeoides* (L.) Benth. *Indian Journal of Plant Protection*, *31*, 6–11.

Mallikarjuna, N., & Moss, J. P. (1995). Production of hybrids between *Cajanus platycarpus* and *Cajanus cajan*. *Euphytica*, *83*(1), 43–46.

Mallikarjuna, N., & Saxena, K. B. (2002). Production of hybrids between *Cajanus acutifolius* and *C.cajan*. *Euphytica*, *124*(1), 107–110.

Mallikarjuna, N., & Saxena, K. B. (2005). A new cytoplasmic male-sterility system derived from cultivated pigeonpea cytoplasm. *Euphytica*, *142*(1–2), 143–148.

Mallikarjuna, N., Sharma, H. C., & Upadhyaya, H. D. (2007). Exploitation of wild relatives of pigeonpea and chickpea for resistance to *Helicoverpa armigera*. *SAT eJournal*, *3*(1), 4.

Malviya, N., & Yadav, D. (2010). RAPD analysis among pigeonpea [*Cajanus cajan* (L.).] cultivars for their genetic diversity. *Genetic Engineering and Biotechnology Journal*, *GEBJ-1*, 1–9.

Odeny, D. A. (2007). The potential of pigeonpea (*Cajanus cajan* (L.) Millsp.) in Africa. *Natural Resources Forum*, 297–305.

Odeny, D. A., Jayashree, B., Ferguson, M., Hoisington, D., Crouch, J., & Gebhardt, C. (2007). Development characterization and utilization of microsatellite markers in pigeonpea [*Cajanus cajan* (L.) Millsp.]. *Plant Breeding*, *126*, 130–137.

Panguluri, S. K., Janaiah, K., Govil, J. N., Kumar, P. A., & Sharma, P. C. (2006). AFLP finger printing in pigeonpea (*Cajanus cajan* (L.) Millsp.) and its wild relatives. *Genetic Resources and Crop Evolution*, *53*, 523–531.

Plukenet, L. (1692). Phytographia. In F. Singh & D. L. Oswalt (Eds.), *Pigeonpea botany and production practices*. Patancheru, Andhra Pradesh, India: International Crops Research Institute for the Semi-Arid Tropics. Skills Development Series No. 9 (1992).

Raju, N. L., Gnanesh, B. N., Lekha, P. T., Balaji, J., Pande, S., Byregowda, M., et al. (2010). The first set of EST resource for gene discovery and marker development in pigeonpea (*Cajanus cajan* L.). *BMC Plant Biology*, *10*, 45.

Rao, N. K., Reddy, L. J., & Bramel, P. J. (2003). Potential of wild species for genetic enhancement of some semi-arid food crops. *Genetic Resources and Crop Evolution*, *50*, 707–721.

Reddy, K. N., Upadhyaya, H. D., Reddy, L. J., & Gowda, C. L. L. (2006). Evaluation of pollination control methods for pigeonpea (*Cajanus cajan* (L.) Millsp.) germplasm regeneration. *International Chickpea and Pigeonpea Newsletter*, *13*, 35–38.

Reddy, L. J. (1973). Interrelationships of *Cajanus* and *Atylosia* species as revealed by hybridization and pachytene analysis. *PhD Thesis*. Kharagpur, India: Indian Institute of Technology.

Reddy, L. J., Kameswara Rao, N., & Saxena, K. B. (2001). Production and characterization of hybrids between *Cajanus cajan* × *C. reticulatus* var. *grandifolius*. *Euphytica*, *121*, 93–98.

Reddy, L. J., Saxena, K. B., Jain, K. C., Singh, U., Green, J. M., Sharma, D., et al. (1997). Registration of high-protein pigeonpea elite germplasm ICPL 87162. *Crop Science*, *37*, 294.

Reddy, L. J., Upadhyaya, H. D., Gowda, C. L. L., & Singh, S. (2005). Development of core collection in pigeonpea [*Cajanuscajan* (L.) Millsp.] using geographical and qualitative morphological descriptors. *Genetic Resources and Crop Evolution, 52*, 1049–1056.

Remanandan, P., Sastry, D. V. S. S. R., & Mengesha Melak, H. (1988). *ICRISAT Pigeonpea germplasm catalog: Evaluation and analysis*. Patancheru, A.P., India: International Crops Research Institute for the Semi-Arid Tropics.

Royes Vernon, W. (1976). Pigeonpea. In N. W. Simmonds (Ed.), *Evolution of crop plants* (pp. 154–156). London, UK: Longmans.

Rupela, O. P., Gowda, C. L. L., Wani, S. P., & Ranga Rao, G. V. (2004). Lessons from nonchemical input treatments based on scientific and traditional knowledge in a long-term experiment: *Proceedings of the International Conference on Agricultural Heritage of Asia, 6–8 December 2004*. (pp. 184–196). Secunderabad, Andhra Pradesh, India: Asian Agri-History Foundation.

Saxena, K. B. (2005). Pigeonpea (*Cajanus cajan* (L.) Millsp.). In R. J. Singh & P. R. Jauhar (Eds.), *Genetic resources, chromosome engineering and crop improvement* (pp. 86–115). New York: Taylor and Francis.

Saxena, K. B., Faris, D. G., & Kumar, R. V. (1987). Relationship between seed size and protein content in newly developed high protein lines of pigeonpea. *Plant Foods and Human Nutrition, 36*, 335–340.

Saxena, K. B., & Kumar, R. V. (2003). Development of cytoplasmic nuclear male-sterility system in pigeonpea using *C. scarabaeoides* (L.) Thours. *Indian Journal of Genetics, 63*(3), 225–229.

Saxena, K. B., Kumar, R. V., Dalvi, V. A., Mallikarjuna, N., Gowda, C. L. L., Singh, B. B., et al. (2005). Hybrid breeding in grain legumes – a success story of pigeonpea. In M. C. Khairwal & H. K. Jain (Eds.), *Proceedings of the international food legumes research conference,* New Delhi, India.

Saxena, K. B., Kumar, R. V., Reddy, L. J., & Arora, A. (2003). Pigeonpea. In N. C. Singhal (Ed.), *Hybrid seed production in field crops: principles and practices* (pp. 163–181). Ludhiana, India: Kalyani Publishers.

Saxena, K. B., Kumar, R. V., Tikle, A. N., Saxena, M. K., Gautam, V. S., Rao, S. K., et al. (2013). ICPH 2671 – the world's first commercial food legume hybrid: *Plant breeding reviews*. Blackwell Verlag GmbH. pp. 1–7.

Saxena, K. B., Singh, L., & Gupta, M. D. (1990). Variation for natural out-crossing in pigeonpea. *Euphytica, 46*, 143–146.

Saxena, K. B., Sultana, R., Mallikarjuna, N., Saxena, R. K., Kumar, R. V., Sawargaonkar, S. L., et al. (2010). Male-sterility systems in pigeonpea and their role in enhancing yield. *Plant Breeding, 129*, 125–134.

Saxena, R. K., Cui, X., Thakur, V., Walter, B., Close, T. J., & Varshney, R. K. (2011). Single feature polymorphisms (SFPs) for drought tolerance in pigeonpea (*Cajanus* spp.). *Functional Integrative Genomics, 11*, 651–657.

Saxena, R. K., Prathima, C., Saxena, K. B., Hoisington, D. A., Singh, N. K., & Varshney, R. K. (2010). Novel SSR markers for polymorphism detection in pigeonpea (*Cajanus* spp.). *Plant Breeding, 129*, 142–148.

Saxena, R. K., Saxena, K. B., Kumar, R. V., Hoisington, D. A., & Varshney, R. K. (2010). SSR-based diversity in elite pigeonpea genotypes for developing mapping populations to map resistance to *Fusarium* wilt and sterility mosaic disease. *Plant Breeding, 129*, 135–141.

Saxena, R. K., Saxena, K. B., & Varshney, R. K. (2010). Application of SSR markers for molecular characterization of hybrid parents and purity assessment of ICPH 2438 hybrid of pigeonpea [*Cajanus cajan* (L.) Millspaugh]. *Molecular Breeding, 26*, 371–380.

Saxena, R. K., Varma Penmetsa, R., Upadhyaya, H. D., Kumar, A., Carrasquilla-Garcia, N., Schlueter, J. A., et al. (2012). Large-scale development of cost-effective single-nucleotide polymorphism marker assays for genetic mapping in pigeonpea and comparative mapping in legumes. *DNA Research*, *19*, 449–461.

Shannon, C. E., & Weaver, W. (1949). *The mathematical theory of communication*. Urbana: University Illinois Press.

Sharma, D., Kannaiyan, J., & Saxena, K. B. (1987). Sources of resistance to alternaria blight in pigeonpea. *SABRAO Journal*, *19*(2), 109–114.

Sharma, H. C., Pampapathy, G., & Reddy, L. J. (2003). Wild relatives of pigeonpea as a source of resistance to the podfly (*Melalagromyza* obtuse Malloch) and pod wasp (*Tranaostigmodes cajaninae* La Salle). *Genetic Resources and Crop Evolution*, *50*, 817–824.

Sharma, H. C., Sujana, G., & Rao, D. M. (2009). Morphological and chemical components of resistance to pod borer, *Helicoverpa armigera* in wild relatives of pigeonpea. *Arthropod–Plant Interactions*, *3*(3), 151–161.

Sharma, M., Rathore, A., Mangala, U. N., Ghosh, R., Upadhyaya, H. D., Sharma, S., et al. (2012). New sources of resistance to *Fusarium* wilt and sterility mosaic disease in a mini-core collection of pigeonpea germplasm. *European Journal of Plant Pathology*, *133*, 707–714.

Sharma, S. B. (1995). Resistance to *Rotylenchulus reniformis*, *Heterodera cajani* and *Meloidogyne javanica* in accessions of *Cajanus platycarpus*. *Plant Disease*, *79*, 1033–1035.

Sharma, S. B., Remanandan, P., & Jain, K. C. (1993). Resistance to cyst nematode (*Heterodera cajani*) in pigeonpea cultivars and wild relatives of *Cajanus*. *Annals of Applied Biology*, *123*, 75–81.

Sharma, S. B., Remanandan, P., & McDonald, D. (1993). Resistance to *Meloidogyne javanica* and *Rotylenchulus reniformis* in wild relatives of pigeonpea. *Journal in Nematology*, *25*(4S), 824–829.

Singh, U., & Jambunathan, R. (1981). Protease inhibitors and *in vitro* protein digestibility of pigeonpea [*Cajanus cajan* (L.) Millsp.] and its wild relatives. *Journal of Food Science and Technology*, *18*, 246–247.

Sivaramakrishnan, S., Seetha, K., & Reddy, L. J. (2002). Diversity in selected wild and cultivated species of pigeonpea using RFLP of mtDNA. *Euphytica*, *125*, 21–28.

Srivastava, N., Vadez, V., Upadhyaya, H. D., & Saxena, K. B. (2006). Screening for intra and inter specific variability for salinity tolerance in pigeonpea (*Cajanus cajan*) and its related wild species. *Journal of SAT Agricultural Research*, *2*, 12. <http://icrtest:8080/Journal/crop improvement v2il/ v2il screening for. Pdf>.

Subbarao, G. V. (1988). Salinity tolerance in pigeonpea (*Cajanus cajan* (L.) Millsp.) and its wild relatives. *PhD Thesis*. Kharagpur, India: Indian Institute Technology.

Subbarao, G. V., Johansen, C., Jana, M. K., & Rao, J. D. V. K. (1991). Comparative salinity responses among pigeonpea genotype and their wild relatives. *Crop Science*, *31*, 415–418.

Subbarao, G. V., Johansen, C., Kumar Rao, J. V. D. K., & Jana, M. K. (1990). Salinity tolerance in F1 hybrids of pigeonpea and a tolerant wild relative. *Crop Science*, *30*(4), 785–788.

Sujana, G., Sharma, H. C., & Rao, D. M. (2008). Antixenosis and antibiosis components of resistance to pod borer *Helicoverpa armigera* in wild relatives of pigeonpea. *International Journal of Tropical Insect Science*, *28*(4), 191–200.

Thudi, M., Li, Y., Jackson, S. A., May, G. D., & Varshney, R. K. (2012). Current state-of-art of sequencing technologies for plant genomics research. *Briefings in Functional Genomics*, *11*, 3–11.

Tikka, S. B. S., Parmar, L. D., & Chauhan, R. M. (1997). First record of cytoplasmic-genic male sterility system in pigeonpea (*Cajanus cajan* (L.) Millsp.) through wide hybridization. *Gujarat Agricultural University Research Journal*, *22*(2), 160–162.

Upadhyaya, H. D. (2008). Crop germplasm and wild relatives: a source of novel variation for crop improvement. *Korean Journal of Crop Science*, *53*, 12–17. ISSN 0252-9777.

Upadhyaya, H. D., Bhattacharjee, R., Varshney, R. K., Hoisington, D. A., Reddy, K. N., & Singh, S. (2008). Assessment of genetic diversity in pigeonpea using SSR markers. In *Annual Joint Meeting of ASA/CSSA*, Houston, USA, 5–9 October 2008 (Abstracts No. 657-3).

Upadhyaya, H. D., Dronavalli, N., Gowda, C. L. L., & Singh, S. (2012). Mini core collections for enhanced utilization of genetic resources in crop improvement. *Indian Journal of Plant Genetic Resources*, *25*, 111–124. ISSN 0976-1926.

Upadhyaya, H. D., & Gowda, C. L. L. (2009). *Managing and enhancing the use of germplasm – strategies and methodologies*: *Technical Manual No. 10*. Patancheru, India: International Crops Research Institute for the Semi-Arid Tropics.

Upadhyaya, H. D., & Oritz, R. (2001). A mini core subset for capturing diversity and promoting utilization of chickpea genetic resources in crop improvement. *Theoretical and Applied Genetics*, *102*, 1292–1298.

Upadhyaya, H. D., Pundir, R. P. S., Gowda, C. L. L., Reddy, K. N., & Singh, S. (2005). Geographical patterns of diversity for qualitative and quantitative traits in the pigeonpea germplasm collection. *Plant Genetic Resources*, *3*(3), 331–352.

Upadhyaya, H. D., Reddy, L. J., Gowda, C. L. L., Reddy, K. N., & Singh, S. (2006). Development of a mini core subset for enhanced and diversified utilization of pigeonpea germplasm resources. *Crop Science*, *46*, 2127–2132.

Upadhyaya, H. D., Yadav, D., Dronavalli, N., Gowda, C. L. L., & Singh, S. (2010). Mini core germplasm collection for infusing genetic diversity in plant breeding programs. *Electronic Journal of Plant Breeding*, *1*(4), 1294.

van der Maesen, L. J. G. (1980). India is the native home of the pigeonpea. In J. C. Arends, G. Boelema, C. T. de Groot, & A. J. M. Leeuwenberg (Eds.), *Liber gratulatorius in nonerem H.C.D. de Wit. Landbouwhogeschool Miscellaneous Paper no. 19*. Wageningen, The Netherlands: H. Veenman and B.V. Zonen.

van der Maesen, L. J. G. (1990). Pigeonpea origin, history, evolution, and taxonomy. In Y. L. Nene, S. D. Hall, & V. K. Sheila (Eds.), *The pigeonpea* (pp. 15–46). Wallingford, Oxon, UK: CAB International.

Varshney, R. K., Chen, W., Li, Y., Bharti, A. K., Saxena, R. K., Schlueter, J. A., et al. (2012). Draft genome sequence of pigeonpea (*Cajanus cajan*), an orphan legume crop of resource-poor farmers. *Nature Biotechnology*, *30*, 83–89.

Vavilov, N. I. (1951). The origin, variation, immunity and breeding of cultivated plants. *Chronica Botanica*, *13*(1–6), 1–366.

Wanjari, K. B., Patil, A. N., Manapure, P., Manjaya, J. G., & Manish, P. (2001). Cytoplasmic male-sterility with cytoplasm from *Cajanus volubilis*. *Annuals Plant Physiology*, *13*, 170–174.

Yang, S., Pang, W., Ash, G., Harper, J., Carling, J., Wenzl, P., et al. (2006). Low level of genetic diversity in cultivated pigeonpea compared to its wild relatives is revealed by diversity arrays technology. *Theoretical and Applied Genetics*, *113*, 585–595.

Yang, S. Y., Saxena, R. K., Kulwal, P. L., Ash, G. J., Dubey, A., Harpe, J. D. I., et al. (2011). The first genetic map of pigeonpea based on diversity arrays technology (DArT) markers. *Journal of Genetics*, *90*, 103–109.

9 Peanut

H. Thomas Stalker

Department of Crop Science, North Carolina State University, Raleigh, NC

9.1 Introduction

Domesticated peanut (*A. hypogaea* L.), sometimes called groundnut, is an allotetraploid ($2n=4x=40$) species that is widely grown in tropical and subtropical regions of the world. The crop is cultivated in more than 100 countries and has an average production of 35.5 million tonnes annually (FAO, 2009). China is the largest peanut producer, followed by India, United States and Nigeria. The seed is rich in oil (40–60%) and protein (20–40%), which makes it a high-energy seed. Most of the world production is crushed for oil, whereas in the United States more than 60% of production is consumed as edible products. The average yield of the peanut crop ranges from 0.43 t/ha in Africa to 3.54 t/ha in North America, with a world average of 1.35 t/ha (Dwivedi et al., 2007). Disease epidemics and drought are major constraints to peanut production in all production areas. Several species of the genus have been consumed for their seeds, but only *A. hypogaea* is economically important today. However, several wild species (most notably *A. glabrata* and *A. pintoi*) are utilized for grazing (Hernandez-Garay, Sollenberger, Staples, & Pedreria, 2004; Magbanua et al., 2000), and *A. repens* is used as a ground cover in residential areas and roadsides in subtropical and tropical regions. The primary interest in wild species of *Arachis* has been for utilizing sources of disease and insect resistances for crop improvement because of the extremely high levels of resistance in many of the species.

Until the early 1900s, peanut was mostly consumed in the United States in the shell as a roasted product; in most countries this remains the primary method of human consumption. Peanut butter was marketed in the late 1890s as a nutritious and healthy food and by 1899 several brands of peanut butter were marketed (Hammons, 1982). The market further expanded at about the same time with the popularity of peanut candy and penny-in-the-slot peanut machines. Commercialization of peanut products led to mechanical diggers in the early 1900s and once-over combines in the 1940s (Hammons, 1982).

The domesticated peanut is plagued by many disease and insect pests, with early leaf spot (*Cercospora arachidicola* Hori), late leaf spot (*Cercosporidium personatum* (Berk & M.A. Curtis) Deighton) and rust (*Puccinia arachidis* Speg.) being the most widespread and destructive. The three diseases can result in 70% or more yield loss (Subrahmanyam, Williams, McDonald, & Gibbons, 1984). Additional diseases are important on a regional scale and many cause significant yield losses, for example

Genetic and Genomic Resources of Grain Legume Improvement. DOI: http://dx.doi.org/10.1016/B978-0-12-397935-3.00009-8

tomato spotted wilt virus, sclerotinia blight (*Sclerotinia minor* Jagger), southern stem rot (*Sclerotium rolfsii* Sacc.) and Cylindrocladium black rot (*Cylindrocladium crotalariae* (Loos) Bell and Sobers). The most important insect problems in peanut on a global scale are aphids, thrips, jassids and *Spodoptera* species (Isleib, Wynne, & Nigam, 1994). Other insects are more regional, such as termites, millipedes, ants and white grubs.

Although not resulting in yield losses, aflatoxin (caused by *Aspergillus* spp.) is a serious problem due to human health issues. Aflatoxin is most prevalent during periods of drought stress, which occurs often in most production areas. Allergens also are a major commercialization problem because of the increasing percentage of the population that has anaphylactic reactions after consuming peanuts. Peanut allergens are caused by 2S, 7S and 11S protein families that comprise the seed storage proteins. Unfortunately, all peanut products with the exception of very highly purified oil will cause allergic reactions in susceptible individuals.

9.2 Origin, Distribution, Diversity and Taxonomy

9.2.1 Arachis *Species*

Arachis species are distinguished from most other taxa by having a peg and geocarpic reproductive development. As opposed to other Papilionoid legumes, the ovary is at the base of the hypanthium rather than being enclosed by the petals. After fertilization there are three to four cell divisions and then the embryo is quiescent until after it is carried into the soil by a peg. The embryo reinitiates development after the pod is formed. In wild species, the peg can grow to more than a meter in length and individual pods are usually separated along the peg. This specialized type of reproductive development has led to seed survival because they are planted in the soil, but at the same time, dispersal is restricted to a few meters. Species in different sections of the genus also have evolved mechanisms to survive in harsh environments, for example tuberoid roots, tuberiform hypocotyls or rhizomes. Wild peanut species are adapted to a wide range of environments from xerophytic forests, to partially flooded areas, to grasslands and subtropical forests. They grow from sea level in Brazil to about 1450m in elevation in the foothills of the Andes Mountains in Argentina. However, they are most frequently associated with savannah-like regions. Most *Arachis* species have a spreading habit, but a few grow upright (e.g. *A. paraguariensis*).

Eighty species have been described in *Arachis* (Krapovickas & Gregory, 1994; Valls & Simpson, 2005) (Table 9.1), and they are divided into nine sections based on morphology and cross-compatibility relationships (Figure 9.1). Additional species are expected to be named as new materials and are collected in South America. Many species in different sections have overlapping distributions, but strong hybridization barriers have evolved to reproductively isolate taxa.

The earliest reports of chromosome numbers in *Arachis* were by Kawakami (1930) who reported that *A. hypogaea* is tetraploid ($2n=4x=40$). A few years later

Table 9.1 *Arachis* Species Identities

Section and Species	2*n*	Type specimen Collector[a]	No.
Section *Arachis*			
batizocoi Krapov. & W.C. Gregory	20	K	9505
benensis Krapov., W.C. Gregory & C.E. Simpson	20	KGSPSc	35005
cardenasii Krapov. & W.C. Gregory	20	KSSc	36015
correntina (Burkart) Krapov. & W.C. Gregory	20	Clos	5930
cruziana Krapov., W.C. Gregory & C.E. Simpson	20	KSSc	36024
decora Krapov., W.C. Gregory & Valls	18	VSW	9955
diogoi Hoehne	20	Diogo	317
duranensis Krapov. & W.C. Gregory	20	K	8010
glandulifera Stalker	20	St	90-40
gregoryi C.E. Simpson, Krapov. & Valls	20	VS	14960
helodes Martius ex Krapov. & Rigoni	20	Manso	588
herzogii Krapov., W.C. Gregory & C.E. Simpson	20	KSSc	36030
hoehnei Krapov. & W.C. Gregory	20	KG	30006
hypogaea L.	40	Linn.	9091
ipaensis Krapov. & W.C. Gregory	20	KMrFr	19455
kempff-mercadoi Krapov., W.C. Gregory & C.E. Simpson	20	KGPBSSc	30085
krapovickasii C.E. Simpson, D.E. Williams, Valls & I.G. Vargas	20	WiSVa	1291
kuhlmannii Krapov. & W.C. Gregory	20	KG	30034
linearifolia Valls, Krapov & C.E. Simpson	20	VPoBi	9401
magna Krapov., W.C. Gregory & C.E. Simpson	20	KGSSc	30097
microsperma Krapov., W.C. Gregory & Valls	20	VKRSv	7681
monticola Krapov. & Rigoni	40	K	8012
palustris Krapov., W.C. Gregory & Valls	18	VKRSv	6536
praecox Krapov., W.C. Gregory & Valls	18	VS	6416
schininii Valls & C.E. Simpson	20	VSW	9923
simpsonii Krapov. & W.C. Gregory	20	KSSc	36009
stenosperma Krapov. & W.C. Gregory	20	HLK	410
trinitensis Krapov. & W.C. Gregory	20	Wi	866
valida Krapov. & W.C. Gregory	20	KG	30011
villosa Benth.	20	Tweedi	1837
williamsii Krapov. & W.C. Gregory	20	WiCl	1118
Section *Caulorrhizae*			
pintoi Krapov. & W.C. Gregory	20	GK	12787
repens Handro	20	Otero	2999
Section *Erectoides*			
archeri Krapov. & W.C. Gregory	20	KCr	34340
benthamii Handro	20	Handro	682

(Continued)

Table 9.1 (Continued)

Section and Species	2*n*	Type specimen Collector[a]	No.
brevipetiolata Krapov. & W.C. Gregory	20	GKP	10138
cryptopotamica Krapov. & W.C. Gregory	20	KG	30026
douradiana Krapov. & W.C. Gregory	20	GK	10556
gracilis Krapov. & W.C. Gregory	20	GKP	9788
hatschbachii Krapov. & W.C. Gregory	20	GKP	9848
hermannii Krapov. & W.C. Gregory	20	GKP	9841
major Krapov. & W.C. Gregory	20	Otero	423
martii Handro	20	Otero	174
oteroi Krapov. & W.C. Gregory	20	Otero	194
paraguariensis			
ssp. *paraguariensis* Chodat & Hassl.	20	Hassler	6358
ssp. *capibarensis* Krapov. & W.C. Gregory	20	HLKHe	565
porphyrocalyx Valls & C.E. Simpson	18	VSPtWiSv	13271
stenophylla Krapov. & W.C. Gregory	20	KHe	572
Section *Extranervosae*			
burchellii Krapov. & W.C. Gregory	20	Irwin, Maxwell & Wasshausen	21163
lutescens Krapov. & Rigoni	20	Stephens	255
macedoi Krapov. & W.C. Gregory	20	GKP	10127
marginata Gardner	20	Gardner	3103
pietrarellii Krapov. & W.C. Gregory	20	GKP	9923
prostrata Benth.	20	Pohl	1836
retusa Krapov., W.C. Gregory & Valls	20	VPtSv	12883
setinervosa Krapov. & W.C. Gregory	20	Eiten & Eiten	9904
submarginata Valls, Krapov. & C.E. Simpson	20	SiW	3729
villosulicarpa Hoehne	20	Gehrt	SP47535
Section *Heteranthae*			
dardani Krapov. & W.C. Gregory	20	GK	12946
giacomettii Krapov., W.C. Gregory, Valls & C.E. Simpson	20	VPzV1W	13202
interrupta Valls & C.E. Simpson	20	VPiFaSv	13082
pusilla Benth.	20	Blanchet	2669
seridoensis Valls, C.E. Simpson, Krapov & R. Veiga	20	VRSv	10969
sylvestris (A. Chev.) A. Chev.	20	Chevalier	486
Section *Procumbentes*			
appressipila Krapov. & W.C. Gregory	20	GKP	9990
chiquitana Krapov., W.C. Gregory & C.E. Simpson	20	KSSc	36027
hassleri Valls & C.E. Simpson	20	SvPiHn	3818
kretschmeri Krapov. & W.C. Gregory	20	KrRa	2273

(*Continued*)

Table 9.1 (Continued)

Section and Species	2*n*	Type specimen	
		Collector[a]	No.
lignosa (Chodat and Hassl.) Krapov. & W.C. Gregory	20	Hassler	7476
matiensis Krapov., W.C. Gregory & C.E. Simpson	20	KSSc	36014
pflugeae C.E. Simpson, Krapov & Valls	20	VOlSiS	13589
rigonii Krapov. & W.C. Gregory	20	K	9459
subcoriacea Krapov. & W.C. Gregory	20	KG	30037
vallsii Krapov. & W.C. Gregory	20	VRGeSv	7635
Section *Rhizomatosae*			
Ser. *Prorhizomatosae*			
burkartii Handro	20	Archer	4439
Ser. *Rhizomatosae*			
glabrata	40		
var. *glabrata* Benth.		Riedel	1837
var. *hagenbeckii* Benth. (Harms ex. Kuntze) F.J. Herm.		Hagenbeck	2255
nitida Valls, Krapov & C.E. Simpson	40	VMPiW	14040
pseudovillosa (Chodat & Hassl.) Krapov. & W.C. Gregory	40	Hassler	5069
Section *Trierectoides*			
guaranitica Chodat & Hassl.	20	Hassler	4975
tuberosa Bong. ex Benth	20	Riedel	605
Section *Triseminatae*			
triseminata Krapov. & W.C. Gregory	20	GK	12881

Source: From Krapovickas and Gregory (1994); Upadhyaya et al., (2005).
[a]Collectors: B, Banks; Bi, Bianchetti; Cl, Claure; Cr, Cristobal; Fa, Faraco; Fr, Fernandez; G, Gregory; Ge, Gerin; H, Hammons; He, Hemsy; Hy, Hn, Heyn; K, Krapovickas; Kr, Kretchmere; L, Langford; M, Moss; Mr, Mroginski; Ol, Oliveira; P, Pietrarelli; Pi, Pizarro; Po, Pott; Pt, Pittman; R, Rao; Ra, Raymon; S, Simpson; Sc, Schinini; Si, Singh; St, Stalker; Sv, Silva; V, Valls; Va, Vargas; Ve, Veiga; Vl, Valente; W, Werneck; Wi, Williams. Others, as listed.

the chromosome behaviour and morphology were reported by Husted (1936). Gregory (1946) reported the first chromosome number of a wild species (*A. glabrata*) as $2n=4x=40$ and also observed diploid species ($2n=2x=20$). Not until 2005 were species having 18 chromosomes discovered (Penaloza & Valls, 2005). Most species in the genus are diploid, but tetraploids exist in sections *Arachis* and *Rhizomatosae*, and several species in sections *Arachis* and *Erectoides* are aneuploid ($2n=2x=18$). Polyploidy is believed to have evolved independently in sections *Arachis* and *Rhizomatosae* (Smartt & Stalker, 1982), and Nelson, Samuel, Tucker, Jackson, & Stahlecker-Roberson (2006) concluded that polyploidy evolved multiple times within section *Rhizomatosae*. Tallury et al. (2005) reported molecular evidence that indicates the diploid section *Rhizomatosae* species (only one known) did not give rise to the tetraploids. Because *A. glabrata* will hybridize with species of both

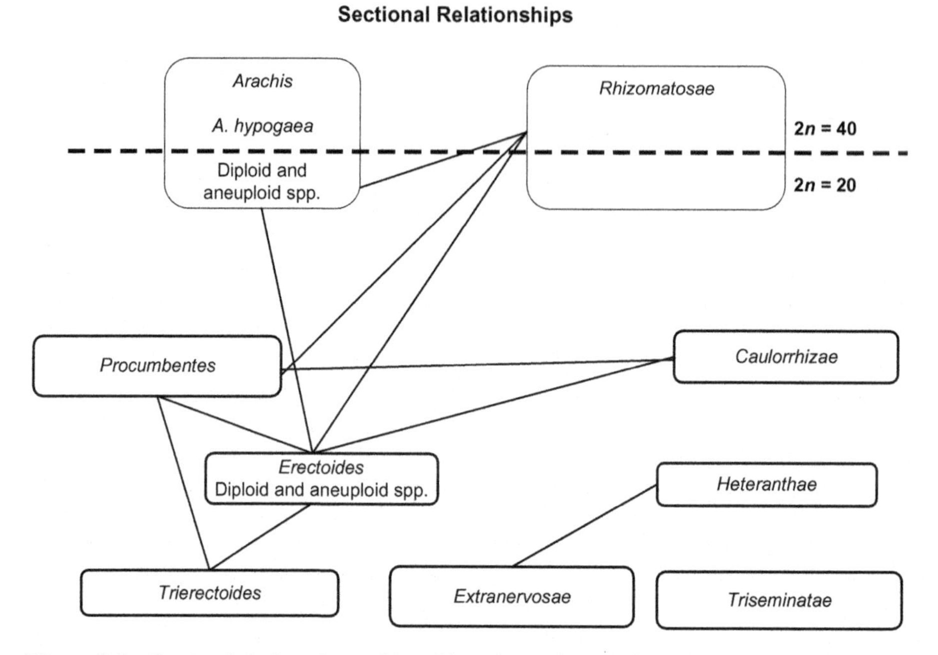

Figure 9.1 Sectional designations of *Arachis* and crossing relationships.
Source: After Krapovickas and Gregory (1994).

sections *Erectoides* and *Arachis*, Smartt and Stalker (1982) concluded that two diploids from sections *Erectoides* and *Arachis* likely hybridized and spontaneously doubled in chromosome number.

Krapovickas and Gregory (1994) concluded that *Erectoides*, *Extranervosae*, *Heteranthae*, *Trierectoides* and *Triseminatae* are 'older' sections, while *Arachis*, *Caulorrhizae*, *Procumbentes*, and *Rhizomatosae* are more 'recent' in origin. The largest group is section *Arachis*, which includes the cultivated species, one other tetraploid (*A. monticola*), 26 diploid ($2n=2x=20$) and three aneuploid ($2n=2x=18$) species.

9.2.2 *Arachis hypogaea*

Cultivated peanut is a New World crop that was widely distributed throughout much of South America in pre-Columbian times. *A. hypogaea* evolved from two diploid species of section *Arachis* approximately 3500 years ago in the southern Bolivia to northern Argentina region of South America (Gregory, W.C., Gregory, M.P., Krapovickas, Smith, & Yarbrough, 1973). Because of the narrow genetic base of the domesticated species, it most likely evolved from a single hybridization event, and the genome has been highly conserved (Young, Weeden, & Kochert, 1996). Domesticated peanut is taxonomically a member of section *Arachis* and will hybridize with other species in the group, with the possible exceptions of *A. glandulifera* (D genome) and the aneuploid ($2n=2x=18$) species. The species is

Table 9.2 *Arachis hypogaea* Subspecific and Varietal Classification

Botanical Variety	Market Type	Location	Traits
hypogaea		Bolivia, Amazon	No flowers on the main stem; alternating pairs of floral and reproductive nodes on lateral branches; branches short; relatively few trichomes
	Virginia		Large seeds; less hairy
	Runner		Small seeds; less hairy
hirsuta	Peruvian runner	Peru	More hairy
fastigiata			Flowers on the main stem; sequential pairs of floral and vegetative axes on branches
	Valencia	Brazil Guaranian Goias Minas Gerais Paraguay Peru Uruguay	Little branched; curved branches
peruviana		Peru, NW Bolivia	Less hairy, deep pod reticulation
aequatoriana		Ecuador	Very hairy, deep pod reticulation; purple stems, more branched, erect
vulgaris	Spanish	Brazil Guaranian Goias Minas Gerais Paraguay Uruguay	More branched; upright branches

Source: After Stalker and Simpson (1995).

highly diplodized, although multivalents occur at a low frequency (Stalker, 1985). At least five secondary constriction types are found among different varieties of the species (Stalker & Dalmacio, 1986), which indicates that chromosome evolution has occurred. *A. duranensis* (A genome) and *A. ipaensis* (B genome) are believed to be the diploid progenitors of the cultivated peanut (Calbrix, Beilinson, Stalker, & Neilson, 2012; Jung et al., 2003; Kochert et al., 1996; Seijo et al., 2004). Further, according to an analysis of cytoplasmic genes *A. duranensis* was the female parent in the original hybrid (Hilu & Stalker, 1995). Secondary centres of diversity developed in South America and tertiary centres in Africa (Gregory et al., 1973; Smartt & Stalker, 1982). The species has evolved into two subspecies and six botanical varieties (Table 9.2). The subspecies are in large part separated morphologically based on the presence or absence of flowers on main stem and regularly alternating vegetative and reproductive nodes on branches.

The tetraploids of section *Arachis* (*A. hypogaea* and *A. monticola*) are completely cross compatible and belong to the same biological species. Whether *A. monticola* is a progenitor or wild escape from cultivation has not been resolved, but cytologically it is more similar to the Spanish types which are more advanced in evolutionary terms than other *A. hypogaea* types (Stalker & Dalmacio, 1986; Stalker & Simpson, 1995).

The cultivated peanut has a more upright growth habit, shorter branches, suppressed hypanthium length, stronger and shorter pegs and pods with the internode between seeds that is suppressed when compared to wild species of the genus (Stalker & Simpson, 1995). The most primitive *A. hypogaea* types have alternating inflorescences, main stems without flowers, prostrate growth habits and long lateral branches, are late maturing and have hairy leaves, two-seeded pods with a beak and small seeds with a long dormancy period (Stalker & Simpson, 1995). Standardized descriptor criteria have been published in the United States (Pittman, 1995) and at the International Crops Research Institute for the Semi-Arid Tropics (ICRISAT) (IBPGR and ICRISAT, 1992). These descriptors have been used to evaluate most of the ICRISAT collection, whereas less than 20% of the US core collection has been assessed.

Spanish and Portuguese explorers carried the peanut to Africa, the Pacific Islands and Asia. There is also evidence that Chinese explorers carried peanut to Asia in pre-Columbian times from the coast of Peru (Mathews, 1983). Peanut was most likely introduced to the United States with the slave trade when ships stopped on the northern coast of Brazil to take on supplies before their voyage north. A small-seeded peanut with a runner habit was the first type successfully cultivated in the southeast United States (Hammons, 1982). The centre of origin for Spanish types is the Guarani region of Argentina, Paraguay and southern Brazil (Hammons, 1982). This type was introduced into the United States from Spain in the early 1870s. The Valencia type spreads from Paraguay and central Brazil (Krapovickas, 1969) and was apparently introduced from Spain to the United States from Valencia, Spain; the name continues to be used for the botanical and market types. The origin of the large-seeded Virginia peanut is not clear, but Gregory, Krapovickas, and Gregory (1980) associated it to the Bolivian and Amazonian geographical regions. The currently grown Virginia-type peanut is believed to be a chance hybrid between a runner type (that was typical of peanuts introduced from Africa into the southeast United States) and an unidentified large-seeded genotype (Hammons, 1982). Seeds of varieties *hypogaea*, *vulgaris* and *fastigiata* have been exchanged widely by peanut breeders across continents, but other varieties have rather limited distributions.

Four market types have been designated in the United States as follows:

1. Runner (subspecies *hypogaea* var. *hypogaea*), with small to medium seeds that range from 550 to 650 mg/seed. The runner market class has become the dominant type grown in the United States, with about 80% of the total production. They have a long growing season of 120 or more days and have a highly indeterminate growth habit. In general, the runners are higher yielding than other market types.
2. The Virginia market class (subspecies *hypogaea* var. *hypogaea*) has large to very large seeds. They have a long growing season and require more soil calcium than the other types

of peanut. A premium is paid for the large seeds in the marketplace and they are generally consumed as in-shell or salted products. Virginia peanuts account for about 15% of the US production.

3. Spanish (subspecies *fastigiata* var. *vulgaris*) peanuts have a similar seed size to runner types, but yields are generally lower and they only account for 4% of the US market. However, they are the preferred type on a global scale, where mechanization is not available for harvest because of their short growing season and bunch growth habit. Spanish types are mostly consumed as peanut candy or salted nuts.
4. Valencia (subspecies *fastigiata* var. *fastigiata*) types usually have three or more seeds and are sold in the shell. They are very sweet as compared to other varieties. However, as a group they are highly susceptible to leaf spots, and yields can be greatly suppressed by diseases. Isleib, Holbrook, and Gorbet (2001) conducted a pedigree analysis of US cultivars and illustrated that the germplasms from both *A. hypogaea* subspecies are in the lineage of most modern cultivars.

Although none of the early molecular marker studies with *A. hypogaea* were very informative (Bertioli et al., 2011 for review), simple sequence repeat (SSR) markers have promise to investigate variation within the cultivated species. Several thousands of microsatellite markers have been developed (Barkley et al., 2007; Krishna et al., 2004; Nagy et al., 2012; Tang et al., 2007; Varshney et al., 2009) and have been used to group the varieties (Jiang et al., 2007; Kottapalli, Burow, M., Burow, G., Burke, & Puppala, 2007). The molecular studies generally confirmed the morphological divisions of varieties in the species. However, varieties *peruviana* and *aequatoriana* accessions grouped more closely with the subspecies *hypogaea*, which conflicts with their placement into subspecies *fastigiata* (Cunha et al., 2008; Freitas, Moretzsohn, & Valls, 2007). Only a few accessions of *peruviana* and *aequatoriana* were available for study, and the results may be an artifact of sample size (Bertioli et al., 2011).

9.3 Genomic Affinities and Speciation

The first published attempt at interspecific hybridization in the genus was between the two tetraploids *A. hypogaea* (section *Arachis*) and *A. glabrata* (section *Rhizomatosae*) (Hull & Carver, 1938), but no hybrids were obtained. Krapovickas and Rigoni (1951) later hybridized *A. hypogaea* with *A. villosa* var. *correntina* and the F_1s were vigorous but sterile. The cultivated peanut has since been hybridized with most species in section *Arachis*. Similar to other genera which have polyploid series, crosses are usually more successful when *A. hypogaea* is used as the female parent. The triploid interspecific hybrids usually have 10 bivalents and 10 univalents, but trivalents are also observed, which indicates that some chromosome homology exists between the A and B genomes. Earlier cytological research identified one significantly smaller chromosome (termed 'A' chromosome) in species of section *Arachis* and a unique chromosome that had a large secondary constriction (termed 'B' chromosome) in the species *A. batizocoi* (Husted, 1936). Hybridization between diploid species was first reported between *A. duranensis* and *A. villosa* var. *correntina* (Raman & Kesavan, 1962) and meiosis was regular. Later studies indicated that hybrids between the species having the small chromosome pair were partially fertile to fertile and most will produce F_2 seeds; however, hybrids between the species with the small chromosome and *A. batizocoi* are sterile (Stalker & Simpson, 1995). Thus, the terminology 'A' and 'B' genome was used in peanut to describe the two genomes. Because the cultivated peanut has one significantly smaller chromosome and a chromosome with a secondary constriction, it was described as an allotetraploid with AABB genomes. Stalker, Dhesi, Parry, and Hahn (1991) crossed a series of species designated as having the A genome with *A. batizocoi* and found that F_1s had many univalents, and bivalents were loosely associated. Hybrids between either A or B genome species with *A. glandulifera* (D genome) also have many univalents and are sterile (Stalker et al., 1991). Thus, there is a considerable amount of cytological differentiation between the three genomes. Gregory, M.P. and Gregory, W.C. (1979) conducted an extensive hybridization programme using 91 *Arachis* collections and reported cross-compatibility relationships among species. Their results indicated that hybridization between species in the same section is more successful than crosses between sections, and F_1s of intersectional crosses were highly sterile. To overcome crossing barriers, complex hybrids have been attempted (Gregory, M.P. & Gregory, W.C. 1979; Stalker, 1981), but fertility was not restored. Thus, introgression to *A. hypogaea* by conventional hybridization is believed to be restricted to members of section *Arachis*. Even within section *Arachis* there can be difficulties obtaining interspecific hybrids due to genomic and/or ploidy differences.

Based on cross-compatibility data, Smartt and Stalker (1982) and Stalker (1991) concluded that genomic groups have evolved in the genus that mostly follow sectional designations (Am – *Ambinervosae*, T – *Triseminatae*, C – *Caulorrhizae*, EX – *Extranervosae*, and E – *Erectoides*, R – *Rhizomatosae*, and A, B and D – *Arachis*). The B genome was recently divided into B, F and K genomes by Seijo et al. (2004) and Robledo and Seijo (2010). Based on rDNA loci and chromosomes with centromeric heterochromatin, Robledo, Lavia, and Seijo (2009) described three karyolotypic subgroups within the A genome and grouped the cultivated peanut with *A. duranensis*, *A. villosa*, *A. schininii* and *A. correntina*. Other studies support placing *A. hypogaea* closely with *A. duranensis* (Bravo, Hoshino, Angelici, Lopes, & Gimenes, 2006; Calbrix et al., 2012; Cuc et al., 2008; Koppolu, Upadhyaya, Dwivedi, Hoisington, & Varshney, 2010; Milla, Isleib, & Stalker, 2005; Moretzsohn et al., 2004). The chromosomes of species with a B genome are karyologically more diverse than those with an A genome (Fernandez & Krapovickas, 1994; Seijo et al., 2004). The B genome does not have centromeric heterochromatin and includes *A. ipaensis* (the B component of *A. hypogaea*), *A. magna*, *A. gregoryi*, *A. valida* and *A. williamsii* (Robledo & Seijo, 2010; Seijo et al., 2004). The D genome species is more distantly removed from *A. hypogaea* than other species of section *Arachis*. Also, molecular analysis indicated that the aneuploids in section *Arachis* are more closely related to the B and D genome species than to A genome species (Tallury et al., 2005). Evolution is apparently continuing in section *Arachis* at a rapid pace and multiple translocations have been observed in diploid accessions of *A. duranensis* (Stalker, Dhesi, & Kochert, 1995) and *A. batizocoi* (Guo et al., 2012; Stalker et al., 1991). At least five different secondary constriction types have been observed in *A. hypogaea*, which were most likely from translocation events (Stalker & Dalmacio, 1986), and this species is also evolving cytologically. Analyses of species in sections other than section *Arachis* have been infrequent. Stalker (1985) reported that the two diploid section *Erectoides* species *A. rigonii* × *A. paraguariensis* hybrids had many univalents, and Krapovickas and Gregory (1994) later placed these species in different sections. Intersectional hybrids were reported by Mallikarjuna (2005), who used *in vitro* techniques, but the hybrids have not been used for cultivar development.

In addition to morphological and cross-compatibility studies, molecular investigations have been used to better clarify the understanding of phylogenetic relationships among peanut species. Most of these investigations have involved species in section *Arachis* because of the importance of *A. hypogaea*. Many molecular systems have been utilized, including isozymes (Lu & Pickersgill, 1993; Stalker, Phillips, Murphy, & Jones, 1994), seed storage proteins (Bianchi-Hall, Keys, & Stalker, 1993; Liang, Luo, Holbrook, & Guo, 2006; Singh, Krishnan, Mengesha, & Ramaiah, 1991), restriction fragment length polymorphisms (RFLPs) (Kochert, Halward, Branch, & Simpson, 1991; Paik-Ro, Smith, & Knauft, 1992), amplified fragment length polymorphisms (AFLPs) (Milla et al., 2005); SSRs (He et al., 2005; Hong et al., 2010; Hopkins et al., 1999; Nagy et al., 2012), randomly amplified polymorphic DNA (RAPDs) (Halward, Stalker, Larue, & Kochert, 1992; Hilu & Stalker, 1995; Lanham, Fennell, Moss, & Powell, 1992) and *in situ* hybridization (Raina & Mukai, 1999). All of the studies have indicated that the cultivated peanut has significantly less molecular variation than diploid species, which supports the hypothesis that *A. hypogaea* originated from a single hybridization event. Additionally, there has been little or no apparent introgression from the diploid species to *A. hypogaea* (Kochert et al., 1996).

As opposed to the cultivated species, large amounts of molecular variation have been documented among wild species of the genus. Although there have been differences observed among marker systems regarding species relationships, and there remain questions about species positions in sectional groupings (Friend, Quandt, Tallury, Stalker, & Hilu, 2010), the molecular data generally fits the sectional relationship model proposed by Krapovickas and Gregory (1994). However, questions remain about several sections. For example, Hoshino et al. (2006) used microsatellites to evaluate species in the nine peanut accessions, and while most species grouped as expected, several species in the *Procumbentes* grouped with species from section *Erectoides*, and others clustered into sections *Trierectoides* and *Heteranthae*. Galgaro, Lopes, Gimenes, Valls, and Kochert (1998) also indicated that species in section *Heteranthae* did not group together. Friend et al. (2010) conducted a more comprehensive investigation of *Arachis* species and also found that sections *Extranervosae*, *Triseminatae* and *Caulorrhizae* each separated into distinct groups based on trnT-trnF sequences; but species in sections *Erectoides*, *Heteranthae*, *Procumbentes*, *Rhizomatosae* and *Trierectoides* formed a major lineage. Species in section *Arachis* grouped into two major clades, with the B and D genome species plus 18 chromosome aneuploids being in one group and the A genome species in the other.

9.4 Erosion of Genetic Diversity from the Traditional Areas

Genetic diversity in *A. hypogaea* has dramatically decreased in most areas where peanut is cultivated because improved cultivars are replacing landraces (Williams, 2001). The trend has accelerated since Williams's (2001) review of peanut genetic conservation efforts. Wild species diversity also continues to decline as native habitats are destroyed at a rapid pace due to urbanization, farmers opening new areas for cultivation, excessive grazing and other human activities. Genetic losses are most dramatic in Brazil and Bolivia, but occur in all areas where *Arachis* species are found.

In an analysis of the distribution of *Arachis* species using 2175 observations of wild species locations in conjunction with modeling based on climatic adaptation to extrapolate geographical distributions of species, Jarvis et al. (2003) concluded that wild peanut species potentially inhabit 5 million km^2 (with 364,000 km^2 having four or more species growing sympatrically). Like many other genera, most of the species accessions acquired were found along roads, which leaves vast areas of South America unexplored for peanut germplasm. The authors predicted gaps in collections and investigated species distributions and land use. Based on restricted ranges of individual species and land use pressures by human activities, several species were identified as being under threat of extinction, including *A. archeri*, *A. setinervosa*, *A. marginata*, *A. hatschbachii*, *A. appressipila*, *A. villosa*, *A. cryptopotamica*, *A. helodes*, *A. margna* and *A. magna*. Other species are poorly represented in collections (i.e. where only one or a few accessions are maintained), for example *A. monticola*, *A. ipaensis*, *A. cruziana*, *A. williamsii*, *A. martii*, *A. pietrarelli* and *A. vallsii*. Other species, such as *A. burkartii*, *A. triseminata*, *A. tuberosa* and *A. dardani*, have experienced significant reductions in range due to agriculture land use (Jarvis et al., 2003). Their study suggested that priority for *ex situ* conservation efforts should be in areas southeast of Cuiaba, Brazil and around San Jose de Chiquitos in Bolivia.

9.5 Status of Germplasm Resources Conservation

Peanut germplasm has been collected in South America for a long time. There was a concentrated effort beginning in the 1950s to systematically acquire *Arachis* genetic resources. The first major collection trips were in 1959 and 1960 by W.C. Gregory (North Carolina State University), A. Krapovickas (Instituto de Botánica del Nordeste, Argentina) and J.R. Pietrarelli (Estación Experimental INTA Manfredi, Argentina), followed by two additional expeditions by W.C. Gregory during 1961 and 1967, and then one in 1968 by R.O. Hammons (USDA, GA) and W.R. Langford (USDA, GA). Thirty-five additional collection trips were made to collect both cultivated and wild peanuts between 1976 and 1992 (Stalker & Simpson, 1995) and several more since that time. National scientists in Argentina and Brazil have greatly expanded their national collections since 2000, but materials have remained in country. *In situ* conservation of genetic resources was not a high priority in peanut during the twentieth century because *ex situ* conservation was well funded and a large number of collection trips to South America resulted in many hundreds of new accessions of both cultivated and wild peanut species. However, *in situ* conservation efforts have increased in Brazil during recent years (J.F.M. Valls, personal communication).

Collection and exchange of unimproved peanut germplasm was unrestricted prior to implementation of the Convention on Biological Diversity in 1993. The Convention was ratified by 179 countries; since then, laws restricting access to genetic resources have been widely implemented (Williams, K.A. & Williams, D.E. 2001). This is important to peanut because the nations in South America, where much of the diversity for cultivated peanut and all of the wild species exist, have restricted collection and export of peanut. The Andean Pact also was implemented in 1996 whereby five countries (Bolivia, Colombia, Ecuador, Peru and Venezuela) established strict provisions to restrict germplasm access. This pact has had a significant negative impact on conservation of peanut genetic resources (Williams, K.A. & Williams, D.E. 2001). Although not an Andean Pact nation, Brazil also implemented very strict constraints for collecting and exchanging germplasm for both the international and national Brazilian scientists. Since the pacts were signed in the 1990s, germplasm exchange from South America has been very limited. The exception is cultivated landraces in Ecuador, which were obtained during the late 1990s (Williams, K.A. & Williams, D.E. 2001).

A memorandum of understanding was signed by the USDA and ICRISAT to facilitate germplasm exchange (Shands & Bertram, 2000), whereby both institutions agreed not to claim ownership or intellectual property rights on exchanged germplasm. This is important because ICRISAT is the international centre for peanut genetic resources. Likewise, when germplasm is passed through the USDA to state or private institutions the same policy applies (Williams, K.A. & Williams, D.E. 2001).

Priorities for future collection of *A. hypogaea* are the landraces found in Central and South America, Africa, Asia and China, where the primitive types are being replaced by elite cultivars (Stalker & Simpson, 1995). More specific collection priorities were presented by Valls, Ramanatha, Simpson, and Krapovickas (1985) for *Arachis* species in Brazil (which are still valid today), including (i) the northwest state of Mato Grosso; (ii) the states of Acre, Rondonia, Maranhao, Ceara, Rio Grande do Norte and Paraiba, the northwest region of Goias and the northern region of Piaui and (iii) the southeast Amazon region of Brazil. Collection in Uruguay is also a priority. In addition, areas such as eastern Bolivia and northwestern Paraguay are undercollected for *Arachis* species (Williams, 2001). For cultivated peanut, the northern and western areas of Brazil, Colombia, Venezuela and the Guyanas have not been systematically collected, and many areas in Mexico, Bolivia and Ecuador are undercollected (Williams, 2001). Accessions of varieties *hirsuta*, *peruviana* and *aequatoriana* are poorly represented in germplasm collections and priority needs to be placed to obtain additional materials of these types. Both India and China have excellent plant improvement programmes; improved cultivars have taken over most of the production areas while at the same time replacing traditional cultivars. Much of the traditional genetic diversity in Asia has already been lost (Williams, 2001).

A. hypogaea genetic resources are preserved at multiple locations; Pandey et al. (2012) summarized information about these collections. The largest single collection is at ICRISAT, where 15,445 accessions are held from 93 countries (Upadhyaya, Ferguson, & Bramel, 2001). Other large collections are held by the National Bureau of Plant Genetic Resources (NBPGR) in India (14,585 accessions); the Directorate of Groundnut Research Junagarh (DGRJ) in India, where 9024 accessions are maintained; the Oil Crops Research Institute (OCRI) in China (8083 accessions) and the US Department of Agriculture with 9917 accessions, of which approximately half are unimproved landraces collected in South America (Holbrook, 2001). Additional collections are held by other institutions in the United States, Brazil and Argentina. There is significant duplication of accessions among all of the above-mentioned institutions because germplasm exchange has been extensive since the mid 1970s.

Collection priorities of *A. hypogaea* at ICRISAT are based on the numbers of accessions collected in a particular region, combined with diversity studies of morphological and molecular data (Upadhyaya, Ferguson, et al., 2001). Although the primary centre of diversity of cultivated peanut is northern Argentina and southern Bolivia, the regions are represented by only 368 and 444 accessions, respectively, in the ICRISAT collection (Upadhyaya, Ferguson, et al., 2001). A large part of their collection was obtained from the Indian subcontinent and several African countries (Upadhyaya, Ferguson, et al., 2001); there remain significant gaps in the collection in Asia and Africa. Priority areas designated by ICRISAT include Bolivia, Argentina, Brazil, Paraguay, Peru, Uruguay, Ecuador, Laos, China, Angola, Madagascar, Namibia and South Africa. Also, the varieties *aequatoriana*, *hirsute* and *peruviana* are under-represented at ICRISAT.

By the year 2000, more than 3400 *Arachis* species accessions were documented as seeds, plants or herbaria specimens (Stalker, Beute, Shew, & Isleib, 2002). New species have been discovered and preserved in germplasm collections in Argentina, Brazil, United States, ICRISAT and the International Centre for Tropical Agriculture (CIAT) (Simpson, 1991; Valls et al., 1985). With a few exceptions (e.g. Argentina and Brazil), conservation within the country has not been a priority (Williams, 2001). Presently, about 1300 *Arachis* species accessions are available in germplasm collections as plants or seeds (Stalker et al., 2002). The largest wild species collections are located at Embrapa Recursos Genéticos e Biotecnologia, Brazil (1200 accessions), Texas A&M University (1200 accessions); the USDA (607 accessions); ICRISAT (477 accessions); IBONE (472 accessions) in Argentina and at North Carolina State University (428 accessions) (Pandey et al., 2012). Duplication also exists among collections for *Arachis* species, and approximately 800 entries are maintained in the United States (Stalker & Simpson, 1995).

9.6 Germplasm Maintenance and Evaluation

9.6.1 Maintenance

9.6.1.1 Arachis hypogaea

Maintenance of the domesticated collection is rather straightforward except for handling large numbers of accessions. However, many of the accessions are susceptible to tomato spotted wilt virus and other diseases, and seed regeneration is highly problematic in areas where there is a prevalence of diseases that kill susceptible genotypes. When accessions are introduced into the United States, they are quarantined in the greenhouse before being taken to the field for larger increases. Accessions are stored at −18°C in vacuum sealed packets for 15–20 years before regeneration. Many accessions in the US collection were introduced as seed mixtures, so large plots are needed to maintain variation at the original gene frequencies. Although peanut is classified as a self-pollinated species, outcrossing occurs where bees are prevalent. Virginia types have lower outcrossing rates (1–3%) than Valencia types, which can be as high as 8% (Knauft, Chiyembekeza, & Gorbet, 1992). This can be problematic in breeding or seed-increase nurseries.

9.6.1.2 Arachis Species

Maintenance of the *Arachis* species is more difficult than for *A. hypogaea* and is accomplished either in the greenhouse or field. Stalker and Simpson (1995) reported that about 28% of the accessions in cultivation are maintained vegetatively because of poor seed set and nearly 25% of the species from which seed can be obtained under nursery conditions have fewer than 50 seeds in storage. The situation has not significantly changed since 1995. Especially problematic for long-term preservation are perennial accessions that produce rhizomes or tubers, which include all the species in sections *Rhizomatosae* and *Extranervosae* because they produce very few seeds under cultivation. Other species such as *A. guaranitica* and *A. tuberosa* go into a permanent dormancy when seeds are dried, but seeds of *A. tuberosa* have been maintained for nearly 2 years when stored in moist sphagnum moss at room temperature (Stalker & Simpson, 1995). Light quality and day length also have significant effects on reproduction and seed development in peanut.

The field nursery system used for the *Arachis* species at North Carolina State University is to initially germinate seeds in the greenhouse and then transplant accessions into small blocks where peanut has not previously been grown. Plant blocks are separated by 5–10 m in all directions, and cross-compatible types are not planted in adjacent plots within or between rows. The planting scheme also avoids the problem of pegs growing into plots of other accessions. Harvest is completed by sifting the soil in plots to recover pods. In large part because regeneration of *Arachis* species requires a large amount of land, very sandy soil and intensive labour, very few investigators regenerate the *Arachis* species collection in the field even though many more seeds can be obtained than in greenhouses. Because of the difficulties associated with propagating many of the *Arachis* species, either in the field or in greenhouses, many accessions have been lost in collections. Thus, it is critical that multiple locations be used to maintain the wild species of peanut to assure preservation of the genetic resources.

9.6.2 Evaluation

Several review articles have been published that summarize genetic resources of the domesticated peanut and related *Arachis* species (Dwivedi et al., 2003; Dwivedi et al., 2007; Holbrook & Stalker, 2003; Isleib & Wynne, 1992; Singh & Simpson, 1994; Stalker & Simpson, 1995; and Tillman & Stalker, 2009), so only a brief review will be presented in this chapter. Standards for evaluation of peanut also have been published by IBPGR and ICRISAT (1992) and the USDA (Pittman, 1995).

Most of the US *A. hypogaea* collection has been evaluated for resistance to early and late leaf spots and rust (see Holbrook & Stalker, 2003 for review) and few other traits. Moderate levels of resistance have been identified in the *A. hypogaea* collection, but extremely high levels of resistance apparently do not exist in the germplasm collection for most of the important peanut pathogens (Stalker & Moss, 1987; Stalker & Simpson, 1995). However, extremely high levels of resistance to both diseases and insects have been identified in *Arachis* species (see Stalker & Moss, 1987 for review). Although large numbers of accessions have been evaluated for agronomically useful traits in the USDA and ICRISAT collections, relatively few accessions have been utilized by breeders for cultivar development in the United States (Isleib et al., 2001).

The greatest evaluation efforts of peanut have been at ICRISAT and Dwivedi et al. (2007) summarized their research at ICRISAT. One hundred forty-three accessions were found resistant to peanut rust (Mehan, Reddy, Vidyasagar Rao, & McDonald, 1994); 54 were resistant to late leaf spot (Subrahmanyam et al., 1995); 10 were resistant to *Aspergillus* flavus infection and two accessions did not produce aflatoxin after infection (Mehan, 1989; Mehan, McDonald, Ranakrishna, & Williams, 1986) and 154 were resistant to groundnut rosette virus (Subrahmanyam, Anaidu, Reddy, Kumar, & Ferguson, 2001). Mehan et al. (1986) also identified four *Arachis* species that are resistant to aflatoxin production. No resistance was identified for peanut strip virus (PStV) in the cultivated collection (Prasad Rao et al., 1991). Subrahmanyam et al. (2001) found 12 *Arachis* species accessions to be immune to groundnut rosette virus. *A. diogoi* was the only species identified with no infection to peanut bud necrosis virus (Subrahmanyam et al., 1995); this species is also the only one with immunity to tomato spotted wilt virus. None of 7000 accessions screened for peanut clump virus (PCV) had useful resistance, whereas four *Arachis* accessions of *A. kuhlmannii*, *A. duranensis* and *A. ipaensis* were immune. ICRISAT scientists also have evaluated *Arachis* species for late and early leaf spots and they identified highly resistant materials (Upadhyaya, Ferguson, et al., 2001).

Insect resistance was identified for jassids (*Empoasca kerri* Pruthi), thrips (*Thrips palmi* Karny), aphids (*Aphis craccivora* Koch), leaf minor (*Aproaerema modicella* Deventer) and termites (*Odontotermes* spp.). Several accessions were identified with multiple resistances for insects (Upadhyaya, Ferguson, et al., 2001). Wrightman and Ranga Rao (1994) reported several *Arachis* species with high levels of resistance to pests, including entries in *A. duranensis*, *A. cardenasii*, *A. paraguariensis* and *A. pusilla*. Researchers at ICRISAT evaluated about 8000 accessions for oil content and 5501 accessions for protein content and found 66 lines with more than 50% oil and 125 lines with more than 30% protein (Upadhyaya, Ferguson, et al., 2001). Nageswara Rao, Udaykumar, Farquhar, Talwar, and Prasad (1995) evaluated crop growth rate, water use efficiency and assimilate partitioning. Because of the large sizes of the collections, core collections have been developed to facilitate evaluation for diseases and other agronomic traits. Holbrook, Anderson, and Pittman (1993) analysed morphological and geographical distributions of the USDA germplasm collection and developed a core collection represented by 831 accessions. This core collection has four of the six varieties of *A. hypogaea*; the remaining two (*peruviana* and *aequatoriana*) need to be added. The core collection has been evaluated for all of the US descriptors. Upadhyaya, Ortiz, Bramel, and Singh (2001) developed a larger core collection from the ICRISAT germplasm, comprising 1704 accessions. Jiang et al. (2008) also developed a Chinese core collection with 576 accessions. Mini-core collections, representing approximately 10% of the US and ICRISAT core collections, were then developed as a subset to facilitate evaluation research. There are 112 accessions in the US mini-core (Holbrook & Dong, 2005) and 184 in the ICRISAT mini-core (Upadhyaya, Bramel, Ortiz, & Singh, 2002). Jiang et al. (2010) developed a mini-core of the Chinese core collection with 298 accessions. Dwivedi, Puppala, Upadyaya, Manivannan, and Singh (2008) also developed a core collection for the Valencia-type peanuts. Evaluations of the US core collection have identified new sources of resistance for Cylindrocladium black rot and early leaf spot (Isleib, Beute, Rice, & Hollowell, 1995), tomato spotted wilt virus (Anderson, Holbrook, & Culbreath, 1996), root-knot nematode (*Meloidogyne arenaria* (Neal) Chitwood) and preharvest aflatoxin contamination (Holbrook, Bruniard, Moore, & Knauft, 1998), rhizoctonia limb rot (*Rhizoctonia solani* Kuhn) (Franke, Brenneman, & Holbrook, 1999), Sclerotinia blight (*Sclerotinia minor* Jagger) and pepper spot (*Leptosphaerulina crassiasca* (Sechet) Jackson and Bell) (Damicone, Jackson, Dashiell, Melouk, & Holbrook, 2003). In addition, the mini-core has been evaluated for traits that are expensive to analyse such as for microsatellite markers (Kottapalli et al., 2007; Wang et al., 2011) and oil content (Wang, Barkley, Chinnan, Stalker, & Pittman, 2010). Germplasm evaluations of the core accessions at ICRISAT identified accessions with early maturity (Upadhyaya, Reddy, Gowda, & Singh, 2006), tolerance to low temperatures (Upadhyaya, Ortiz, et al., 2001) and drought tolerance (Upadhyaya, Mallikarjuna Swamy, Goudar, Kullaiswaym, & Singh, 2005). Importantly, the US core collection was evaluated for usefulness for identifying additional germplasm in the entire collection by extrapolating core collection data for late leaf spot to the entire collection. Holbrook and Anderson (1995) found that evaluating the core collection is a good indicator of late leaf spot resistance in the entire collection.

9.7 Use of Germplasm in Crop Improvement

Plant introductions have been important to peanut production, in large part for resistance to diseases such as Sclerotinia blight, root-knot nematode and tomato spotted wilt virus (Isleib et al., 2001). Most of the runner market types can be traced back to four ancestors that were used in early breeding programmes, including the two variety *hypogaea* lines Dixie Giant and Basse and the two variety *vulgaris* lines Small White Spanish and Spanish 18-38 (Isleib et al., 2001). The ancestry of the Virginia market class included those four lines and a large-seeded selection of Jenkins Jumbo in the Florida programme. Basse (PI 203396) was introduced from Gambia and is in 32 of 41 runner-type cultivars (as of 2000) and was the source of late leaf spot, tomato spotted wilt virus and southern stem rot (*Sclerotium rolfsii* Sacc.) resistance in runner-type cultivars. PI 109839 was collected from Venezuela in 1935 and is the source of early leaf spot resistance in most cultivars. In all, 13 plant introductions were in the pedigrees of most US cultivars before 2000 (Isleib et al., 2001). Seven introductions serve as the basis of Spanish-type cultivars (Isleib et al., 2001). Most runner- and Virginia-type peanuts have a mixture of subspecies *hypogaea* and variety *vulgaris* and to a lesser extent from variety *fastigiata*.

Only 119 cultivars were released in theUnited States (276 worldwide) before 2000 (Isleib et al., 2001; Paterson et al., 2004). Hammons (1976) and Knauft and Gorbet (1989) characterized the peanut crop as being genetically vulnerable to disease and insect pests. Historically, only a few cultivars have dominated the production areas, especially in the southeast (Isleib et al., 2001). For example, during the 2012 growing season, there were 13 runner, 2 Spanish, 11 Virginia and 1 Valencia market type cultivars grown in the United States. However, Georgia-06G accounted for 65.6% of the runner production and 50.7% of the total US peanut production area. One Spanish cultivar accounted for 80% of this market type production. Four cultivars had more than 10% of the production area of the Virginia market type, with Bailey having 30.5%. Thus, the US germplasm base remains rather narrow. In other countries there is also a predominance of one or a few cultivars being grown across large production regions, and many of these are replacing lower yielding landraces.

9.8 Limitations in Germplasm Use

Peanut breeding is largely accomplished by the public sector breeders, which have relatively small programmes as compared to large, privately owned seed companies. Hybridization is a laborious process because individual flowers need to be emasculated and then hand-pollinated, after which one or sometimes two seeds are produced. High temperatures or low humidity can significantly decrease fertilization percentages. The peanut has a long generation time (120–150 or more days), so at most there can be two plant generations per year in a breeding programme. Utilizing genetic materials in different market classes can result in poor quality or unacceptable seed or pod traits at the breeder's location. For example, the Spanish types have a shorter growing season, which is important in areas where early frost will damage the crop, but they are lower yielding than materials in the runner and Virginia market classes; Spanish peanuts are also more susceptible to tomato spotted wilt virus, leaf spots and other diseases. Crosses among the market classes can increase diversity but also cause problems with market quality and reduce yield potential.

The most important limitation to germplasm use in peanut is identifying lines with sufficiently high levels of resistance to utilize for crop improvement. Land resources and personnel have not been available for systematic evaluation of the US germplasm collection for many disease and insect resistances or other agronomic traits, so a limited number of genotypes have been utilized in breeding programmes. Further, most 'resistant' *A. hypogaea* accessions have only moderate levels of resistance, many of which express multigenic inheritance and are difficult to incorporate into elite breeding materials. Other sources of resistance are extremely difficult to evaluate in field plots (such as aflatoxins and diseases caused by soil-borne fungi). Although high levels of resistance to immunity have been identified in wild species, only members of section *Arachis* will hybridize with *A. hypogaea*, and even in this group there are barriers to germplasm use.

9.9 Germplasm Enhancement Through Wide Crosses

Because the domesticated peanut is an allotetraploid with two genomes and the species being utilized for introgression are diploids, sterility barriers result from ploidy differences and genomic incompatibilities between the species. Traits of interest from *Arachis* species have been difficult to follow in progenies of interspecific hybrids because of low population sizes and high sterility levels in progenies. Utilizing molecular markers associated with traits of interest may help overcome many of these problems, but unfortunately only few molecular markers have been available to enhance selection efficiency. Introgression from *Arachis* species to *A. hypogaea* appears to be in large blocks (Garcia, Stalker, & Kochert, 1995; Nagy et al., 2012) rather than as single genes or small chromosome segments. Thus, linkage drag of undesirable traits can restrict the use of genetic resources.

The first peanut cultivar released from interspecific hybridization was from a cross between *A. hypogaea* and the second tetraploid species in section *Arachis* (*A. monticola* Krapov. & Rigoni). Biologically, *A. monticola* could be considered a weedy subspecies of *A. hypogaea*. Spancross was released by Hammons (1970); Tamnut 74 was later released by Simpson and Smith (1975). Neither of these cultivars had phenotypic characters that could be identified as being derived from the wild species, which is not surprising because *A. monticola* has most of the same disease and insect problems as found in *A. hypogaea*.

Several methods have been utilized to create populations of fertile *A. hypogaea* interspecific hybrids and restore plants to the tetraploid level. First, hybrids can be made by crossing *A. hypogaea* with diploids to produce triploid ($3x=30$) F_1s, after which cuttings can be colchicine-treated to restore fertility at the hexaploid ($6x=60$) level. Many triploids will also produce a few seeds through the fusion of unreduced gametes, especially if they are placed in the field for long periods of time. Backcrossing the hexaploids with *A. hypogaea* results in pentaploids ($5x=50$) that are usually vigorous but partially sterile. Additionally, they produce few flowers and are difficult to use in the crossing programmes, but they sometimes yield a few seeds and the ploidy level stabilizes at the tetraploid level. A major problem with this scheme has been the few seeds produced at the hexaploid and pentaploid levels; the lack of selection methods during the semi-sterile generations for traits of interest has resulted in tetraploid lines without traits of interest for crop improvement. Hundreds of tetraploid progenies have been recovered with many diploid species, but to date, no useful germplasm has resulted from backcrossing hexaploids with *A. hypogaea*. Although backcrossing hexaploids with diploids will theoretically drop the chromosome number to the tetraploid level in one generation, these $6x \times 2x$ crosses (or reciprocals) have not produced viable progenies.

An alternative method to backcrossing hexaploids with the cultivated species is to allow $6x$ plants to self-pollinate and, by selecting fertile progenies, a few plants may spontaneously lose chromosomes and stabilize at the 40-chromosome level. The loss of chromosomes appears to be infrequent and random, but the advantage of this procedure is associating chromosomes in different species at a high ploidy level which can increase the frequency of recombination. For example, after *A. hypogaea* × *A. cardenasii* hexaploids were selfed for five generations they produced 40-chromosome progenies that were highly variable for seed size, colour and other morphological traits (Company, Stalker, & Wynne, 1982). Garcia et al. (1995) analysed introgression from *A. cardenasii* to *A. hypogaea* with RFLPs and found wild species-specific markers on 10 of 11 linkage groups on the diploid RFLP map developed by Halward, Stalker, and Kochert (1993). Most of the introgression (88%) was apparently in the A genome of *A. hypogaea*, with the remaining 12% in the B genome. Germplasm lines have been released from this cross with resistance to early leaf spot, nematodes and several insect pests (Stalker et al., 2002; Stalker & Lynch, 2002; Isleib et al., 2006). The cultivar Bailey was released after utilizing these lines as sources of multiple disease resistances (Isleib et al., 2010).

A second method to introgress germplasm from diploid species to *A. hypogaea* is to first double the chromosome number of the diploid species to the tetraploid level. This method has the advantage of avoiding several generations of mostly sterile hybrids. Further, recovering tetraploids is much faster than by going through the triploid–hexaploid procedure; autotetraploids generally have low vigour and when annual species are used they are short-lived. Ideally, A and B genome species would be hybridized at the diploid level and then the chromosomes doubled to produce AABB genome allopolyploids to be crossed with the cultivated species. However, chromosome doubling of the sterile AB genome diploids can be problematic. Examples of success with this methodology are TxAG-6 and TxAG-7 (Simpson, Nelson, Starr, Woodard, & Smith, 1993) which originated from the complex hybrid 4×[*A. batizocoi* (B genome)×(*A. cardenasii* (A genome)×*A. diogoi* (A genome))]. TxAG-6 had very good nematode resistance, but also significant linkage drag, which resulted in low yields and poor seed and pod quality. RFLP markers linked to the nematode gene conferring resistance were used to select favourable genotypes. The nematode-resistant cultivars COAN (Simpson & Starr, 2001) and NemaTAM (Simpson, Burrow, Patterson, Starr, & Church, 2003) were released by introgressing genes from TxAG-6. By using SSR markers, Nagy et al. (2012) showed that recombination was greatly reduced in the chromosome area where the nematode-resistant gene is located, due to a large introgressed segment from the wild species that comprised one-third to one-half of a chromosome in hybrids. The same procedure resulted in release of the nematode-resistant cultivar Tifguard (Holbrook, Timper, Culbreath, & Kvien, 2008), but it was highly susceptible to tomato spotted wilt virus and production has been limited.

9.10 Peanut Genomic Resources

9.10.1 Tool Development

Molecular research with peanut began in the 1980s with the analysis of proteins and isozymes variation in *A. hypogaea*, but there was little variation observed. In contrast, large amounts of variation exist in *Arachis* species for these marker systems. The same trend was found for RFLPs, RAPDs and AFLPs (see Stalker, Weissinger, Milla-Lewis, & Holbrook, 2009 for review). Prior to 2005, there were only a few hundred markers available for peanut. AFLPs were the first molecular marker system used to differentiate closely related peanut cultivars (Herselman, 2003), and Moretzsohn et al. (2005) later used SSR markers to separate the cultivated lines. Large-scale SSR marker development was initiated in Asia and the United States, and more than 6000 SSR markers are now available (see Pandey et al., 2012 for review). Pandey et al. (2011) also developed a set of 199 highly informative SSR markers that should be widely used in breeding programmes. A 20 SSR marker set was developed to analyse 300 cultivated accessions by Upadhyaya et al. (2002), which should serve as a useful reference for future molecular research with peanut.

DArT markers were developed in a cooperative programme between researchers in Australia, India, France and Brazil with about 15,000 markers (Pandey et al., 2012). Analysis of diploid and tetraploid species indicated that there was a moderate level of polymorphism in the diploids (Kilian, 2008; Varshney, Glaszmann, Leung, & Ribaut, 2010), but they are not highly useful for analysing the tetraploid genome. Thus, like other types of markers, they may be useful for gene introgression research but not for cultivar development. More than 2000 SNPs have been discovered at the University of Georgia (Pandey et al., 2011; Guo et al., 2012), and Illumina GoldenGate SNP arrays have been developed for diploid peanuts. Unfortunately, because of homology between the A and B genomes in diploids, the arrays may not be highly useful for analysis of *A. hypogaea* (Pandey et al., 2011).

9.10.2 Molecular Maps of Peanut

Several molecular maps have been produced in peanut with different marker systems. The first map used RFLPs and utilized variation between the diploid species *A. stenosperma* × *A. cardenasii*, where a total of 117 RFLP markers were mapped into 11 linkage groups (Halward et al., 1993). Moretzsohn et al. (2009) and Guo et al. (2012) compared linkages on A and B genome maps and found a considerable amount of homology. The first tetraploid map was created by using progenies of a cross between *A. hypogaea* and the interspecific hybrid TxAG-6, where 383 markers were mapped (Burow, Starr, Simpson, & Paterson 1996). A summary of the 22 published maps to date using RFLPs, AFLPs, SSRs, SNPs, SCAR and CAPS markers can be found in Pandey et al. (2012). All but one of the maps has fewer than 400 markers (range 12–449). Nagy et al. (2012) published a more saturated map having 2319 SNP, SSR and single-stranded DNA conformation polymorphism (SSCP) markers for a cross between two *A. duranensis* accessions. Guo et al. (2012) then compared high-density A and B genome maps and found a large amount of synteny, but also several inversions, translocations and other chromosomal structural changes. The data also has been used to develop markers closely associated with a nematode resistance gene (Nagy et al., 2010). For the domesticated peanut, several maps have been developed using SSRs (see Pandey et al., 2011) and most recently Qin et al. (2012) made an integrated *A. hypogaea* map.

9.10.3 Transcriptome Resources

Expressed sequence tag (EST) development is an alternative for developing gene-based markers and for identifying genes for expression of traits. To date, 246,733 ESTs are in the public domain at the National Center for Biotechnology Information (http://www.ncbi.nlm.nih.gov/). The ESTs have been developed from a variety of plant tissues including seeds (57.6%), leaves (19.0%) and roots (23.2%), and range in length from 37 to 2038 bp (Pandey et al., 2012). Non-public ESTs also have been developed, most notably at the University of Georgia, where more than 350MB of transcript sequences from 17 *A. hypogaea* genotypes resulted in about one million ESTs. A consensus transcription assembly was developed with 211,244 contigs (Pandey et al., 2012). Guimarães et al. (2011) developed 743,232 additional ESTs by using the diploid species *A. duranensis* and *A. stenosperma*, which were placed under stresses due to *C. personatum* and water deficit. From these ESTs they produced 39,626 unigenes that were annotated for the species, and since the parents were the same ones as used for a reference map by Moretzsohn et al. (2005) it is a highly valuable genetic resource.

9.10.4 Whole Genome Sequences

The peanut genome is complex due to its allopolyploid nature. The genome size is large, with 2.9 pg DNA per haploid genome, about 27% highly repetitive DNA and 37% middle-repetitive DNA (Paterson et al., 2004). A Peanut Genome Consortium (PGC) (http://www.peanutbioscience.com/peanutgenomeproject.html) has been formed to address the technical problems and to develop a strategy to sequence the peanut genome. Parallel sequencing of diploid progenitor species will be necessary to sort out the A and B genomes present in *A. hypogaea*. This is a multinational effort not only to develop a sequenced genome, but also to develop tools that can be utilized by breeders for cultivar development. It is anticipated that the domesticated peanut will be sequenced and the information made available within a few years.

9.10.5 Linking Agronomic Traits with Markers

To date, the number of genes associated with molecular markers in peanut is small, but the large number of molecular markers becoming available has great potential for utilizing in a crop improvement programme. Bertioli et al. (2003) described numerous linkages of resistant genes in peanut. Pandey et al. (2012) listed quantitative trait loci (QTLs) for some of the important traits found in the cultivated peanut. Chu et al. (2011) outlined a breeding scheme to utilize marker-assisted selection (MAS) to pyramid nematode resistance and the high oleic acid trait in peanut cultivars, and the system has greatly increased efficiency for developing breeding lines. In addition to markers being useful for associating with specific traits, they also may be useful for following introgression from *Arachis* species to *A. hypogaea*. This is important because recombination between the cultivated genomes and those of other species is rare, thus restricting selection for desired traits in interspecific hybrid derivatives (Holbrook & Stalker, 2003). Guimarães et al. (2010, 2011) identified eight genes in *A. stenosperma* roots that provide resistance to *M. arenaria*; QTLs for resistance to late leaf spot also have been identified (Leal-Bertioli et al., 2009) in section *Arachis* species.

9.11 New Sources of Genetic Diversity

9.11.1 Targeting Induced Local Lesions in Genomes

Targeting Induced Local Lesions in Genomes (TILLING) is a method developed to find genes of interest in a mutant population of a species through reverse genetics. By screening DNA sequence changes in the gene of interest, mutants can be detected and evaluated for its effect on phenotypes. To discover genes influencing peanut allergens a TILLING population was developed from the cultivar Tifrunner (Knoll et al., 2011). Gene knockouts for genes encoding *Ara h* 1 and *Ara h 2* (seed storage protein genes) and for the *FAD2* gene that is involved in conversion of oleic to linoleic acid were discovered. However, each of the 2S, 7S and 11S seed storage proteins of peanut are produced by gene families; Calbrix et al. (2012) reported 10, 13 and 10 subgroups, respectively, for the three protein classes. Thus, knockouts of single genes may reduce allergen problems, but will not eliminate them.

9.11.2 Peanut Transformation

Ozias-Akins et al. (1993) reported the first successful transformed peanut plant. Micro-bombardment has been the technique most commonly used in peanut, and several genes have been transferred conferring disease resistance (Dar, Reddy, Gowda, & Ramesh, 2006; Higgins, Hall, Mitter, Cruickshank, & Dietzgen, 2004; Magbanua et al., 2000; Ozias-Akins & Gill, 2001; Yang, Singsit, Wang, Gonsalves, & Ozias-Akins, 1998). However, efficiency levels are low and the process takes many months to obtain a mature plant (Egnin, Mora, & Prakash, 1998). Cheng, Jarret, Li, Aiqiu Xing, and Demski (1996) used Agrobacterium-mediated transformation on a Valencia-type peanut, but the technique apparently does not work on other genotypes. To date, biolistic methodologies are the most reliable in peanut, and single constructs can be inserted into the peanut genome. Improved lines with tomato spotted wilt virus (Yang et al., 1998) and Sclerotinia (Chenault, Maas, Damicone, Payton, & Melouk, 2009) resistances have been produced, but the regulatory process of germplasm release for consumption has thus far prevented commercialization.

9.12 Conclusions

A large amount of peanut germplasm of both cultivated and wild species has been collected and is being maintained at multiple international locations. Gaps exist in the collection for cultivated materials as well as wild species. Newly discovered materials are currently unavailable to the international community because of germplasm collection and distribution restrictions imposed by countries where peanut is found. Eighty species have been named, and additional ones will be described in the future. Improved cultivars are replacing landraces, and the genetic variability of cultivated peanut is rather narrow in most production regions.

Crossing relationships are generally known in *Arachis*, although there remain questions about the origins of the tetraploid species in section *Rhizomatosae* and biosystematic relationships of species in several other sections. Although more than 15,000 lines of *A. hypogaea* are in germplasm collections, relatively few entries are in the pedigrees of improved cultivars. Wild species have great potential for improving disease and insect resistance, though genomic and ploidy level differences cause sterility problems when hybridizing with *A. hypogaea*. Progress has been made to incorporate genes from *Arachis* species into improved cultivars. Molecular research has lagged behind many other crop species, in large part due to a lack of significant amounts of variation in the cultivated species, but new marker systems such as simple sequence repeats have promise for enhancing genetic resources.

References

Anderson, W. F., Holbrook, C. C., & Culbreath, A. K. (1996). Screening the peanut core collection for resistance to tomato spotted wilt virus. *Peanut Science*, *23*, 57–61.

Barkley, N. A., Dean, R. E., Pittman, R. N., Wang, M. L., Holbrook, C. C., & Pederson, G. A. (2007). Genetic diversity of cultivated and wild-type peanuts evaluated with M13-tailed SSR markers and sequencing. *Genetics Research Cambridge*, *89*, 93–106.

Bertioli, D. J., Leal-Bertioli, S. C., Lion, M. B., Santos, V. L., Pappas, G., Jr., et al., Cannon, S. B., et al. (2003). A large scale analysis of resistance gene homologues in *Arachis*. *Molecular Genetics and Genomics*, *270*, 34–45.

Bertioli, D. J., Seijo, G., Freitas, F. O., Valls, J. F. M., Leal-Bertioli, S. C. M., & Moretzsohn, M. C. (2011). An overview of peanut and its wild relatives. *Plant Genetic Resources: Characterization and Utilization*, *9*, 134–149.

Bianchi-Hall, C., Keys, R. D., & Stalker, H. T. (1993). Diversity of seed storage proteins in wild peanuts (*Arachis* species). *Plant Systematics and Evolution*, *186*, 1–15.

Bravo, J. P., Hoshino, A. A., Angelici, C. M. L. C. D., Lopes, C. R., & Gimenes, M. A. (2006). Transferability and use of microsatellite markers for the genetic analysis of the germplasm of some *Arachis* section species of the genus *Arachis*. *Genetics and Molecular Biology*, *29*, 516–524.

Burow, M. D., Starr, J. L., Simpson, C. E., & Paterson, A. H. (1996). Identification of RAPD markers in peanut (*Arachis hypogaea*) associated with root-knot nematode resistance derived from *A. cardenasii*. *Molecular Breeding*, *2*, 307–319.

Calbrix, R. G., Beilinson, V., Stalker, H. T., & Neilson, N. C. (2012). Diversity of seed storage proteins of *Arachis hypogaea* and related species. *Crop Science*, *52*, 1676–1688.

Chenault, K. D., Maas, A. L., Damicone, J. P., Payton, M. E., & Melouk, H. A. (2009). Discovery and characterization of a molecular marker for *Sclerotinia minor* (Jagger) resistance in peanut. *Euphytica*, *166*, 357–365.

Cheng, M., Jarret, R. L., Li, Z., Aiqiu Xing, A., & Demski, J. W. (1996). Production of fertile transgenic peanut (*Arachis hypogaea* L.) plants using *Agrobacterium tumefaciens*. *Plant Cell Reports*, *15*, 653–657.

Chu, Y., Wu, C. L., Holbrook, C. C., Tillman, B. L., Person, G., & Ozias-Akins, P (2011). Marker-assisted selection to pyramid nematode resistance and the high oleic trait in peanut. *Plant Genome*, *4*, 110–117.

Company, M., Stalker, H. T., & Wynne, J. C. (1982). Cytology and leafspot resistance in *Arachis hypogaea* × wild species hybrids. *Euphytica*, *31*, 885–893.

Cuc, L. M., Mace, E. S., Crouch, J. H., Quang, V. D., Long, T. D., & Varshney, R. K. (2008). Isolation and characterization of novel microsatellite markers and their application for diversity assessment in cultivated groundnut (*Arachis hypogaea*). *BMC Plant Biology*, *8*, 55.

Cunha, F. B., Nobile, P. M., Hoshino, A. A., Moretzsohn, M. C., Lopes, C. R., & Gimens, M. A. (2008). Genetic relationships among *Arachis hypogaea* L. (AABB) and diploid *Arachis* species with AA and BB genomes. *Genetic Resources and Crop Evolution*, *55*, 15–20.

Damicone, J. P., Jackson, K. E., Dashiell, K. E., Melouk, H. A., & Holbrook, C. C. (2003). Reaction of the peanut core to sclerotinia blight and pepperspot. *Proceedings of American Peanut Research and Education Society*, *35*, 55. (Abstract).

Dar, W. D., Reddy, B. V. S., Gowda, C. L. L., & Ramesh, S. (2006). Genetic resources enhancement of ICRISAT-mandate crops. *Current Science*, *91*, 880–884.

Dwivedi, S. L., Bertioli, D. J., Crouch, J. H., Valls, J. F., Upadhyaya, H. D., Favero, A., et al. (2007). Peanut. In C. Kole (Ed.), *Genome mapping and molecular breeding in plants: Vol. 2. Oilseeds* (pp. 115–151). Berlin: Springer-Verlag.

Dwivedi, S. L., Gurtu, S., Nigam, S. N., Ferguson, M. E., & Paterson, A. H. (2003). Molecular breeding of groundnut for enhanced productivity and food security in the semi-arid tropics: Opportunities and challenges. *Advances in Agronomy*, *80*, 153–221.

Dwivedi, S. L., Puppala, N., Upadyaya, H. D., Manivannan, N., & Singh, S. (2008). Developing a core collection of peanut specific to Valencia market type. *Crop Science*, *48*, 625–632.

Egnin, M., Mora, A., & Prakash, C. S. (1998). Factors enhancing *Agrobacterium tumefaciens*-mediated gene transfer in peanut (*Arachis hypogaea* L.). *In Vitro Cellular and Developmental Biology – Plant*, *34*, 310–318.

FAOSTAT. Food and Agricultural Organization of the United Nation. FAO statistical database. (2009). <http://faostat.fao.org/faostat/collections?subset=agriculture> (accessed 7.11.2012).

Fernandez, A., & Krapovickas, A. (1994). Cromosomas y evolucion en *Arachis* (Leguminosae). *Bonplandia*, *8*, 1064–1070.

Franke, M. D., Brenneman, T. B., & Holbrook, C. C. (1999). Identification of resistance to *Rhizoctonia* limb rot in a core collection of peanut germplasm. *Plant Disease*, *83*, 944–948.

Freitas, F. O., Moretzsohn, M. C., & Valls, J. F. M. (2007). Genetic variability of Brazilian Indian landraces of *Arachis hypogaea* L.. *Genetics and Molecular Research*, *6*, 675–684.

Friend, S. A., Quandt, D., Tallury, S. P., Stalker, H. T., & Hilu, K. W. (2010). Species, genomes and section relationships in genus *Arachis* (Fabaceae): A molecular phylogeny. *Plant Systematics and Evolution*, *290*, 185–199.

Galgaro, L., Lopes, C. R., Gimenes, M., Valls, J. F. M., & Kochert, G. (1998). Genetic variation between several species of sections extranervosae, caulorrhizae, heteranthae, and triseminatae (genus *Arachis*) estimated by DNA polymorphism. *Genome*, *41*, 445–454.

Garcia, G. M., Stalker, H. T., & Kochert, G. A. (1995). Introgression analysis of an interspecific hybrid population in peanuts (*Arachis hypogaea* L.) using RFLP and RAPD markers. *Genome*, *38*, 166–176.

Gregory, M. P., & Gregory, W. C. (1979). Exotic germplasm of *Arachis* L. interspecific hybrids. *Journal of Heredity*, *70*, 185–193.

Gregory, W. C. (1946). Peanut breeding program underway. Research and farming. *Annual Report, North Carolina Agricultural Experiment Station*, *69*, 42–44.

Gregory, W. C., Gregory, M. P., Krapovickas, A., Smith, B. W., & Yarbrough, J. A. (1973). Structures and genetic resources of peanuts: *Peanuts – culture and uses* (pp. 47–133). Stillwater, OK: American Peanut Research and Education Society.

Gregory, W. C., Krapovickas, A., & Gregory, M. P. (1980). Structure, variation, evolution and classification in *Arachis*. In R. J. Summerfield & A. H. Bunting (Eds.), *Advances in legume sciences* (pp. 469–481). Kew, UK: Royal Botanic Gardens.

Guimarães, P. M., Brasileiro, A. C. M., Araújo, A. C. G., Leal-Bertioli, S. C. M., da Silva, F. R., & Morgante, C. V. (2010). A study of gene expression in the nematode resistant wild peanut relative, *Arachis stenosperma*, in response to challenge with *Meloidogyne arenaria*. *Tropical Plant Biology*, doi:10.1007/s12042-010-9056-z.

Guimarães, P. M., Brasileiro, A. C. M., Leal-Bertioli, S. C. M., Pappas, G., Togawa, R., & Bonfim, O. (2011). Comparative 454 pyrosequencing of transcripts from two wild *Arachis* genotypes under biotic and abiotic stress. In *Proceedings: Plant and animal genome XIX conference*, January 15–19, 2011, San Diego, CA.

Guo, Y., Sameer Khanal, G. S., Tang, S., Bowers, J. E., Heesacker, A. F., Khalilian, N., et al. (2012). Comparative mapping in intraspecific populations uncovers a high degree of macrosynteny between A and B genome diploid species of peanut. *BMC Genomics, 13*, 608.

Halward, T. M., Stalker, H. T., & Kochert, G. (1993). Development of an RFLP map in diploid peanut species. *Theoretical and Applied Genetics*, *87*, 379–384.

Halward, T. M., Stalker, H. T., Larue, E., & Kochert, G. (1992). Use of single-primer DNA amplifications in genetic studies of peanut (*Arachis hypogaea* L.). *Plant Molecular Biology*, *18*, 315–325.

Hammons, R. O. (1970). Registration of Spancross peanut. *Crop Science*, *10*, 459.

Hammons, R. O. (1976). Peanuts: Genetic vulnerability and breeding strategy. *Crop Science*, *16*, 527–530.

Hammons, R. O. (1982). Origin and early history of the peanut. In H. E. Pattee & C. T. Young (Eds.), *Peanut science and technology* (pp. 1–20). Yoakum, TX: American Peanut Research and Education Society.

He, G. H., Meng, R., Gao, H., Guo, B., Gao, G., Newman, M., et al. (2005). Simple sequence repeat markers for botanical varieties of cultivated peanut (*Arachis hypogaea* L.). *Euphytica*, *142*, 131–136.

Hernandez-Garay, A., Sollenberger, L. E., Staples, C. R., & Pedreria, C. G. S. (2004). 'Florigraze' and 'Arbrook' rhizome peanut as pasture for growing Holstein heifers. *Crop Science*, *44*, 1355–1360.

Herselman, L. (2003). Genetic variation among southern African cultivated peanut (*Arachis hypogaea* L.) genotypes as revealed by AFLP analysis. *Euphytica*, *133*, 319–327.

Higgins, C., Hall, R., Mitter, N., Cruickshank, A., & Dietzgen, R. (2004). Peanut stripe potyvirus resistance in peanut (*Arachis hypogaea* L.) plants carrying viral coat protein gene sequences. *Transgenic Research*, *13*, 59–67.

Hilu, K., & Stalker, H. T. (1995). Genetic relationships between peanut and wild species of *Arachis* section *Arachis* (Fabaceae): Evidence from RAPDs. *Plant Systematics and Evolution*, *188*, 167–178.

Holbrook, C. C. (2001). Status of the *Arachis* germplasm collection in the United States. *Peanut Science, 28*, 84–89.

Holbrook, C. C., & Anderson, W. F. (1995). Evaluation of a core collection to identify resistance to late leafspot in peanut. *Crop Science, 35*, 1700–1702.

Holbrook, C. C., Anderson, W. F., & Pittman, R. N. (1993). Selection of a core collection from the U.S. germplasm collection of peanut. *Crop Science, 33*, 859–861.

Holbrook, C. C., Bruniard, J., Moore, K. M., & Knauft, D. A. (1998). Evaluation of the peanut core collection for oil content. *Agronomy Abstracts*, 159. (Abstract).

Holbrook, C. C., & Dong, W. (2005). Development and evaluation of a mini core collection for the U. S. peanut germplasm collection. *Crop Science, 45*, 1540–1544.

Holbrook, C. C., & Stalker, H. T. (2003). Peanut breeding and genetic resources. *Plant Breeding Reviews, 22*, 297–356.

Holbrook, C. C., Timper, P., Culbreath, A. K., & Kvien, C. K. (2008). Registration of 'Tifguard' peanut. *Journal of Plant Registration, 2*, 92–94.

Hong, B., Chen, X. P., Liang, X. O., Liu, H. Y., Zhou, G. Y., Li, S. X., et al. (2010). A SSR-based composite genetic linkage map for cultivated peanut (*Arachis hypogaea* L.) genome. *BMC Plant Biology, 10*, 17.

Hopkins, M. S., Casa, A. M., Wang, T., Mitchell, S. E., Dean, R. E., Kochert, G. D., et al. (1999). Discovery and characterization of polymorphic simple sequence repeats (SSRs) in cultivated peanut (*Arachis hypogaea* L.). *Crop Science, 39*, 1243–1247.

Hoshino, A. A., Bravo, J. P., Angelici, C. M. L. C. D., Barbosa, A. V. G., Lopes, C. R., & Gimenes, M. A. (2006). Heterologous microsatellite primer pairs informative for the whole genus *Arachis*. *Genetics and Molecular Biology, 29*, 665–675.

Hull, F. H., & Carver, W. A. (1938). Peanut improvement. *Annual report, Florida agricultural experiment station* (pp. 39–40).

Husted, L. (1936). Cytological studies of the peanut *Arachis*. II. Chromosome number, morphology and behavior and their application to the origin of cultivated forms. *Cytologia, 7*, 396–423.

IBPGR & ICRISAT. (1992). *Descriptors for groundnut* (p. 125). International Board Plant Genetic Resources, Rome, Italy, and International Crops Research Institute for the Semi-Arid Tropics, Patancheru, Andhra Pradesh, India.

Isleib, T. G., Beute, M. K., Rice, P. W., & Hollowell, J. E. (1995). Screening the peanut core collection for resistance to cylindrocladium black rot and early leaf spot. *Proceeding of American Peanut Research and Education Society, 27*, 25. (abstr.).

Isleib, T. G., Holbrook, C. C., & Gorbet, D. W. (2001). Use of *Arachis* spp. plant introductions in peanut cultivar development. *Peanut Science, 28*, 96–113.

Isleib, T. G., Milla-Lewis, S. R., Pattee, H. E., Copeland, S. C., Zuleta, M. C., Shew, B. B., et al. (2010). Registration of 'Bailey' peanut. *Journal of Plant Registrations, 5*, 27–39.

Isleib, T. G., Rice, P. W., Mozingo, R. W., II, Copeland, S. C., Graeber, J. B., & Stalker, H. T. (2006). Registration of N96076L peanut germplasm. *Crop Science, 46*, 2329–2330.

Isleib, T. G., & Wynne, J. C. (1992). Use of plant introductions in peanut improvement: *USE of plant introductions in cultivar development, Part 2* (pp. 75–116). Madison, WI: CSSA Special Publication.

Isleib, T. G., Wynne, J. C., & Nigam, S. N. (1994). Groundnut breeding. In J. Smartt (Ed.), *The groundnut crop: A scientific basis for improvement* (pp. 552–623). London: Chapman & Hall.

Jarvis, A., Ferguson, M. E., Williams, D. E., Guarino, L., Jones, P. G., Stalker, H. T., et al. (2003). Biogeography of wild *Arachis*: Assessing conservation status and setting future priorities. *Crop Science, 43*, 1100–1108.

Jiang, H., Liao, B., Ren, X., Lei, Y., Mace, E., Fu, T., et al. (2007). Comparative assessment of genetic diversity of peanut (*Arachis hypogaea* L.) genotypes with various levels of resistance to bacterial wilt through SSR and AFLP analyses. *Journal of Genetics and Genomics*, *34*, 544–554.

Jiang, H. F., Ren, X. P., Liao, B. S., Huang, J. Q., Lei, Y., Chen, B. Y., et al. (2008). Peanut core collection established in China and compared with ICRISAT mini core collection. *Acta Agronomica Sinica*, *34*, 25–30.

Jiang, H. F., Ren, X. P., Zhang, X. J., Huang, J. Q., Lei, Y., Yan, L. Y., et al. (2010). Comparison of genetic diversity based on SSR markers between peanut mini core collections from China and ICRISAT. *Acta Agronomica Sinica*, *36*, 1084–1091.

Jung, S., Tate, P. L., Horn, R., Kochert, G., Moore, K., & Abbott, A. G. (2003). The phylogenetic relationship of possible progenitors of the cultivated peanut. *Journal of Heredity*, *94*, 334–340.

Kawakami, J. (1930). Chromosome numbers in Leguminosae. *Botanical Magazine (Tokyo)*, *44*, 319–328.

Kilian, A. (2008). DArT-based whole genome profiling and novel information technologies in support system of modern breeding of groundnut. In *Proceedings of the 3rd international conference for peanut genomics and biotechnology on advances in Arachis through genomics and biotechnology (AAGB)*, November 4–8, 2008, Hyderabad, India.

Knauft, D. A., Chiyembekeza, A. J., & Gorbet, D. W. (1992). Possible reproductive factors contributing to outcrossing in peanut. *Peanut Science*, *19*, 29–31.

Knauft, D. A., & Gorbet, D. W. (1989). Genetic diversity among peanut cultivars. *Crop Science*, *29*, 1417–1422.

Knoll, J. E., Ramos, M. L., Zeng, Y., Holbrook, C. C., Chow, M., Chen, S., et al. (2011). TILLING for allergen reduction and improvement of quality traits in peanut (*Arachis hypogaea* L.). *BMC Plant Biology*, *11*, 81.

Kochert, G., Halward, T., Branch, W. D., & Simpson, C. E. (1991). RFLP variability in peanut (*Arachis hypogaea* L.) cultivars and wild species. *Theoretical and Applied Genetics*, *81*, 565–570.

Kochert, G., Stalker, H. T., Gimenes, M., Galgaro, L., Romero Lopes, C., & Moore, K. (1996). RFLP and cytogenetic evidence on the origin and evolution of allotetraploid domesticated peanut, *Arachis hypogaea* (Leguminosae). *American Journal of Botany*, *83*, 1282–1291.

Koppolu, R., Upadhyaya, H. D., Dwivedi, S. L., Hoisington, D. A., & Varshney, R. K. (2010). Genetic relationships among seven sections of the genus *Arachis* studied by using SSR markers. *BMC Plant Biology*, *10*, 15.

Kottapalli, K. R., Burow, M., Burow, G., Burke, J., & Puppala, N. (2007). Molecular characterization of the U.S. peanut mini core collection using microsatellite markers. *Crop Science*, *47*, 1718–1727.

Krapovickas, A. (1969). The origin, variability and spread of the groundnut (*Arachis hypogaea*). In P. J. Ucko & G. W. Dimbleby (Eds.), *The domestication and exploration of plants and animals* (pp. 427–441). London: Duckworth Publishers.

Krapovickas, A., & Gregory, W. C. (1994). Taxonomy of the genus *Arachis* (Leguminosae). *Bonplandia*, *8*, 1–186.

Krapovickas, A., & Rigoni, V. A. (1951). Estudios citologicos en el genero *Arachis*. *Revista de Investigaciones Agricolas*, *5*, 289–294.

Krishna, G. K., Zhang, J., Burow, M., Pittman, R. N., Delikostadinov, S. G., Lu, Y., et al. (2004). Genetic diversity analysis in Valencia peanut (*Arachis hypogaea* L.) using microsatellite markers. *Cellular and Molecular Biology Letters*, *9*, 111–120.

Lanham, P. G., Fennell, S., Moss, J. P, & Powell, W. (1992). Detection of polymorphic loci in *Arachis* germplasm using random amplified polymorphic DNAs. *Genome, 35*, 885–889.

Leal-Bertioli, S. C., Jose, A. C., Alves-Freitas, D. M., Moretzsohn, M. C., Guimaraes, P. M., Nielen, S., et al. (2009). Identification of candidate genome regions controlling disease resistance in *Arachis*. *BMC Plant Biology, 9*, 112.

Liang, X. Q., Luo, M., Holbrook, C. C., & Guo, B. Z. (2006). Storage protein profiles in Spanish and runner market type peanuts and potential markers. *BMC Plant Biology, 6*, 24.

Lu, J., & Pickersgill, B. (1993). Isozyme variation and species relationships in peanut and its wild relatives (*Arachis* L. – Leguminosae). *Theoretical and Applied Genetics, 85*, 550–560.

Magbanua, Z. V., Willed, H. D., Roberts, J. K., Chowdhury, K., Abad, J., Moyer, J. W., et al. (2000). Field resistance to tomato spotted wilt virus in transgenic peanut (*Arachis hypogaea* L.) expressing an antisense nucleocapsid gene sequence. *Molecular Breeding, 6*, 227–236.

Mallikarjuna, N. (2005). Production of hybrids between *Arachis hypogaea* and *A. chiquitana* (section *Procumbentes*). *Peanut Science, 32*, 148–152.

Mathews, B. W., Carpenter, J. R., Cleveland, E., Gibson, Z., & Niino-Duponte, R. Y. (2000). Perennial forage peanut (*Arachis pintoi*) in pastures for raising replacement heifers/stocker steers in Hawaii. *Journal Hawaiian Pacific Agriculture, 11*, 1–10.

Mathews, J. (1983). Theory takes root in archaeologist's peanut discovery. *Philadelphia Inquirer, 4*.

Mehan, V. K. (1989). Screening of groundnuts for resistance to seed invasion by *Aspergillus flavus* and to aflatoxin production. In *Aflatoxin contamination of groundnut: Proceedings of an international workshop*, October 6–9, 1989 (pp. 323–334). ICRISAT Center, India, Patancheru, Andhra Pradesh, India.

Mehan, V. K., McDonald, D., Ranakrishna, N., & Williams, J. H. (1986). Effect of genotype and date of harvest on infection of peanut seed by *Aspergillus flavus* and subsequent contamination with aflatoxin. *Peanut Science, 13*, 46–50.

Mehan, V. K., Reddy, P. M., Vidyasagar Rao, K., & McDonald, D. (1994). Components of rust resistance in peanut genotypes. *Phytopathology, 84*, 1421–1426.

Milla, S. R., Isleib, T. G., Stalker, H. T., & Scoles, G. J. (2005). Taxonomic relationships among *Arachis* sect. *Arachis* species as revealed by AFLP markers. *Genome, 48*, 1–11.

Moretzsohn, M. C., Barbosa, A. V., Alves-Freitas, D. M., Teixeira, C., Leal-Bertioli, S. C., Guimaraes, P. M., et al. (2009). A linkage map for the B-genome of *Arachis* (Fabaceae) and its synteny to the A-genome. *BMC Plant Biology, 9*, 40.

Moretzsohn, M. C., Hopkins, M. S., Mitchell, S. E., Kresovich, S., Valls, J. F. M., & Ferreira, M. E. (2004). Genetic diversity of peanut (*Arachis hypogaea* L.) and its wild relatives based on the analysis of hypervariable regions of the genome. *BMC Plant Biology, 4*, 11.

Moretzsohn, M. C., Leoi, L., Proite, K., Guimarães, P. M., Leal-Bertioli, S. C. M., Gimenes, M. A., et al. (2005). A microsatellite-based, gene-rich linkage map for the AA genome of *Arachis* (Fabaceae). *Theoretical and Applied Genetics, 111*, 1060–1071.

Nageswara Rao, R. C., Udaykumar, M., Farquhar, G. D., Talwar, H. S., & Prasad, T. G. (1995). Variation in carbon isotope discrimination and its relationship to specific leaf area and ribulose-1,5-bisphosphate carboxylase content in groundnut genotypes. *Australian Journal of Plant Physiology, 22*, 545–551.

Nagy, E. D., Chu, Y., Guo, Y., Khanal, S., Tang, S., Li, Y., et al. (2010). Recombination is suppressed in an alien introgression in peanut harboring *Rma*, a dominant root-knot nematode resistance gene. *Molecular Breeding, 26*, 357–370.

Nagy, E. D., Guo, Y., Tang, S., Bowers, J. E., Okashah, R. A., Taylor, C. A., et al. (2012). A high-density genetic map of *Arachis duranensis*, a diploid ancestor of cultivated peanut. *BMC Genomics*, *13*, 469.

Nelson, A., Samuel, D. M., Tucker, J., Jackson, C., & Stahlecker-Roberson, A. (2006). Assessment of genetic diversity and sectional boundaries in tetraploid peanuts (*Arachis*). *Peanut Science*, *33*, 64–67.

Ozias-Akins, P., & Gill, R. (2001). Progress in the development of tissue culture and transformation methods applicable to the production of transgenic peanut. *Peanut Science*, *28*, 123–131.

Ozias-Akins, P., Schnall, J. A., Anderson, W. F., Singsit, C., Clemente, T. E., Adang, M. J., et al. (1993). Regeneration of transgenic peanut plants from stably transformed embryogenic callus. *Plant Science*, *93*, 185–194.

Paik-Ro, O. G., Smith, R. L., & Knauft, D. A. (1992). Restriction fragment length polymorphism evaluation of six peanut species within the *Arachis* section. *Theoretical and Applied Genetics*, *84*, 201–208.

Pandey, M., Gautami, B., Jayakumar, T., Sriswathi, M., Upahyaya, H. D., Gowda, M. V. C., et al. (2011). Highly informative genic and genomic SSR markers to facilitate molecular breeding in cultivated groundnut (*Arachis hypogaea*). *Plant Breeding*, *131*, 139–147.

Pandey, M. K., Monyo, E., Ozias-Akins, P., Liang, X., Guimarães, P., Nigam, S. N., et al. (2012). Advances in *Arachis* genomics for peanut improvement. *Biotechnology Advances*, *30*, 639–651.

Paterson, A. H., Stalker, H. T., Gallo-Meagher, M., Burow, M. D., Dwivedi, S. L., Crouch, J. H., & Mace, E. S. (2004). Genomics and genetic enhancement of peanut. In R. F. Wilson, H. T. Stalker, & C. E. Brummer (Eds.), *Genomics for legume crops* (pp. 97–109). Champaign, IL: American Oil and Chemical Society Press.

Penaloza, A. P. S., & Valls, J. F. M. (2005). Chromosome number and satellite chromosome morphology of eleven species of *Arachis* (Leguminosae). *Bonplandia*, *14*, 65–72.

Pittman, R. N. (1995). United States peanut descriptors: *USDA-ARS-132*. Washington, DC: U.S. Government Printing Office. 13 pp.

Prasad Rao, R. D. V., Reddy, A. S., Chakrabarty, S. K., Reddy, D. V. R., Rao, V. R., & Moss, J. P. (1991). Identification of peanut stripe virus resistance in wild *Arachis* germplasm. *Peanut Science*, *18*, 1–2.

Qin, H., Feng, S., Chen, C., Guo, Y., Knapp, S., Culbreath, A., et al. (2012). An integrated genetic linkage map of cultivated peanut (*Arachis hypogaea* L.) constructed from two RIL populations. *Theoretical and Applied Genetics*, *124*, 653–664.

Raina, S. N., & Mukai, Y. (1999). Genomic *in situ* hybridization in *Arachis* (Fabaceae) identifies the diploid wild progenitors of cultivated (*A. hypogaea*) and related wild (*A. monticola*) peanut species. *Plant Systematics and Evolution*, *214*, 251–262.

Raman, V. S., & Kesavan, P. C. (1962). Studies on a diploid interspecific hybrid in *Arachis*. *Nucleus*, *5*, 123–126.

Robledo, G., Lavia, G. I., & Seijo, J. G. (2009). Species relations among wild *Arachis* species with the A genome as revealed by FISH mapping of rDNA loci and heterochromatin detection. *Theoretical and Applied Genetics*, *118*, 1295–1307.

Robledo, G., & Seijo, G. (2010). Species relationships among the wild B genome of *Arachis* species (section Arachis) based on FISH mapping of rDNA loci and heterochromatin detection: A new proposal for genome arrangement. *Theoretical and Applied Genetics*, *121*, 1033–1046.

Seijo, J. G., Lavia, G. I., Fernandez, A., Krapovickas, A., Ducasse, D., & Moscone, E. A. (2004). Physical mapping of the 5S and 18S–25S rRNA genes by FISH as evidence that *Arachis duranensis* and *A. ipaensis* are the wild diploid progenitors of *A. hypogaea* (Leguminosae). *American Journal of Botany, 91*, 1294–1303.

Shands, H. L., & Bertram, R. (2000). Access to plant germplasm in the CGIAR centers: An update. *Crop Science-Soil Science-Agronomy News, March*, 8.

Simpson, C. E. (1991). Pathways for introgression of pest resistance into *Arachis hypogaea* L.. *Peanut Science, 18*, 22–26.

Simpson, C. E., Burrow, M. D., Patterson, A. H., Starr, J. L., & Church, G. T. (2003). Registration of 'NemaTAM' peanut. *Crop Science, 43*, 1561.

Simpson, C. E., Nelson, S. C., Starr, J. L., Woodard, K. E., & Smith, O. D. (1993). Registration of TxAG-6 and TxAG-7 peanut germplasm lines. *Crop Science, 33*, 1418.

Simpson, C. E., & Smith, O. D. (1975). Registration of Tamnut 74 peanut. *Crop Science, 15*, 603–604.

Simpson, C. E., & Starr, J. (2001). Registration of 'COAN' peanut. *Crop Science, 41*, 918.

Singh, A. K., Krishnan, S. S., Mengesha, M. H., & Ramaiah, C. D. (1991). Polygenetic relationships in section *Arachis* based on seed protein profiles. *Theoretical and Applied Genetics, 82*, 593–597.

Singh, A. K., & Simpson, C. E. (1994). Biosystemics and genetic resources. In J. Smartt (Ed.), *The groundnut crop: A scientific basis for improvement* (pp. 96–137). London: Chapman & Hall.

Smartt, J., & Stalker, H. T. (1982). Speciation and cytogenetics in *Arachis*. In H. E. Pattee & C. E. Young (Eds.), *Peanut Science and Technology* (pp. 21–49). Yoakum, TX: American Peanut Research and Education Society.

Stalker, H. T. (1981). Hybrids in the genus *Arachis* between sections *Erectoides* and *Arachis*. *Crop Science, 21*, 359–362.

Stalker, H. T. (1985). Cytotaxonomy of *Arachis*. Hyderabad, India. *International workshop on cytogenetics of Arachis*, International Crops Research Institute for Semi-Arid Tropics (pp. 65–79). Hyderabad, India.

Stalker, H. T. (1991). A new species in section *Arachis* of peanuts with a D genome. *American Journal of Botany, 78*, 630–637.

Stalker, H. T., Beute, M. K, Shew, B. B., & Isleib, T. G. (2002). Registration of five leafspot-resistant peanut germplasm lines. *Crop Science, 42*, 314–316.

Stalker, H. T., & Dalmacio, R. D. (1986). Karyotype analysis and relationships among varieties of *Arachis hypogaea* L. *Cytologia, 58*, 617–629.

Stalker, H. T., Dhesi, J. S., & Kochert, G. D. (1995). Variation within the species *A. duranensis*, a possible progenitor of the cultivated peanut. *Genome, 38*, 1201–1212.

Stalker, H. T., Dhesi, J. S., Parry, D. C., & Hahn, J. H. (1991). Cytological and interfertility relationships of *Arachis* section *Arachis*. *American Journal of Botany, 78*, 238–246.

Stalker, H. T., & Lynch, R. L. (2002). Registration of four insect-resistant peanut germplasm lines. *Crop Science, 42*, 313–314.

Stalker, H. T., & Moss, J. P. (1987). Speciation, cytogenetics, and utilization of *Arachis* species. *Advances in Agronomy, 41*, 1–40.

Stalker, H. T., Phillips, T. D., Murphy, J. P., & Jones, T. M. (1994). Variation of isozyme patterns among *Arachis* species. *Theoretical and Applied Genetics, 87*, 746–755.

Stalker, H. T., & Simpson, C. E. (1995). Germplasm resources in *Arachis*. In H. E. Pattee & H. T. Stalker (Eds.), *Advances in peanut science* (pp. 14–53). Stillwater, OK: American Peanut Research and Education Society.

Stalker, H. T., Weissinger, A. K., Milla-Lewis, S., & Holbrook, C. C. (2009). Genomics – An evolving science in peanut. *Peanut Science, 36*, 2–10.

Subrahmanyam, P., Anaidu, R., Reddy, L. J., Kumar, P. L., & Ferguson, M. E. (2001). Resistance to groundnut rosette disease in wild *Arachis* species. *Annals of Applied Biology*, *139*, 45–50.

Subrahmanyam, P., McDonald, D., Waliyar, F., Reddy, L. J., Nigam, S. N., Gibbons, R. W., et al. (1995). Screening methods and sources of resistance to rust and late leaf spot of groundnut. *ICRISAT Information Bulletin No. 47*. International Crops Research Institute for Semi-Arid Tropics, Hyderabad, India.

Subrahmanyam, P., Williams, J. H., McDonald, D., & Gibbons, R. W. (1984). The influence of foliar diseases and their control by selective fungicides on a range of groundnut (*Arachis hypogaea* L.). *Annuals of Applied Biology*, *104*, 467–476.

Tallury, S. P., Hilu, K. W., Milla, S. R., Friend, S. A., Alsaghir, M., Stalker, H. T., et al. (2005). Genomic affinities in *Arachis* section *Arachis* (Fabaceae): Molecular and cytogenetic evidence. *Theoretical and Applied Genetics*, *111*, 1229–1237.

Tang, R., Gao, G., He, L., Han, Z., Shan, S., Zhong, R., et al. (2007). Genetic diversity in cultivated groundnut based on SSR markers. *Journal of Genetics and Genomics*, *34*, 449–459.

Tillman, B. L., & Stalker, H. T. (2009). Peanut. In J. Vollmann & I. Rajcan (Eds.), *Oil crops, handbook of plant breeding* (Vol. 4, pp. 297–315). New York, NY: Springer Science Press.

Upadhyaya, H. D., Bramel, P. J., Ortiz, R., & Singh, S. (2002). Developing a mini core of peanut for utilization of genetic resources. *Crop Science*, *42*, 599–600.

Upadhyaya, H. D., Ferguson, M. E., & Bramel, P. J. (2001). Status of the *Arachis* germplasm collection at ICRISAT. *Peanut Science*, *28*, 89–96.

Upadhyaya, H. D., Mallikarjuna Swamy, B. P., Goudar, P. V. K, Kullaiswaym, B. Y., & Singh, S. (2005). Identification of diverse accessions of groundnut through multienvironmental evaluation of core collection for Asia. *Field Crops Research*, *93*, 293–299.

Upadhyaya, H. D., Ortiz, R., Bramel, P. J., & Singh, S. (2001). Development of a groundnut core collection using taxonomical, geographical and morphological descriptors. *Genetic Resources and Crop Evolution*, 139–148.

Upadhyaya, H. D., Reddy, L. J., Gowda, C. L. L., & Singh, S. (2006). Identification of diverse groundnut germplasm: Sources of early maturity in a core collection. *Field Crops Research*, *97*, 261–267.

Valls, J. F. M., Ramanatha Rao, V., Simpson, C. E., & Krapovickas, A. (1985). Current status of collection and conservation of South American groundnut germplasm with emphasis on wild species of *Arachis*. In J.P. Moss (ed.), *Proceedings of an international workshop on cytogenetics of Arachis*, October 31–November 2, 1983. International Crops Research Institute for Semi-Arid Tropics (pp. 15–35). Patancheru, Andhra Pradesh, India.

Valls, J. F. M., & Simpson, C. E. (2005). New species of *Arachis* (Leguminosae) from Brazil, Paraguay and Bolivia. *Bonplandia*, *14*, 35–63.

Varshney, R. K., Bertioli, D. J., Moretzsohn, M. C., Vedez, V., Krishnamurthy, L., Aruna, R., et al. (2009).). The first SSR-based genetic linkage map for cultivated groundnut (*Arachis hypogaea* L.). *Theoretical and Applied Genetics*, *118*, 729–739.

Varshney, R. K., Glaszmann, J. C., Leung, H., & Ribaut, J. M. (2010). More genomic resources for less-studied crops. *Trends in Biotechnology*, *28*, 452–460.

Wang, M. L., Barkley, N. A., Chinnan, M., Stalker, H. T., & Pittman, R. N. (2010). Oil content and fatty acid composition variability in wild peanut species. *Plant Genetic Resources: Characterization and Utilization*, *8*, 232–234.

Wang, M. L., Sukumaran, S., Barkley, N. A., Chen, Z., Chen, C. Y., Guo, B., et al. (2011). Population structure and marker-trait association analysis of the U.S. peanut (*Arachis hypogaea* L.) mini-core collection. *Theoretical and Applied Genetics*, *123*, 1307–1317.

Williams, D. E. (2001). New directions for collecting and conserving peanut genetic diversity. *Peanut Science*, *28*, 135–140.

Williams, K. A., & Williams, D. E. (2001). Evolving political issues affecting international exchange of *Arachis* genetic resources. *Peanut Science*, *28*, 132–135.

Wrightman, J. A., & Ranga Rao, G. V. (1994). Groundnut pests. In J. Smartt (Ed.), *The groundnut crop – A scientific basis for improvement* (pp. 395–479). London: Chapman & Hall.

Yang, H., Singsit, C., Wang, A., Gonsalves, D., & Ozias-Akins, P. (1998). Transgenic peanut plants containing a nucleocapsid protein gene of tomato spotted wilt virus show divergent levels of gene expression. *Plant Cell Report*, *17*, 693–699.

Young, N. D., Weeden, N. F., & Kochert, G. (1996). Genome mapping in legumes (family Fabaceae). In A. H. Paterson (Ed.), *Genome Mapping in Plants* (pp. 211–277). Texas: Landes Biomedical Press.

10 Asian *Vigna*

Ishwari Singh Bisht and Mohar Singh

National Bureau of Plant Genetic Resources, Pusa, New Delhi, India

10.1 Introduction

The genus *Vigna* is a large pantropical genus with 82 described species distributed among 7 subgenera, namely *Ceratotropis*, *Haydonia*, *Lasiospron*, *Macrorhyncha*, *Plectotropis*, *Sigmoidotropis* and *Vigna*, and 150 species (Maréchal, Mascherpa, & Stainer, 1978; Tomooka, Vaughan, & Moss, 2002). Among pulse crops, *Vigna* is a large genus that belongs to tribe *Phaseolae* of the family Papilionaceae. Among the subgenera of the genus *Vigna* only the subgenus *Ceratotropis* has its centre of species diversity in Asia. The subgenus *Ceratotropis* currently consists of 16 (Verdcourt, 1970) to 17 (Maréchal et al., 1978; Tateishi, 1985) recognized species, which are naturally distributed across Asia and thus are often called Asiatic or Asian *Vigna* (Singh et al., 2006). Tomooka et al. (2002) describes 21 species of Asian *Vigna*, 8 of which are used for human food or animal feed. This is in contrast to the African *Vigna* (the subgenus *Vigna*) of which, of the 36 species, only 2 species have been domesticated (Maréchal et al., 1978) and the closely related genus *Phaseolus* of the New World that consists of about 50 species, of which only 5 are cultivated (Debouck, 2000). The Asian *Vigna* were initially classified as the genus *Phaseolus* by de Candolle (1825). Later on, Verdcourt (1970) limited the use of *Phaseolus* exclusively to those American species that have a tightly coiled style and pollen grain lacking coarse reticulation. As a consequence, Asian *Vigna* was classified as a subgenus, *Ceratotropis* (Maréchal et al., 1978). The comparative taxonomic system of Maréchal et al. (1978) and that of Tateishi (1985) are shown in Table 10.1. The revision of subgenus *Ceratotropis* by Tateishi (1985) is the most comprehensive one to date (Tomooka, Egawa, & Kaga, 2000). The eight cultivated species of the subgenus *Ceratotropis* as described by Tomooka et al. (2002) are *Vigna radiata* (green gram or mung bean), *V. mungo* (black gram or urd bean), *V. angularis* (small red bean or azuki/adzuki bean), *V. umbellata* (rice bean or red bean), *V. aconitifolia* (moth bean), *V. reflexopiloxa* var. *glabra* (Creole bean), *V. trilobata* (wild bean) and *V. trinervia* (Tooapée).

The last three species are of minor importance and *V. trilobata* can perhaps be regarded as a semidomesticate. The Asian *Vigna* are considered to be recently evolved and morphological differentiation between taxa is limited (Baudoin & Maréchal, 1988).The five main cultivated species of Asian *Vigna* belonging to the subgenus *Ceratotropis* are closely related, characteristically small-seeded and distinguished on the basis of seedling type. The first and second leaves are sessile in

Genetic and Genomic Resources of Grain Legume Improvement. DOI: http://dx.doi.org/10.1016/B978-0-12-397935-3.00010-4

Table 10.1 Species and Infraspecific Taxa of the Asian *Vigna*, Subgenus *Ceratotropis*, Recognized by Maréchal et al. (1978) and Tateishi (1985)

S. No.	Maréchal et al. (1978)	S. No.	Tateishi (1985)
1.	*V. aconitifolia* (Jacquin) Maréchal	1.	*V. aconitifolia* (Jacquin) Maréchal
2.	*V. angularis* (Willdenow) Ohwi & Ohashi var. *angularis* var. *nipponensis* (Ohwi) Ohwi & Ohashi	2.	*V. angularis* (Willdenow) Ohwi & Ohashi var. *angularis* var. *nipponensis* (Ohwi) Ohwi & Ohashi
3.	*V. dalzelliana* (O. Kuntze) Verdcourt	3.	*V. dalzelliana* (O. Kuntze) Verdcourt
		4.	*V. exilis* Tateishi
		5.	*V. grandiflora* (Prain) Tateishi
4.	*V. hirtella* Ridley	6.	*V. hirtella* Ridley
5.	*V. khandalensis* (Santapau) Raghavan & Wadhwa	7.	*V. khandalensis* (Santapau) Raghavan & Wadhwa
6.	*V. minima* (Roxburgh) Ohwi & Ohashi	8.	*V. minima* (Roxburgh) Ohwi & Ohashi ssp. *gracilis* (Prain) Tateishi ssp. *minima* var. *minima* var. *minor* (Matsumura) Tateishi ssp. *nakashime* (Ohwi) Tateishi
7.	*V. rikiuensis* (Ohwi) Ohwi & Ohashi		
8.	*V. nakashimae* (Ohwi) Ohwi & Ohashi		
9.	*V. mungo* (L.) Hepper var. *mungo* var. *silvestris* Lukoki, Maréchal & Otoul	9.	*V. mungo* (L.) Hepper var. *mungo* var. *silvestris* Lukoki, Maréchal & Otoul
		10.	*V. nepalensis* Tateishi
10.	*V. radiata* (L.) Wilczek var. *radiata* var. *sublobata* (Roxburgh) Verdcourt var. *setulosa* (Dalzell) Ohwi & Ohashi	11.	*V. radiata* (L.) Wilczek var. *radiata* var. *sublobata* (Roxburgh) Tateishi
11.	*V. glabrescence* Maréchal, Mesherpa & Stainier		
12.	*V. reflexo-pilosa* Hayata	12.	*V. reflex-pilosa* Hayata subsp. *glabra* (Roxburgh) Tateishi subsp. *reflexo-pilosa*
		13.	*V. stipulacea* (Lamarck) Tateishi
		14.	*V. subramaniana* (Babuex Raizada) Tateishi
13.	*V. trilobata* (L.) Verdcourt	15.	*V. trilobata* (L.) Verdcourt
14.	*V. bourneae* Gamble	16.	*V. trinervia* (Heyne ex Wight et Arnott) Tateishi var. *trinervia* var. *bourneae* (Gamble) Tateishi

(*Continued*)

Table 10.1 (Continued)

S. No.	Maréchal et al. (1978)	S. No.	Tateishi (1985)
15.	*V. umbellata* (Thunberg) Ohwi & Ohashi var. *umbellata* var. *gracilis* (Prain) Maréchal, Mesherpa and Stainier	17.	*V. umbellata* (Thunberg) Ohwi and Ohashi
16	*V. malayana* M.R. Henderson		
17.	*V. popuana* Baker F.		

Source: Tomooka et al. (2000) and Singh et al. (2006).

Table 10.2 Domesticated Asian *Vigna* Species and their Wild Relatives

Species	Cultigen	Wild form	Distribution of wild forms
V. angularis	var. *angularis*	var. *nipponensis*	Himalayas, northern Myanmar, China, Korea, Japan
V. mungo	var. *mungo*	var. *silvestsis*	India, Myanmar
V. radiata	var. *radiata*	var. *sublobata*	East Africa, Madagascar, Asia, New Guinea, Australia
V. umbellata	var. *umbellata*	var. *gracilis*	East India, Thailand, Indochina, China
V. aconitifolia	var. *aconitifolia*	var. *silvestris*	Pakistan, India

Source: Data from Tomooka et al. (2000) and Singh et al. (2006).

V. radiata and *V. mungo* and have epigeal cotyledons, whereas *V. angularis* and *V. umbellata* have petiolate and hypogeal cotyledons (Maekawa, 1995). *V. aconitifolia* has intermediate seedling type with epigeal cotyledons and petiolate first and second leaves (Baudet, 1974). Tateishi (1996) used seedling characteristics to recognize three subgroups *s. str.* in the subgenus *Ceratotropis*, namely (i) the mung bean group *s. str.*, (ii) the azuki bean group *s. str.* and (iii) *V. aconitifolia* group with intermediate seedling characteristics. Taxonomically, cultigens and their conspecific wild forms are recognized in all the species except *V. aconitifolia* (Table 10.2) (Lukoki, Maréchal, & Otoul, 1980; Maréchal et al., 1978; Tomooka et al., 2000). However, Dana (1998) reported a wild form of *V. aconitifolia* (*V. aconitifolia* var. *silvestris*) also. Pods of domesticated species of *Vigna* have reduced dehiscence and seeds are nondormant. Both day-neutral and short-day genotypes are found in all the species. Chromosome complements in *Vigna* species are $2n=2x=22$ with exception of *V. glabrescens* ($2n=4x=44$). Chromosome rearrangements play a significant role in the genetic differentiation of Asian *Vigna* species (Biswas & Dana, 1975, 1976; Dana 1966a, 1966b; Machado, Tai, & Baker, 1982; Satyan, Mahishi, & Shivashankar, 1982; Sen & Ghosh, 1961). Even the two closest relatives, *V. radiata* and *V. mungo*, have some structural differentiation of their genomes. The wild related species and other cultigens of *Vigna* do not form a particularly extensive or accessible gene pool

(Smartt, 1990). Lawn (1995) proposed that the Asian *Vigna* consists of three more or less isolated genepools, based on cross-compatibility studies which correspond with groups based on seedling characters proposed by Tateishi (1996), as follows:

Gene pool (Lawn, 1995): angularis-umbellata; radiata-mungo; aconitifolia-trilobata
Group (Tateishi, 1996): Azuki bean *s.str.*; mung bean *s. str.*; *V. aconitifolia.*

10.2 Origin, Distribution and Diversity

The Asian *Vigna* have been domesticated in Asia from the Indian subcontinent to the Far East (Smartt, 1990). Records of Asian *Vigna* from 3500 to 3000 BC were found in archaeological sites at Navdatoli in Central India (Jain & Mehra, 1980). Mung bean or green gram (*Vigna radiata* syn. *Phaseolus aureus* Roxb.; *P. radiatus* L.) has been considered to have been domesticated in India (Vavilov, 1926). Other authors have supported his theory based on the morphological diversity (Singh, Joshi, Chandel, Pant, & Saxena, 1974), existence of wild and weedy types (Chandel, 1984; Paroda & Thomas, 1988), and archaeological remains (Jain & Mehra, 1980) of mung bean in India. Wild forms of mung bean, *V. radiata* var. *sublobata*, show a wide area of distribution stretching from Central and East Africa, Madagascar, through Asia and New Guinea, to northern and eastern Australia (Tateishi, 1996). Mung bean is the most widely distributed among the six Asian *Vigna* species. It is of immense importance because of its adaptation to a short growing season, low water supply and poor soil fertility conditions. It is widely cultivated throughout South and Southeast Asia, including India, Pakistan, Bangladesh, Sri Lanka, Myanmar, Thailand, Philippines, Laos, Cambodia, Vietnam, Indonesia, Malaysia, South China and Taiwan. It is also grown to a lesser extent in many parts of Africa, the United States especially in Oklahoma and has been recently introduced in parts of Australia. In India green gram is mainly grown in the states of Orissa, Andhra Pradesh and Maharashtra. It is a photo- and thermosensitive crop. The best temperature for its cultivation is 30–35 °C with good atmospheric humidity. The wild *V. radiata* var. *sublobata* occurs in the Tarai Mountains, sub-Himalayan tract and sporadically in the western and eastern peninsular tracts of India (Arora & Nayar, 1984). Jain and Mehra (1980) indicated two races of *V. sublobata* (L.) Philipp, one closer to *mungo*, the other to *V. radiata*. Reciprocal differences were found in most of the interspecific crosses. Tomooka, Lairungreang, Nakeeraks, Egawa, and Thavarasook (1991, 1992) revealed the geographical distribution of growth types, seed characters and protein types in mung bean landraces collected from throughout Asia. In South and West Asia, mung bean strains characterized by small seeds with various seed colours, including black, brown and green mottled with black and showing diverse growth habit and protein types were distributed. In the Southeast Asian countries, mung bean strains characterized by various sizes of seed with shiny green seed testa were distributed, showing tall plants with a high branching habit, late maturity and simple protein type composition. In East Asia, mung bean strains characterized by medium-sized dull green seed testa were distributed, showing short plants with an early maturity, low-branching habit and relatively diverse (similar to that of West Asia) protein types.

Black gram or urd bean (*Vigna mungo* syn. *Phaseolus mungo*) is also an important pulse crop of India. Black gram is widely adapted both to semi-arid and subtropical areas. The primary centre of origin of urd bean is India, with a secondary centre in Central Asia. Reference to it in Vedic texts such as Kautilya's 'Arthasasthra' and in 'Charak Samhita' lends support to the presumption of its origin in India. India is the largest producer and consumer of black gram in the world. It has spread to other tropical areas in Asia, Africa and America. Distribution of black gram is comparatively restricted to wet tropics. It is abundantly grown in India, Pakistan, Sri Lanka, Myanmar and some parts of Southeast Asia, parts of Africa and America. In the West Indies it is grown mainly as a green manure crop. Black gram is a rich protein food, containing about 26% protein, or almost three times that of cereals. Black gram supplies a major share of the protein requirement of the vegetarian population in the country. The biological value improves greatly when wheat or rice is combined with black gram because of the complementary relationship of the essential amino acids, such as arginine, leucine, lysine, isoleucine and valine phenylalanine. In addition to being an important source of human food and animal feed, it also plays an important role in sustaining soil fertility by improving soil physical properties and fixing atmospheric nitrogen. Also, it being a drought-resistant crop, it is suitable for dryland farming and predominantly used as an intercrop with other crops.The wild form of *V. mungo* is *V. mungo* var. *silvestris* (Chandel, 1984; Smartt, 1990). In India, wild forms are widely distributed in the Konkan belt of the Western Ghats and in Khandala. Black gram is basically a tropical crop, but it is grown in both winter and summer in India.

Moth bean (*Vigna aconitifolia* syn. *Phaselous aconitifolius* Jacq.) is an important legume crop of arid and semi-arid regions. Its wild form has been designated as *V. aconitifolia* var. *silvestris* (Dana, 1998). Moth bean is found growing wild in Pakistan, India and Myanmar and from the Himalayas in the north to Sri Lanka in the south. On this basis, it is considered to be native to India, Pakistan and Myanmar (Rachie & Roberts, 1974). In cultivated form it has spread to China, Indonesia, Malaysia, Africa and the southern United States. It is widely grown in the Indian subcontinent, Japan, Malaysia, Hong Kong, Singapore and particularly Thailand. In India it is grown in Rajasthan, western Uttar Pradesh, Punjab, Gujarat, Madhya Pradesh, Maharashtra and Karnataka. It is the most important pulse crop in Rajasthan. Amongst *Vigna* species, *V. aconitifolia* is undoubtedly the most drought tolerant and is commonly grown in arid and semi-arid regions, especially in the northwestern desert region of the Indian subcontinent.

Rice bean [*Vigna umbellata* syn. *Phaseolus calcaratus* Roxb., *Vigna calcarata* (Roxb.) Kurz., *Azuki a umbellata* (Thund.) Owhi.], also known as red bean, climbing mountain bean and oriental bean, is considered to have originated in South and Southeast Asia, where it is a multipurpose crop. Its wild form (*V. umbellata* var. *gracilis*) is found in the Himalayas and central China to Malaysia. The cultivated forms seem to have originated from the wild populations occurring in the Indian subcontinent (Chandel & Pant, 1982). It is cultivated in India, China, Korea, Japan, Myanmar, Malaysia, Indonesia, Philippines, Indonesia, Mauritius, Fiji, Bangladesh,

Sri Lanka and Nepal (Purseglove, 1974; Rachie & Roberts, 1974). Rice bean is a short-day plant in India, grown predominantly in the northeastern region, particularly in Assam, Meghalaya, Manipur and to a limited scale in the eastern peninsular tract (Chhota Nagpur region, Bihar and parts of Orissa), western peninsular tracts (particularly the southern hills) and the subtemperate hilly region (Himachal Pradesh and Uttarakhand). Azuki or adzuki bean [*Vigna angularis* syn. *Phaseolus mungo* L. var. *angularis*, *Phaseolus radiatus* L. var. *aurea* Prain, *Phaseolus angularis* W.F. Wight, *Dolichos angularis* Wild. *Azukia subtrilobata* (Fr. et Sav.) Y. Takah., *Azukia angularis* (W.) Ohwi.], also known as small red bean, is a multipurpose food legume. The origin of azuki bean is not clear, but it probably originated in Asia. Its wild types (*V. angularis* var. *nipponensis*) have been found from northern Honshu in Japan to Nepal. In the southern latitudes, *V. angularis* var. *nipponensis* occurs in mountain areas. It is recorded in China, India, Korea, Myanmar and Taiwan. It is cultivated in China, Korea, Japan, Taiwan and eastern Russia for human food. It was introduced into the United States, Angola, India, Kenya, New Zealand, Zaire, Belgium and Argentina. Azuki bean is reported to be a short-day plant that performs best under warm and dry conditions. Wild *V. trilobata* is found from the Himalayas to Sri Lanka, and also in Myanmar, Malaysia, China, Pakistan, Afghanistan and Ethiopia. But in India it is cultivated only as a cover crop and for fodder. The tribal people of India eat seeds gathered from wild plants.

10.3 Genetic Resource Management

10.3.1 Exploration and Collection

In India, the systematic plant exploration and collection work was initiated as early as the 1940s with the establishment of the Plant Introduction unit in the Division of Botany, Indian Agriculture Research Institute (IARI), New Delhi. This ultimately developed into the National Bureau of Plant Genetic Resources (NBPGR) which gave great impetus to germplasm augmentation and conservation of *Vigna* species. The earliest efforts to collect mung bean landraces from all over India and Myanmar were made in 1925 (Bose, 1939). Concerted efforts to collect all *Vigna* species were made to collect the available landraces from various states of India during the late 1960s and early 1970s under the PL-480 project operative at IARI. About 2600 germplasm accessions have been collected from different agro-ecological areas (Malik et al., 2001). The areas explored include the whole of Gujarat and Rajasthan; parts of Bihar, Punjab, Madhya Pradesh, Uttar Pradesh, Himachal Pradesh, Andhra Pradesh, Orissa, Jammu and Kashmir, Haryana except the central zone; the northeastern, western and coastal regions of Maharashtra; southern districts of Karnataka and eastern Kerala; and major areas of Tamil Nadu. Most of the accessions assembled have the determinate growth habit. Wide variation was observed in seed size. The accessions with bold seeds were mainly collected from Maharashtra and Madhya Pradesh and those with more number of seeds per pod from Haryana, Rajasthan Bihar, Gujarat and Jammu and Kashmir. Likewise, in black gram, about 3000 germplasm accessions were collected (Malik et al., 2001). Areas surveyed include Uttar Pradesh, northern and southern Bihar, Madhya Pradesh except southeastern parts; Vidharba, southern and interior coastal Maharashtra, and parts of Himachal Pradesh, Punjab and Rajasthan. Other areas explored include Mysore, Tamil Nadu, Jammu & Kashmir, Goa, Karnataka, Manipur, Tripura, Nagaland, Kerala, Orissa, Rajasthan, Bihar, Haryana, Sikkim, West Bengal and Lakshadweep. Desired plant types were found in collections from Maharashtra, Gujarat, the southern and western parts of Orissa and Madhya Pradesh, while collection from western Rajasthan, the northeastern tract of Sikkim and the eastern Himalayan region generally have bold seeds.

In moth bean, a total of 1956 accessions have been assembled by NBPGR. The collection represents material largely from the states of Rajasthan, Gujarat, Maharashtra, Karnataka, western Uttar Pradesh, Punjab, Haryana and Madhya Pradesh. Moth bean collections possessed variation in growth habit, leaf location and pod and seed colour. The collections made from Rajasthan and Gujarat seems to be more promising. In India, intensive collection efforts from 1971 until now resulted in the assembly of 983 accessions of rice bean at NBPGR. These include primitive cultivars/landraces primarily from the northeastern region and parts of Orissa, Bihar and West Bengal. Since 1971, several wild forms were also collected from the Khasi and Jaintia hills, Meghalaya; Shimla hills and Chamba in Himachal Pradesh, and Western Ghats (Arora, Chandel, Pant, & Joshi, 1980; Chandel, 1981). The exploration and collection efforts were undertaken in several countries in South and Southeast Asia and sizeable germplasm collections have been assembled particularly in Philippines, Indonesia and China. Attempts have also been made to build up the germplasm collection in Japan, Nepal and Sri Lanka (Singh et al., 2006).

Limited exploration and germplasm collection efforts have been made for azuki bean. Systematic collections need to be undertaken to assemble the entire diversity available in the region, particularly from Korea, China and Japan. Only 151 accessions comprising both indigenous and exotic germplasm have been collected by NBPGR. In India, the areas surveyed are parts of Himachal Pradesh and Uttar Pradesh. Intensive efforts to collect wild forms of rice bean were initiated during 1971 (Arora et al., 1980; Chandel, 1981). Between 1974 and 1994 intensive surveys were made in 33 districts of 7 states, namely Gujarat, Rajasthan, Maharashtra, Madhya Pradesh, Bihar, Orissa and West Bengal, and wild *Vigna* species, namely *V. aconitifolia* var. *silvestris*, *V. dalzelliana*, *V. hainiana* Babu, Gopin. & S.K. Sharma, *V. khandalensis*, *V. mungo* var. *silvestris*, *V. radiata* var. *setulosa*, *V. radiata* var. *sublobata* and *V. trilobata* were collected (Dana, 1998). More explorations were conducted by NBPGR and other collaborators in parts of the Eastern and Western Ghats, central tracts of Orissa, Maharashtra, Rajasthan, Kerala, Tamil Nadu, Khasi hills, Himachal Pradesh hills, Uttarakhand hills, Jammu & Kashmir and the northeastern hill region (Bisht et al., 2005). The collected germplasm comprised *Vigna trilobata*, *V. radiata* var. *sublobata*, *V. vexillata* (L.) A. Rich., *V. mungo* var. *silvestris*, *V. dalzelliana*, *V. capensis* (L.) Walp., *V. khandalensis* (*V. grandis*), *V. pilosa* Baker, *V. wightii* Benth. Ex Bedd., *V. bourneae*, etc. About 200 accessions of wild *Vigna* species are presently being maintained at NBPGR Regional Station, Thrissur (Kerala), India. Under international programmes, grain legume crops including *Vigna* species were also collected from Russia, Mali, Nigeria, Malawi and Zambia during 1977–1980. Systematic exploration and collection of wild Asian *Vigna* species have been conducted by Japan in collaborations with Thailand, Sri Lanka and Vietnam since 1989. Accessions of wild species collected are maintained in the gene bank of the Ministry of Agriculture, Forestry and Fisheries, Japan (Singh et al., 2006).

10.3.2 Germplasm Introduction

The NBPGR has introduced substantial germplasm accessions, including trial material (Singh, Chand et al., 2001) of pulse crops from over 50 countries in the last two to three decades. In mung bean, several promising accessions were introduced with high grain yield, uniform and synchronized maturity, long pod with shiny seed coat, large seeds, resistance to biotic and abiotic stresses, and adaptation and appropriate maturity for different seasons from AVRDC (Taiwan), Thailand and Bangladesh. Some of the useful exotic germplasm accessions of black gram include promising genetic stocks with high yield potential, resistance to diseases and high value for quality traits introduced from AVRDC. Further, some promising introductions were made in rice bean from the United States (Gautam et al., 2000). Some accessions of rice bean procured from Taiwan had high yield, wide adaptability and drought resistance. In azuki bean, a few accessions were introduced which possessed long pods and high grain yield. Emphasis was also given to the introduction of wild and related species of *Vigna* from France, Germany, Italy, Japan, Nigeria and the United States (Gautam et al., 2000). In general, South and Southeast Asia are very rich in genetic diversity of Asian *Vigna* species. India has also supplied the germplasm of *Vigna* species to various countries for research purposes.

Various international and national gene banks maintaining wide diversity of *Vigna* species are: International Center for Agriculture Research in the Dry Areas (ICARDA), Aleppo, Syria; Asian Vegetable Research and Development Centre (AVRDC), Shanhua, Taiwan; International Institute for Tropical Agriculture (IITA), Ibadan, Nigeria; Bogor Research Institute for Food Crops (BORIF), Bogor, Indonesia; Commonwealth Scientific and Industrial Organization (CSIRO), Canberra, Australia; Malang Research Institute for Food Crops (MARIF), Malang, Indonesia; National Plant Genetic Research Institute (NPGRI), University of Philippines, Los Banos, Philippines, and US Department of Agriculture (USDA), Southern Regional Plant Introduction Station, Georgia, and NBPGR.

10.3.3 Germplasm Evaluation

The germplasm accessions of mung bean maintained at NBPGR, New Delhi, have been systematically characterized and information is well documented for both qualitative and quantitative sets of descriptors (Bisht, Mahajan, & Kawalkar, 1998; Kawalkar et al., 1996). Variation in different qualitative traits, namely growth habit, branching pattern, twining tendency, raceme position, pod pubescence and seed colour have been documented. The range of variation observed in major agronomic traits is given in Table 10.3. A representative core set of 152 accessions has

Table 10.3 Mean Range for Quantitative Traits of *Vigna* Species

Character	Mung bean	Urd bean	Moth bean	Rice bean
Days to 50% flowering	44.51 (33.00–78.00)	55.00 (41.00–73.00)	63.73 (32.00–84.00)	89.26 (62.00–123.00)
Day to 80% maturity	77.37 (53.00–104.00)	93.00 (67.00–130.00)	82.84 (57.00–105.00)	148.05 (95.00–180.00)
Plant height (cm)	84.10 (17.50–115.20)	94.47 (34.90–157.62)	26.45 (11.68–49.30)	147.46 (62.40–372.50)
Primary branches	3.14 (1.00–7.00)	4.12 (2.8–7.7)	6.09 (1.58–12.00)	5.15 (3.00–13.00)
Clusters/ plant (no.)	8.53 (3.00–28.00)	20.98 (9.00–73.67)	20.98 (9.00–73.67)	36.37 (9.00–124.00)
Post/ cluster (no.)	3.87 (2.58–7.5)	3.34 (2.18–6.97)	2.89 (1.49–7.67)	3.80 (2.00–7.00)
Pod plant (no.)	14.7 (2.80–50.10)	27.05 (2.18–76.95)	49.52 (24.83–128.33)	113.07 (38.00–296.00)
Pod length (cm)	6.59 (3.70–10.00)	4.71 (3.34–6.22)	3.95 (2.20–5.30)	10.11 (7.60–12.80)
Seeds per pod (no.)	10.86 (2.20–14.60)	6.51 (4.29–12.44)	6.75 (2.00–10.00)	9.11 (7.00–12.00)
100-seed weight (g)	3.15 (2.00–5.20)	3.03 (1.87–4.90)	2.90 (1.50–4.60)	8.35 (4.70–21.40)
Yield/plant (g)	3.55 (0.50–8.50)	3.75 (0.90–9.10)	6.17 (1.55–19.08)	22.35 (2.70–73.00)

Source: Data from Singh et al. (2006).

been developed from 1532 well-characterized Indian mung bean accessions, with the primary objective of effective germplasm utilization (Bisht, Mahajan, & Patel, 1998). This set has also been used for genetic enhancement in mung bean as the initial starting material (Bisht et al., 2004). Improved mung bean cultivars have a narrow genetic base that limits yield potential and are poorly adapted to varying growth conditions in different agro-ecological conditions. The genetic potential of landrace germplasm accessions in the gene banks therefore needs better exploitation. At NBPGR, genetic enhancement/pre-breeding studies in mung bean have been initiated involving diverse parents mainly from the cultivated gene pool, using the Bureau's core collection as starting material. Germplasm enhancement aims at widening the genetic base of breeding materials by transferring desired genes from unimproved germplasm into enhanced varieties. Mild and decentralized selected material is maintained in target sites across the country. A total of 102 progenies were finally advanced to F_5 for further selection and use by the breeders in National Agricultural Research System. The genetic potential of a few selected enhanced progenies with desired plant types and better yield traits were reported by Bisht et al. (2004). The study clearly demonstrates the potential of germplasm accessions conserved in gene banks for use in large-scale base-broadening efforts in mung bean. The AVRDC in Taiwan maintains 5616 accessions of mung bean. These accessions are characterized and available for exchange and utilization. The accessions have diversity in morphological traits in relation to geographical regions (Chen, Cheng, Jen, & Tsou, 1999). Taiwan is also in the process of developing representative core collections of mung bean germplasm.

About 400 accessions of urd bean were characterized and evaluated at NBPGR. A high range of variability was observed for different agro-morphological traits (Singh, Kumar et al., 2001). Most of the accessions have semi-erect growth habit, medium terminal leaf length, medium petiole length, green and pubescent leaves, determinate growth pattern, black seed colour and oval seed shape. Wide variability was also observed in quantitative characters like days to 50% flowering, days to maturity, number of pods per cluster, number of pods per plant, number of seeds per pod, 100-seed weight and yield (Table 10.3).

About 2000 accessions of moth bean were characterized and evaluated at NBPGR Regional Station, Jodhpur (Singh, Kumar et al., 2001). A wide range of variation was observed for yield and other growth characters in moth bean, as given in Table 10.3. The varieties showed a wide variation in nodulation and nitrogenase activity (Rao, Venkateswarlu, & Henry, 1984). Varietal differences exist for resistance to insect pests and diseases (Dabi & Gour, 1988). A wide range of variation was also observed for quality characters (Singh et al., 1974). Three promising accessions, namely PLMO39, PLMO55 and IC8851, have been identified amongst germplasm evaluated at NBPGR, Regional Station, Jodhpur. A wide range of variation for different agro-morphological attributes, biochemical constituents, and disease and pest reaction was reported among 690 accessions of rice bean studied for 36 descriptors (Arora, 1986; Arora et al., 1980; Chandel, Joshi, Arora, & Pant, 1978; Chandel, Joshi, & Pant, 1982; Negi et al., 1998; Singh et al., 1974). Accessions of rice bean exhibited wide variation for important agronomic traits (Table 10.3). Early maturing types were available in the material collected from Assam, India. The taller genotypes were observed from Orissa, profusely branched types from Mizoram and Manipur, types with more number of seeds per pod from Meghalaya and Mizoram, and collections with higher number of pods per peduncle, bold seeds and high grain yield from Manipur. The germplasm from Manipur possessed genotypes with several specific desirable traits. A high degree of polymorphism was noticed in the seed colour of rice bean. Several landraces had black, red, cream, violet, purple, maroon, brown, chocolate or mottled grains with greenish, brownish or ash grey background. A rare uniform light green colour occurred in a few accessions from the Mao hills bordering Manipur and Nagaland (Chandel, Arora, & Pant, 1988; Negi et al., 1998; Sarma, Singh, Gupta, Singh, & Srivastava, 1995).

Distinct seed coat colour groups in rice bean germplasm had considerable variation in epicotyl colour, shape and size of leaves, plant height, flowering time, flower colour, seed weight and protein content among the groups (Sastrapradja & Sutarno, 1977). However, there was not much variation in characters occurring within each group of seed coat colour. Evaluation of azuki bean germplasm was undertaken in Korea from 1985 to 1987, and 800 accessions were evaluated for 68 descriptors. Variability has been recorded for growth habit, time of maturity and seed colour. Early maturing cultivars are strictly bushy and mostly erect, while late maturing types are highly viny and branched, and some are decumbent.

Diversity in morphological characters of 206 accessions of 14 wild *Vigna* species from India was assessed. Of these, 12 species belonged to Asian *Vigna* in the subgenus *Ceratotropis* and two were *V. vexillata* and *V. pilosa*, belonging to subgenus *Plectotropis* and *Dolichovigna*, respectively (Bisht et al., 2005). Data on 71 morphological traits, both qualitative and quantitative, were recorded. Data on 45 qualitative and quantitative traits exhibiting higher variation were subjected to multivariate analysis for establishing species relationships and assessing the pattern of intraspecific variation. Of the three easily distinguishable groups in the subgenus *Ceratotropis*, all the species in the *mungo-radiata* group except *V. khandalensis*, namely *V. radiata* var. *sublobata*, *V. radiata* var. *setulosa*, *V. mungo* var. *silvestris* and *V. hainiana*, showed greater homology in vegetative morphology and growth habit. The species, however, differed in other plant, flower, pod and seed characteristics. Intraspecific variation was higher in *V. mungo* var. *silvestris* populations and three distinct clusters could be identified in multivariate analysis. *V. umbellata* showed more similarity to *V. dalzelliana* than *V. bourneae* and *V. minima* in the *angularis-umbellata* (azuki bean) group. Intraspecific variation was higher in *V. umbellata* than other species in the group. In the *aconitifolia-trilobata* (moth bean) group, *V. trilobata* populations were more diverse than *V. aconitifolia*. The cultigens of the conspecific wild species were more robust in growth, with large vegetative parts and often of erect growth with a three- to fivefold increase in seed size and seed weight, except *V. aconitifolia*, which has retained the wild-type morphology to a greater extent. More intensive collection, characterization and conservation of species diversity and intraspecific variations, particularly of the close wild relatives of Asian *Vigna* with valuable characters, such as resistance to biotic/abiotic stresses and more pod-bearing clusters per plant, assumes great priority in crop improvement programmes.

10.3.4 Germplasm Conservation

Vigna species have orthodox seeds that can be dried and stored for a long period with minimum loss of viability. A total of 10,551 accessions of various *Vigna* species comprising mung bean (3704), urd bean (3131), moth bean (1486), rice bean (2045) and azuki bean (185) have been stored at −18°C in the long-term repository of the national gene bank at NBPGR, New Delhi. The world germplasm collections of mung bean, urd bean, moth bean, rice bean, azuki bean are maintained at various institutes worldwide (Table 10.4). Green gram germplasm accessions are maintained by more than 35 institutions globally, with a total holding of more than 25,000 accessions. AVRDC, Taiwan, maintains 5616 accessions of mung bean. Limited germplasm accessions of moth bean are also available in countries, such as Bangladesh, Belgium and Kenya. Active collections are being maintained at NBPGR, mung bean and urd bean at Akola (Maharastra), moth bean at Jodhpur (Rajasthan), rice bean at Bhowali (Uttarakhand), Shillong (Meghalaya) and Shimla (Himachal Pradesh). The working collections of *Vigna* species are also maintained at the Indian Institution of Pulse Research (IIPR), Kanpur, India and its coordinating centres (Asthana, 1998).

10.3.5 Germplasm Registration

Four accessions each of mung bean and urd bean and two accessions each of moth bean and rice bean have been registered and maintained at NBPGR (Table 10.5).

10.3.6 Germplasm Documentation

NBPGR has done characterization and preliminary evaluation of over 5000 accessions of *Vigna* species and published seven catalogues or monographs on moth bean (3), mung bean (2) and rice bean (2). The ICAR Research Complex for Northeastern Region, Meghalaya, India has also brought out a bulletin on rice bean (Sarma et al., 1995). A catalogue describing the mung bean and urd bean collection maintained by the Division of Tropical Crops and Pasture, CSIRO, Australia, has been published by Imrie, Beech, and Thomas (1981). Subsequently, a germplasm catalogue of 6093 AVRDC accessions of mung bean and other *Vigna* species according to various descriptors has been published (Tay, Huang, & Chen, 1989). To generate a standardized and uniform database, a minimal set of descriptors of agri-horticultural crops has been proposed by NBPGR (Mahajan, Sapra, Srivastava, Singh, & Sharma, 2000) to characterize and evaluate the accessions of mung bean, urd bean, moth bean, rice bean and azuki bean.

10.4 Germplasm Utilization

Genetic resources are the basic raw material for crop improvement. Before 1960, most of the improved varieties were direct selections of germplasm collected from different agro-climatic regions within and outside the country. Several high-yielding varieties conferring resistance/tolerance to biotic and abiotic stresses have been developed

Table 10.4 Registered Germplasm of *Vigna* Species in the National Gene Bank at the NBPGR

S. No.	National identity	Donor identity	INGR No.	Year	Pedigree	Developing Institute	Novel Unique Features
Mung bean							
1	IC296679	Pentafoliate	97003	1997	LM 696×ML33	CCSHAU, Hisar	Pentafoliate with five small leaflets
2	IC296771	BSN-1	00011	2000	Nagpuri Local	OUAT, Bhubaneswar	High seed weight, extra long pod and high protein content
3	IC0589309	IPM 205-7	11043	2011	IPM 02-1×EC398889	IIPR, Kanpur, Uttar Pradesh	For super early maturity
4	IC0589310	IPM 409-4	11044	2011	PDM 288×IPM 03-1	IIPR, Kanpur, Uttar Pradesh	Extra early maturity in different genetic background
Urd bean							
1	IC553269	NA	07028	2007	Pant Urd-30	SVBPUA&T, Meerut	Brown pod and yellow seed
2	IC296878	Amp 36-13	02008	2002	Amphidiplod of K 851×MCK-2/ interspecific hybrid	CCSHAU, Hisar	Dwarf semi-erect with ground pod-bearing habit
3	NA	VBG-09-012	11045	2011	*V. mungo* ADT 3×*V. mungo* var. *silvestris*	NPRC, Vamban, Pudukkottai, Tamil Nadu	Multipod formation at base of peduncle, leaf axils and base of clusters
4	NA	VBG-04-014	11046	2011	Vamban 1×*V. mungo* var. *silvestris*	NPRC, Vamban, Pudukkottai, Tamil Nadu	Unique plant type
Moth bean							
1	IC296803	CZM-32	01024	2001	Mutant of moth bean variety Jadia	CAZRI, Jodhpur	Drought tolerant
2	IC432859	RMM-12	04095	2004	RMO-40	CAZRI, Jodhpur	Single stem, early maturity, high influx of sodium ions in root from soil
Rice bean							
1	IC0589127	PRR 2007-1	11020	2011	Naini X PRR 9402	GBPUA&T, Ranichauri, Uttarakhand	Narrow leaf, early maturity, determinate growth habit
2	IC0589128	PPR 2007-2	11021	2011	PRR 2×PRR 9301	GBPUA&T, Ranichauri, Uttarakhand	Narrow leaf, early maturity

Table 10.5 Germplasm Holding of Asian *Vigna* at Main Centres Worldwide

Country	Institute/Centre	Accessions
Green gram		
Bangladesh	Bangladesh Agricultural Research Institute (BARI), Gazipur	498
	Plant Genetic Resources Centre, Bangladesh Agricultural Research Institute, Gazipur	85
Colombia	Corporación Colombiana de Investigación Agropecuaria (CORPOICA), Palmira, Valle	135
Denmark	Plant Genetic Resources Unit, Crop Improvement Division, Ministry of Agriculture, Kabul	280
Germany	Gene Bank, Institute for Plant Genetics and Crop Plant Research (IPK), Gatersleben	61
India	National Bureau of Plant Genetic Resources (NBPGR), New Delhi	3704
Indonesia	Centre for Biology, Indonesian Institute of Sciences, Research and Development, Bogor Research Institute for Legumes and Tuber Crops (RILET), Malang	100 867
Japan	Department of Genetic Resources, National Institute of Agrobioliological Resources (NIAB), Tsukuba-gun, Ibaraki-ken	124
Kenya	National Gene Bank of Kenya, Crop Plant Genetic Resources, Kikuyu	311
Nepal	Nepal Agricultural Research Council (NARC), Lalitpur, Kathmandu	56
Nigeria	International Institute of Tropical Agriculture (IITA), Ibadan	79
Pakistan	Pakistan Agriculture Research Council, PGRI/NARC, Islamabad	754
Philippines	National Plant Genetic Research Institute (NPGRI), IPB/ University of Philippines, Los Banos, Laguna	6869
Russian Federation	N.I. Vavilov Research Institute of Plant Industry (VIR), St Petersburg, Russian Federation	727
Taiwan	Asian Vegetable Research and Development Centre (AVRDC), Shanhua	5616
USA	Southern Regional Plant Introduction Station, USDA-ARS-SAA, Griffin, GA	3891
Vietnam	Food Crops Research Institute, Hai Duong	161
	National Gene Bank Vietnam Agricultural Sciences	200
	Institute of Agriculture Sciences of South Vietnam, Ho Chi Minh City	400
Black Gram		
Bangladesh	Bari, Gazipur	339
	Plant Genetic Resources centre, Bangladesh Institute, Gazipur	106
Colombia	Centro de Investigación La Selva, (CoRPOICA), Rionegro Antioquia	108
India	NBPGR, New Delhi	3131
Nepal	NARC, Lalitpur Kathmandu	83

(*Continued*)

Table 10.5 (Continued)

Country	Institute/Centre	Accessions
Pakistan	Pakistan Agriculture Research Council, PGRI/NARC, Islamabad	693
Russian Federation	VIR, St Petersburg	210
Taiwan	AVRDC, Shanhua	481
USA	Southern Regional Plant Introduction Station, USDA-ARS, Griffin, GA	300
Moth Bean		
India	Indian Grassland and Fodder Research Institute (IGFRI), Jhansi, Uttar Pradesh NBPGR, New Delhi	727
Kenya	National Gene Bank of Kenya, Crop Plant Genetic Resources, Kenya	47
Russian Federation	VIR, St Petersburg	48
Taiwan	AVRDC, Taiwan, Province of China	28
USA	Southern Regional Plant Introduction Station, USDA-ARS, Griffin, GA	57
Rice Bean		
China	Institute of Crop Genetic Resources (CAAS), Beijing, China	1363
India	NBPGR, New Delhi	1486
Indonesia	Centre of Biology, Indonesian Institute of Sciences Research and Development, Bogor	100
Nepal	NARC, Kathmandu	124
Philippines	National Plant Genetic Resources Laboratory, IPB/UPLB College, Laguna	170
Taiwan	AVRDC, Shanhua	72
USA	Southern Regional Plant Introduction Station, USDA-ARS-SAA, Griffin GA	41
Azuki Bean		
Japan	Iwate Agriculture Experiment Station, Morioka-shi, Iwate-ken	214
	Tokachi Agriculture Experiment Station, Tokachi	2500
	Germplasm Storage Centre, NIAB, Tsukuba	142
Taiwan	AVRDC, Shanhua	125
China	Institute of Crop Germplasm Resources, CAAS Beijing	3736
India	NBPGR, New Delhi	185
Korea	Plant Genetic Resources Research Programme	1212
Philippines	National Plant Genetic Resources Laboratory, IPB/UPLB College, Laguna	161
Russian Federation	VIR, St Petersburg	140
USA	Southern Regional Plant Introduction Station, Georgia	301

Sources: FAO (1998), IPGRI Directory of Germplasm Collections, Singh et al. (2006). Figures of NBPGR updated as of 31 March 2012.

by using the germplasm resources. A large number of varieties have been developed in green gram in India. The earlier varieties were developed through selection. Type 1 is the first variety developed through selection from Muzaffarpur (Bihar) in 1948. Shining mung 1, Amrit, Panna, Co 1, Co 2, Khargone 1, Krishna 11 are some of the important varieties developed through this method. Since the 1960s, hybridization has been used to achieve variability. 'Type 44' is the first variety of green gram developed through hybridization (Type 1×Type 49) in Uttar Pradesh, and was released in 1962. Interspecific hybridization of green gram and black gram was attempted in the 1990s to develop early maturing, disease-resistant varieties. Three such varieties were released in India, including Pang Mung 4 (Type 44×UPU 2), HUM 1 (PHUM 1×Pant U 30) and IPM 99–125 (Pant mung 2×AMP 36). Through mutation breeding, over a dozen green gram varieties have been developed. Dhauli is the first mutant variety of green gram released in 1979 from Orissa Agricultural University and Technology. The other varieties include Co 4, Pant Moong 2, TAP 7, BM 4, MUM 2, LGG 407, LGG 450, TARM 1, TARM 2 and TARM 18, etc. The important varieties of green gram and their suitability to different agro-climatic zones and seasons are given in Table 10.6.

Table 10.6 Improved Varieties of Green Gram Recommended for Various Agro-Climatic Zones of India

Zone	Varieties
Northwestern Plains Zone (Punjab, Haryana, Western Uttar Pradesh, Himachal Pradesh, Jammu & Kashmir)	Type 44 (year round), Pusa Baisakhi (Z), PS 16 (Z), PS 7 (Z), Vamban 1 (spring), K 851 (S,Z), SML 32 (Z), Pusa 9072 (Z), PS 10 (Z), SML 668 (Z), Pant Moong 2,ML 267 (K), ML 337 (K), Pant Moong-3 (K), S 8 [Mohini (K)], Ganga 8 (K), Medium & Late: Varsha, Shining moong 1, RS 4, R 288–8, ML 1, ML 5, ML 9, T 51, Early: Pant Moong 1, ML 9, ML 131, Pusa 105
Northeastern Plains Zone (Eastern Uttar Pradesh, Bihar, Orissa, West Bengal, Assam)	Basant [PDM 84-143 (K,S], PDM 11 (Z), K 851 (S,Z), HUM 12 [(Z) Malviya Janchetra], Pusa 9531 (S), PDM 54 (K, Z), TARM 1 (S), PS 16 (K,Z), MG 368 (S), PDM 90239 (Z), Pusa Baisakhi (Z), Sunaina (Z), PDM 199 (Z), Panna [B105 (Z)], PS 10 (Z), PS 7 (Z), PDM 84–139 (Samrat (Z)], ML 337 (K), Pant Mung 4 [UPM 92-1(K)], S 8 (K), Sonali (E), Pant Moong 1 (E), Pant Moong 2 (E), Koperagaon, (M&L), Amrit, BR 2 (M & L), B1 (E)
Central Zone (Madhya Pradesh, Gujarat, Maharashtra)	PDM 11 (S), Pant Mung 5 (Z), Pusa 9531 (Z), HUM-1 (S), HUM 2 (Z), Pusa Baisakhi (R), PS 16 (Z), BM4 (K), PS 16 (K), Mohini (K), Gujarat 2, Sabarmati, Gujarat 12, Khargaon 1, Jalgaon 781, Krishna 11
Peninsular Zone (AP, Tamil Nadu, Karnataka, Kerala)	PDM 84–143 [Basant (K)], ML 337 (K), OUM-11-5 [Kamdeva (K)], PDM 54 (K), Jawahar 5 (K), PS 16 (K), Jawahar 45 (K), K 851 (K), Mohini (K), LGG 456 (R), Pusa 9072 (R), Pusa Baisakhi (R), TARM 1 (S), Malviya Jyoti [HUM 1(S)], Koperagaon, Kondaveedu, KM 1, KM 2, PDM 1, PDM 2, ADT 2, Co 2, Co 4, Co 65, Paiyur 1

Source: www.nsdl.niscair.res.in
S: Spring; Z: *Zaid* (March to June); R: *Rabi* (Winter); K: *Kharif* (Rainy season), E: Early; M & L: Medium & Late.

Mung bean is highly susceptible to yellow mosaic virus (YMV) in the northwest and northeast plain zones, causing a yield loss of about 15–20%. The selection of YMV-resistant varieties is a must for economical green gram cultivation. Some of the resistant varieties include: Pant Mung 1, Pant Mung 2, Pant Mung 3, Pang Mung 4, Narendra Mung 1, PDM 11, PDM 54, PDM 139, M 267, ML 337, ML 613, Basant, Samrat, HUM 1, HUM 2 and Pusa 9531. Powdery mildew (PM) also causes significant yield losses in green gram. TARM 1, TARM 2, TARM 18, CoG 4 are some of the PM-resistant varieties. Pusa 105, Kamdeva, ML 131 are resistant to both PM and YMV. A mung bean variety Keumseongnogdu has been bred in Korea that has multiple disease resistance and high yield potential (Lee et al., 1998). At AVRDC, the major thrust is on the improvement of mung bean germplasm enhanced for quality, including increased sulphur-containing amino acids and high yield under farmers' field conditions. Over 60 improved varieties have been developed in black gram in India since the 1950s. Selection from local material has contributed over 50% of the improved varieties. T9 is the first variety developed from Bareilly local in Uttar Pradesh (1948). Some other varieties developed through selection include T 27, T 65, T 77, Khargone 3, Mash 1-1, Mash 2, Naveen, ADT 1, D 6–7, D 75, Co 2, Co 3, etc. These varieties were later used in hybridization to develop high-yielding and disease-resistant varieties. KM 1 (G 31×Khargone 3) and ADT 2 (AB 1–33×ADT 1) are the first hybrids developed in black gram. Mutation breeding has also been used to develop six varieties in black gram to date. Co 4 is the first mutant black gram developed at Coimbatore in 1978. Other black gram varieties evolved through mutation include Manikya, TAU-1, TAU-2, TAU-4 and TAU-94-2. The important and improved varieties recommended for different agro-climatic zones of India are given in Table 10.7.

Several varieties in the past were developed in moth bean through single plant selections from local material (Singh & Thomas, 1970), which include B18-54 and B15-54 in Rajasthan; Nadiad 8-3-2 and Jagudan 9-2, Yawel 12-1 and Dhulia 3–5 in Maharashtra, and types 4301, 4312 and 4313 in Uttar Pradesh. Kumar and Rodge (2012) listed improved varieties of moth bean for different cropping regions of Rajasthan (Table 10.8).

The improved varieties of rice bean are listed in Table 10.9 along with their salient features. In addition, cultivars K1 and K16 developed in West Bengal having a forage yield of 250–300q/ha were found suitable for growing in West Bengal, Orissa, Tripura, Manipur, Meghalaya, Assam, Arunachal Pradesh, Kerala, Andhra Pradesh and Bihar (Rai & Patil, 1979). Tremendous possibilities exist for developing better cultivars through interspecific hybridization, such as *V. umbellata*×*V. angularis*, using embryo culture (Ahn & Hartmann, 1978) and *V. radiata*×*V. umbellata* (Rushid, Smartt, & Haq, 1987).

Lumpkin and McClary (1994) reviewed the breeding and genetics of azuki bean. In Japan, breeding of azuki bean was initiated as early as 1894 (Konno & Narikawa, 1978; Takahashi, 1917). A significant achievement in azuki bean breeding has also been made in Korean Republic and Taiwan. No improved cultivar so far has been released in India. Wide hybridization has been attempted among *Vigna* spp., aiming to incorporate certain characters, such as mung bean yellow mosaic virus (MYMV)

Table 10.7 Improved Varieties of Black Gram Recommended for Various Agro-Climatic Zones of India

Zone	Varieties	Rabi	Spring
Northwestern Zone (Punjab, Haryana, Rajasthan, Western Uttar Pradesh, Himachal Pradesh, Jammu & Kashmir)	T 9, T 65, PS 1, Pant U 35, Pant U 19, UG 218, Mash 48, Kulu 4, HPU-6, Pusa 1, WBU 108 (Sharda), IPU 94-1 (Uttar), Krishna		PDU 1, KU 300
Northeastern Zone (Eastern Uttar Pradesh, Bihar, West Bengal, Orissa, Assam)	T 9, T 65, PS 1, T 27, T 77, T 22, T 127, Pant U 19, Pant U-30, BR 68, Kalindi (B76), Naveen, Azad Urad 2, Uttar, DPU-88-31(Neelam)		Azad urd 1, UG 606, PDU 1 (Basant Bahar)
Central Zone (Madhya Pradesh, Gujarat, Maharashtra)	T 9, Pusa 1, Khargone 3, Gwalior 2, D 6–7, D 75, Mash 48, Pusa U 30, Ujjain-4, Barka (RBU 38), TPU-4, TU 94-2, VB 3		PDU 1
Peninsular Zone (Andhra Pradesh, Tamil Nadu, Kerala, Karnataka)	T9. WBG 26. Pusa 1, ADT 1, Khargone 3. ADT 2, PDM 2, Co 2 CO 3, Co 4, Co 5, Pant U 30, Mash 35-5, KM 2, Sharda, VB 3, Warangal 26	LBG 17 (Krishnayya), LBG 685, LBG 402, Prabhava), LBG 623, LBG 645	

Source: www.nsdl.niscair.res.in

resistance from *V. mungo* to *radiata*, and disease and insect resistance from *V. umbellata* to *V. radiata* (AVRDC, 1974). Successful hybridization between *V. radiata* and *V. glabrescens* resulted in four pure lines carrying moderate resistance to thrips (AVRDC, 1990). Tomooka et al. (2000) listed useful traits that could be transferred from *V. mungo* to *V. radiata*, such as resistance to diseases and insect pests, tolerance to adverse environments, non-shattering and high methionine content.

10.5 Limitations in Germplasm Use

The Asian *Vigna* species are very sensitive to photoperiod and temperatures, and these two variables have a very high bearing on the plant type and its adaptability in all these crops. Tickoo, Gajraj, and Manji (1994) elaborated this aspect in greater detail in the context of mung bean. Jain (1975) has argued that grain legumes as a group are still undergoing domestication. Not long ago in the cultivation history of

Table 10.8 Improved Varieties of Moth Bean Suitable for Different Cropping Regions in Rajasthan, India

Average Rainfall* (mm)	Region/ District	Cropped Area* (000 ha)	Productivity* (kg/ha)	Varieties (Year of Release)	Maturity	Important Traits (days)
170–200	Churu	**293.00**	470	FMO-96 (1996)	58–59	Erect upright and synchronized growth
	Jaisalmer	170.00	121	CAZRI Moth-3 (2003)	60–62	Erect and synchronized growth, escapes YMV and seed yield 700–800 kg/ha
200–250	Bikaner	283.00	215	RMO-40 (1994)	61–62	Less biomass erect growth and seed yield 600–750 kg/ha
	Barmer	208.00	194	RMO-225 (1999)	64–65	Field tolerance to YMC, synchronized growth and seed yield 650–700 kg/ha
250–300	Ganganagar	0.23	446	CAZRI Moth-3	60–62	–
	Hanumangarh	39.00	417	RMO-435 (2002)	64–65	Leaves dark green and seed yield 600–650 kg/ha
300–350	Jodhpur	159.00	251	CAZRI-Moth-2 (2002)	66–68	First variety from hybridization, dark green colour, seed yield 800–1200 Kg/ha
	Nagaur	215.00	218	RMO-435 RMo-257 (2005)	64–65	–
					63–65	Good for seed and fodder, seed yield 600–650 kg/ha
350–450	Sikar	0.93	289	CAZRI-Moth-1(1990)	73–75	Inputs responsive, natural source of YMV, seed yield 500–550 kg/ha
	Pali	0.32	239			
	Jalore	0.32	470			

Source: Data from Kumar and Rodge (2012).

Table 10.9 Rice Bean Cultivars Released and Notified at the National Level in India

S. No.	Cultivar	Year	Area of Adoption	Salient Features
1.	RBL1	1987	Punjab	Free from storage insects, YMV resistant
2.	RBL6	1991	Northwest and Northeast regions	High yielding, early maturity, resistant to disease and pests, photosensitive or insensitive
3.	PRR1	1996	Uttaranchal hills	High yielding, black seeded, medium maturity
4.	PRR2	1997	Northwest hills	High yielding, yellow seeded, long pods, high protein (20.5%)

Source: Data from Singh et al. (2006).

these crops, and even today in most areas of the growing countries, these crops are being grown under conditions not very much different from those of their wild relatives. Under conditions of low input management, the evolution has been for the survival of the crop species itself rather than for grain yield from the breeders' point of view. The silver lining has been the evolution of the symbiotic relationship of these crops with the nitrogen-fixing *Rhizobia*, and the subsequent high protein content of their seeds. However, it will always be debatable whether the evolution of symbiosis in grain legumes is a curse or a blessing. Other characters, such as indeterminate growth habit, photo- and thermoinsensitivity, low harvest index, shattering of ripe pods, seed hardiness and zero seed dormancy, have all evolved more via natural than human selection (Tickoo et al., 1994).

Further, information on intraspecific diversity, particularly in *mungo-radiata* complex, is lacking. Information on intraspecific diversity is essential for effective use of wild species germplasm in crop improvement programmes. The use of wild relatives as sources of new germplasm is well established in breeding programmes for crop improvement on a worldwide level, yet the efficiency of introduction of useful traits from wild germplasm, such as disease resistance and other agronomic characters into elite cultivars, varies greatly. Wild *Vigna* species have great potential for use in crop improvement programmes. Bruchids are a serious pest of grain legumes during storage. A wild mung bean accession, *V. radiata* var. *sublobata* was reported by AVRDC to be highly resistant to the bruchid *Callosobruchus chinensis* (L.) (Talekar, 1994). MYMV has been a major problem in mung bean. The wild species *V. radiata* var. *sublobata* is an important source for incorporating resistance to MYMV into cultivated varieties (Singh, 1994). In addition to the landraces and cultivars, the wild species therefore need to be collected, characterized and conserved carefully for use in crop improvement programmes.

In common with most grain legume crop species, the wild related species do not form a particularly extensive or accessible genetic resource. Many of the wild related species, such as *V. radiata* var. *sublobata*, *V. mungo* var. *silvestris*, *V. khandalensis*, *V. trilobata*, and *V. hainiana*, are gathered for their ripe seeds, which are boiled and eaten by the tribal/local communities during famines. Their overexploitation has threatened their occurrence in natural habitats. More variability in wild conspecific forms is to be collected, characterized and conserved carefully in addition to the fullest range of landraces and cultivars. Greater exploitation of the conspecific wild species with valuable characters is necessary to make extended cultivation economically attractive (Smartt, 1990). Some populations of *V. mungo* var. *silvestris*, *V. radiata* var. *sublobata* and *V. radiata* var. *setulosa* with valuable characters, such as more pod-bearing clusters and pods per cluster, have great agronomic potential for use in crop improvement programmes beside the resistance/tolerance to biotic stresses (Bisht et al., 2005). Sources of resistance available in *V. radiata* var. *sublobata* (Singh, 1994) need to be exploited more vigorously with help from biotechnological tools.

10.6 *Vigna* Species Genomic Resources

In recent years isozymes, random amplified polymorphic DNA (RAPD), restriction fragment length polymorphism (RFLP), amplified fragment length polymorphism (AFLP) and sequence tagged microsatellite site (STMS) markers have helped to enhance development of genome maps in various pulse crops. The analyses based on isozymes (Jaaska & Jaaska, 1990), four types of proteinase inhibitors (trypsin and chymotrypsin inhibitors, subtilisin and cysteine proteinase inhibitors; Konarev, Tomooka, & Vaughan, 2002), RAPD (Kaga, 1996; Tomooka, Lairungreang, & Egawa, 1996) and RFLP (Kaga, 1996) have confirmed that the azuki bean, mung bean and *aconitifolia* groups are distinct. Using RFLP, bruchid resistance gene has been mapped in a wild relative of *V. radiata* spp. *sublobata* in accession TC 1966 (Young et al., 1992). RAPD, RFLP and AFLP analyses of released cultivars and advanced lines revealed moderate to low levels of polymorphism. Principal component analysis showed a high degree of genetic similarity among the cultivars, due to the high degree of commonality in their pedigree and narrow genetic base (Karihaloo, Bhat, Lakhanpaul, Mahapatra, & Randhawa, 2001).The transformation process is generally reported to be difficult in legumes; however, a highly efficient transformation system has been developed for azuki bean (Sato, 1995). The genome size of species in the subgenus *Ceratotropis* are among the smallest for legumes, ranging from 470 to 560 Mbp for mung bean (Arumuganathan & Earle, 1991).Young, Danesh, Menancio-Hautea, and Kumar (1993) used RFLPs to map genes in mung bean that confer partial resistance to the PM fungus, *Erysiphe polygoni*. The results indicated that putative partial resistance loci for PM in mung bean can be identified with DNA markers, even in a population of modest size analysed at a single location in a single year.

Menancio-Hautea et al. (1993) investigated genome relationships between mung bean (*V. radiata*) and cowpea (*V. unguiculata*) based on the linkage arrangement of RFLP markers. A common set of probes derived from cowpea, common bean (*Phaseolus vulgaris*), mung bean and soybean (*Glycine max*) *Pst*I genomic libraries were used to construct the genetic linkage maps. Results indicated that nucleotide sequences are conserved, but variations in copy number were detected and several rearrangements in linkage orders appeared to have occurred since the divergence of the two species. Entire linkage groups were not conserved, but several large linkage blocks were maintained in both genomes. A genetic linkage map with 86 F_2 plants derived from an interspecific cross between azuki bean (*V. angularis*, $2n=2x=22$) and rice bean (*V. umbellata*, $2n=2x=22$) was developed by Kaga, Ishii, Tsukimoto, Tokoro, & Kamijima (2000). Based on the lineage of the common mapped markers, 7 and 16 conserved linkage blocks were found in the interspecific map of azuki bean×*V. nakashimae* and mung bean map, respectively. Although the present map is not fully saturated, it may facilitate gene tagging, quantitative trait locus (QTL) mapping and further useful gene transfer for azuki bean breeding.

Lambrides, Lawn, Godwin, Manners, and Imrie (2000) reported two genetic linkage maps of mung bean derived from the cross Berken ACC 41. Segregation distortion occurred in each successive generation after F_2. The regions of distortion identified in the Australian maps did not coincide with regions of the Minnesota (USA) map. A simple and rapid method for isolating microsatellite loci in mung bean *V. radiata* based on the 5′-anchored PCR technique revealing 23 microsatellite loci and 6 cryptically simple sequence repeats (SSRs) was reported by Kumar, Tan, Quah, and Yusoff (2002). These markers should prove useful as tools for detecting genetic variation in mung bean varieties for germplasm management and cross-breeding purposes. Humphry, Magner, McIntyre, Aitken, and Liu (2003) identified a major locus conferring resistance to the causal organism of PM, *Erysiphe polygoni* DC, in mung bean (*Vigna radiata* L. Wilczek) using QTL analysis with a population of 147 recombinant inbred individuals. To generate a linkage map, 322 RFLP clones were tested against the two parents and 51 of these were selected to screen the mapping population. The 51 probes generated 52 mapped loci, which were used to construct a linkage map spanning 350 cM of the mung bean genome over 10 linkage groups. Using these markers, a single locus was identified that explained up to a maximum of 86% of the total variation in the resistance response to the pathogen.

Construction of the first mung bean (*V. radiata* L. Wilczek) bacterial artificial chromosome (BAC) libraries was reported by Miyagi et al. (2004). These BAC clones were obtained from two ligations and represent an estimated 3.5 genome equivalents. This correlated well with the screening of nine random single-copy RFLP probes, which detected on average three BACs each. These mung bean clones were successfully used in the development of two PCR-based markers linked closely with a major locus conditioning bruchid (*Callosobruchus chinesis*) resistance. These markers will be invaluable in facilitating the introgression of bruchid resistance into breeding programmes, as well as the further characterization of the resistance locus. Basak, Kundagrami, Ghose, and Pal (2004) developed a YMV-resistance-linked DNA marker in *V. mungo* from populations segregating for YMV reaction. This was the first report of YMV-resistance-linked DNA marker development in any crop species using segregating populations. This YMV-resistance-linked marker is of potential commercial importance in resistance breeding of plants. Han et al. (2005) constructed a genetic linkage map from a backcross population of (*V. nepalensis*×*V. angularis*)×*V. angularis* consisting of 187 individuals. This moderately dense linkage map equipped with many SSR markers will be useful for mapping a range of useful traits, such as those related to domestication and stress resistance. The mapping population will be used to develop advanced backcross lines for high-resolution QTL mapping of these traits.

A black gram linkage map was developed by Chaitieng et al. (2006) and compared with azuki bean. The study suggested that the azuki bean SSR markers can be widely used for Asian *Vigna* species and the black gram genetic linkage map will assist in improvement of this crop. Prakit, Seehalak, and Srinives (2009) reported the development and characterization of genic microsatellite markers for mung bean by mining a sequence database, and the transferability of the markers to Asian *Vigna* species. A total of 157 markers were designed upon searching for SSR in 830 transcript sequences. Cross-species amplification in 19 taxa of Asian *Vigna* using 85 primers showed that amplification rates varied from 80% (*V. aconitifolia*) to 95.3% (*V. reflexo-pilosa*). These mung bean genic microsatellite markers will be useful to study genetic resource and conservation of Asian *Vigna* species. Souframanien and Gopalakrishna (2006) generated a recombinant inbred line mapping population (F_8) by crossing *V. mungo* (cv. TU 94-2) with *V. mungo* var. *silvestris*, and they screened for MYMV resistance. The ISSR marker technique was employed to identify markers linked to the MYMV resistance gene. The $ISSR811_{1357}$ marker was identified and validated using diverse black gram genotypes differing in their MYMV reaction. The marker will be useful for the development of MYMV-resistant genotypes in black gram. Swag et al. (2006) isolated and characterized new polymorphic microsatellites in mung bean (*V. radiata* L.). The newly developed markers are currently utilized for diversity assessment within the mung bean germplasm collection of the Korean gene bank.

Kaga Isemura, Tomooka, and Vaughan (2008) studied the genetics of domestication of azuki bean. Genetic differences between azuki bean (*V. angularis* var. *angularis*) and its presumed wild ancestor (*V. angularis* var. *nipponensis*) were resolved into QTLs for traits associated with adaptation to their respective distinct habits. A genetic linkage map constructed using progenies from a cross between Japanese cultivated and wild azuki beans covers 92.8% of the standard azuki bean linkage map. Domestication of azuki bean has involved a trade-off between seed number and seed size: fewer but longer pods and fewer but larger seeds on plants with shorter stature in cultivated azuki bean being at the expense of overall seed yield. Genes found related to germination and flowering time in cultivated azuki bean may confer a selective advantage to the hybrid derivatives under some ecological conditions and may explain why azuki bean has evolved as a crop complex in Japan. A genetic linkage map of black gram with 428 molecular markers was constructed by Gupta, Souframanien, and Gopalakrishna (2008) using an F_9 recombinant inbred population of 104 individuals. The population was derived from an intersubspecific cross between a black gram cultivar, TU94-2, and a wild genotype, *V. mungo* var. *silvestris*. The current map is the most saturated map for black gram to date and is expected to provide a useful tool for identification of QTLs and for marker-assisted selection of agronomically important characters in black gram. Tuba, Gupta, and Datta (2010) identified markers tightly linked to the genes responsible for resistance which will be useful for marker-assisted breeding for developing MYMV and PM-resistant cultivars in black gram. To make progress in genome analysis of the Asian *Vigna* cultigens, genetic linkage maps for azuki bean (*V. angularis*), mung bean (*V. radiata*), black gram (*V. mungo*) and rice bean (*V. umbellata*), among the fully domesticated *Vigna* species in Asia have been constructed using mapping populations between cultigens and their presumed wild ancestors mainly based on azuki bean genomic SSR markers (Kaga et al., 2010, www.gene.affrc.go.jp/pdf/misc/%20 international-WS_14_33.pdf). Newly developed cowpea genomic SSR markers and soybean EST-SSR markers have been integrated into the mung bean linkage map. Simultaneously, a detailed comparative genome map across four Asian *Vigna* species based on these linkage maps was constructed. Comparison of the order of common azuki bean SSR markers and RFLP markers on the linkage maps allowed detection of high-level macro-synteny among genomes of the four Asian *Vigna* species. The Asian *Vigna* comparative map is being used to develop a comparative map between Asian *Vigna* and soybean. Preliminary comparative approaches using sequence information of azuki bean, cowpea and soybean SSR markers on the mung bean linkage map could suggest presumed syntenic regions between mung bean and soybean. Although much more information is required to test the colinearity of markers, segmentations of soybean linkage block are frequently observed at most mung bean linkage groups. Further efforts are, however, needed to make steady progress on the establishment of a genomic base for the Asian *Vigna* by collaboration in order to utilize the gene and sequence information of soybean in Asian *Vigna* through comparative genome analysis.

Isemura, Kaga, Tabata, Somta, and Srinives (2012) analysed the genetic differences between mung bean and its presumed wild ancestor for domestication-related traits by QTL mapping. A genetic linkage map of mung bean was constructed using 430 SSR and EST-SSR markers from mung bean and its related species, and all these markers were mapped onto 11 linkage groups spanning a total of 727.6 cM. The present mung bean map is the first map where the number of linkage groups coincided with the haploid chromosome number of mung bean. In total 105 QTLs and genes for 38 domestication-related traits were identified. The useful QTLs for seed size, pod dehiscence and pod maturity that have not been found in other Asian *Vigna* species were identified in mung bean, and these QTLs may play an important role as new gene resources for other Asian *Vigna* species. The results provide a foundation that will be useful for improvement of mung bean and related legumes.

10.7 Conclusions

Asian *Vigna* species constitute an economically important group of cultivated and wild species, and a rich diversity occurs in India and other Asian countries. The five species of Asian pulses belonging to genus *Vigna* are closely related and are characteristically small seeded. The green gram is a popular food throughout Asia and other parts of the world, and its level of consumption can be expected to increase. The black gram, although very popular in India, is less likely to generate sufficient demand to stimulate production significantly outside its traditional areas. The azuki bean has generated interest as a pulse outside traditional areas of production and consumption; consumer demand for it could increase in the near future. Perhaps the most interesting future exists for rice bean, which has a high food value and tolerance to biotic and abiotic stresses. It possibly has the highest yielding capacity of any of the Asian *Vigna* and could become a useful crop if a sizeable consumer demand were built up. Moth bean has a future in India as a pulse crop. *Vigna trilobata* is probably most useful as a forage crop in semi-arid conditions. The fullest possible range of landraces and cultivars needs to be collected and conserved together with the conspecific wild related species. The wild germplasm resources have a great potential for widening the genetic base of *Vigna* gene pool by interspecific hybridization. The available genetic resources with valuable characters will therefore be required to make extended cultivation economically attractive.

References

Ahn, C. S., & Hartmann, R. W. (1978). Interspecific hybridization between rice bean [*Vigna umbellata* (Thunb.) Ohwi and Ohashi] and azuki bean [*Vigna angularis* (Wild.) Ohwi and Ohashi]. *Journal American Society Horticultural Science, 103*, 435–438.

Arora, R. K. (1986). Diversity and collection of wild *Vigna* species in India. *FAO/IBPGR Plant Genetic Resources Newsletter, 63*, 6–13.

Arora, R. K., Chandel, K. P. S., Pant, K. C., & Joshi, B. D. (1980). Rice bean – A tribal pulse of eastern India. *Economic Botany, 34*, 260–263.

Arora, R. K., & Nayar, E. R. (1984). *The wild relatives of crop plants in India.* New Delhi, India: National Bureau of Plant Genetic Resources. p. 90.

Arumuganathan, K., & Earle, E. D. (1991). Nuclear DNA content of some important plant species. *Plant Molecular Biology Report, 9*, 208–218.

Asthana, A. N. (1998). Pulse crop research in India. *Indian Journal of Agriculture Sciences, 68*, 448–452.

AVRDC, (1974). *AVRDC progress report 1973.* Tainan, Taiwan: Asian Vegetable Research and Development Centre, Shanhua. p. 86.

AVRDC, (1990). *AVRDC progress report 1989.* Tainan, Taiwan: Asian Vegetable Research and Development Centre Shanhua. p. 350.

Basak, J., Kundagrami, S., Ghose, T., & Pal, A. (2004). Development of yellow mosaic virus (YMV) resistance linked DNA marker in *Vigna mungo* from populations segregating for YMV reaction. *Molecular Breeding, 14*, 375–383.

Baudet, I. C. (1974). Signification taxonomique des caracteres bloastogeniquis dans la trisudes Papilionaceae-Phaseoleae. *Bulletin du Jardin Botanique National de Belgique, 44*, 259–293.

Baudoin, J. P., & Maréchal, R. (1988). Taxonomy and evoluation of the genus Vigna. In S. Shanmugasundaram & B. T. McLean (Eds.), *Mungbean proceedings of second international symposium* (pp. 2–12). Shanhua, Tainan, Taiwan: Asian Vegetable Research and Development Centre.

Bisht, I. S., Bhat, K. V., Lakhanpaul, S., Biswas, B. K., Ram, B., & Tanwar, S. P. S. (2004). The potential of enhanced germplasm for mungbean (*Vigna radiata* (L.) Wilczek) improvement. *Plant Genetic Resources, 2*, 73–80.

Bisht, I. S., Bhat, K. V., Lakhanpaul, S., Latha, M., Jayan, P. K., Biswas, B. K., et al. (2005). Diversity and genetic resources of wild *Vigna* species in India. *Genetic Resources and Crop Evolution*, *52*, 53–68.

Bisht, I. S., Mahajan, R. K., & Kawalkar, T. G. (1998). Genetic diversity in green gram (*Vigna radiata* L. Wilczek) and its use in crop improvement. *Annals of Applied Biology*, *132*, 301–312.

Bisht, I. S., Mahajan, R. K., & Patel, D. P. (1998). The use of characterization data to establish the Indian mung bean core collection and assessment of genetic diversity. *Genetic Resources and Crop Evolution*, *45*, 127–133.

Biswas, M. R., & Dana, S. (1975). Black gram × rice bean cross. *Cytologia*, *40*, 795–797.

Biswas, M. R., & Dana, S. (1976). Phaseplus aconitifolius × phaseolus trilobus. *Indian Journal of Genetics*, *36*, 125–131.

Bose, R. D. (1939). Studies on Indian pulses. IV. Mung or green grams (*Phaseolus aureus* Roxb.). *Indian Journal of Agriculture Sciences*, *2*, 607–724.

Chaitieng, B., Kaga, A., Tomooka, N., Isemura, T., Kuroda, Y., & Vaughan, D. A. (2006). Development of a black gram [*Vigna mungo* (L.) Hepper] linkage map and its comparison with an azuki bean [*Vigna angularis* (Willd.) Ohwi and Ohashi] linkage map. *Theoretical and Applied Genetics*, *113*, 1261–1269.

Chandel, K. P. S. (1981). Wild *Vigna* species in the Himalayas. *Plant Genetic Resources Newsletter*, *45*, 17–19.

Chandel, K. P. S. (1984). Role of wild *Vigna* species in the evolution and improvement of mung [*Vigna radiata* (L.) Wilczek] and urdbean [*V. mungo* (L.) Hepper]. *Annals of Agricultural Research*, *5*, 98–111.

Chandel, K. P. S., Arora, R. K., & Pant, K. C. (1988). Rice bean: *A potential grain legume*: *Science Monograph No. 12*. New Delhi, India: National Bureau of Plant Genetic Resources. p. 60.

Chandel, K. P. S., Joshi, B. S., Arora, R. K., & Pant, K. C. (1978). Rice bean – A new pulse with high potential. *Indian Farming*, *18*, 19–22.

Chandel, K. P. S., Joshi, B. S., & Pant, K. C. (1982). Genetic resources of *Vigna* species in India and their contribution in crop improvement. In R. B. Singh & R. M. Singh (Eds.), *Advances in cytogenetics and crop improvement* (pp. 55–66). New Delhi, India: Kalyani Publishers.

Chandel, K. P. S., & Pant, K. C. (1982). Genetic resources of *Vigna* species in India: Their distribution, diversity and utilization in crop improvement. *Annals of Agricultural Research*, *3*, 19–34.

Chen, S., Cheng, C., Jen, F. M., & Tsou, S. C. S. (1999). Geographical distribution and genetic diversity for morphological characters in a world germplasm collection of mungbean (*Vigna radiata* (L.) Wilczek)]. *China Journal of Agriculture Research*, *47*, 108–124.

Dabi, R. K., & Gour, H. N. (1988). Field screening of moth bean (*Vigna aconitifolia*) for susceptibility to insect pests and diseases. *Indian Journal of Agricultural Sciences*, *58*, 843–844.

Dana, S. (1966a). Species cross between *Phaseolus aureus* Roxb. and *P. trilobus* Ait. *Cytologia*, *3*, 176–187.

Dana, S. (1966b). Interspecific hybrid between *Phaseolus mungo* L. × *P. trilobus* Ait. *Journal of Cytology and Cytogenetics*, *1*, 61–66.

Dana, S. (1998). Collection of wild *Vigna* in seven states in India. *Green Journal*, *1*, 9–12.

Debouck, D. G. (2000). Biodiversity, ecology and genetic resources of Phaseolus beans-seven answered and unanswered questions: *Proceedings of the 7th MAFE international workshop on genetic resources Part 1: wild legumes*. Japan: AFFRC and NIAR. pp. 95–123.

de Candolle, A.P. (1825). Leguminosae. In *Prodromus systematis naturalis regni vegetabilis* 2 (Phaseolus 390–396), Paris, pp. 93–524.

FAO, (1998). *FAOSTAT annual report 1998*. Rome, Italy: *Food and Agricultural Organization of the United Nations*.

Gautam, P. L., Sharma, G. D., Srivastava, U., Kumar, A., Saxena, R. K., & Srinivasan, K. (Eds.), (2000). *20 Glorious years of NBPGR, 1976*. New Delhi, India: National Bureau of Plant Genetic Resources.

Gupta, S. K., Souframanien, J., & Gopalakrishna, T. (2008). Construction of a genetic linkage map of black gram, *Vigna mungo* (L.) Hepper, based on molecular markers and comparative studies. *Genome*, *51*, 628–637.

Han, O. K., Kaga, A., Isemura, T., Wang, X. W., Tomooka, N., & Vaughan, D. A. (2005). A genetic linkage map for azuki bean [*Vigna angularis* (Willd.) Ohwi & Ohashi]. *Theoretical and Applied Genetics*, *111*, 1278–1287.

Humphry, M. E., Magner, T., McIntyre, C. L., Aitken, E. A. B., & Liu, C. J. (2003). Identification of a major locus conferring resistance to powdery mildew (*Erysiphepolygoni* DC) in mungbean (*Vigna radiata* L. Wilczek) by QTL analysis. *Genome*, *46*, 738–744.

Imrie, B. C., Beech, D. B., & Thomas, B. (1981). *A catalogue of the Vigna radiata and V. mungo germplasm collection of division of tropical crops and pastures, genetic resource communication 2*. Canberra, Australia: CSIRO. p. 15.

Isemura, T., Kaga, A., Tabata, S., Somta, P., & Srinives, P. (2012). Construction of a genetic linkage map and genetic analysis of domestication related traits in mungbean (*Vigna radiata*). *PLoS ONE*, *7*(8), e41304. doi:10.1371/journal.pone.0041304.

Jaaska, V., & Jaaska, V. (1990). Isozyme variation in asian beans. *Acta Botanica*, *103*, 281–290.

Jain, H. K. (1975). Breeding for yield and other attributes in grain legumes. *Indian Journal of Genetics and Plant Breeding*, *35*, 169–187.

Jain, H. K., & Mehra, K. L. (1980). Evaluation, adaptation, relationship and cases of the species of *Vigna* cultivation in Asia. In RJ Summerfield & AH Butnting (Eds.), *Advances in legume science* (pp. 459–468). Kew, UK: Royal Botanical Gardens.

Kaga, A. (1996). Construction and application of linkage maps azuki bean (*Vigna angularis*). *Ph.D. Thesis*. Kobe, Japan: Kobe university, p. 210.

Kaga, A., Isemura, T., Shimizu, T., Somta, P., Srinives, P., Tabata, S., et al. (2010). *Asian Vigna*. Genome Research. <www.gene.affrc.go.jp/pdf/%20misc/%20international-WS_14_33.pdf/>.

Kaga, A., Isemura, T., Tomooka, N., & Vaughan, D. A. (2008). The genetics of domestication of the azuki bean (*Vigna angularis*). *Genetics*, *178*, 1013–1036.

Kaga, A., Ishii, T., Tsukimoto, K., Tokoro, E., & Kamijima, O. (2000). Comparative molecular mapping in *Ceratotropis* species using an interspecific cross between azuki bean (*Vigna angularis*) and rice bean (*V. umbellata*). *Theoretical and Applied Genetics*, *100*, 207–213.

Karihaloo, J. L., Bhat, K. V., Lakhanpaul, S., Mahapatra, T., & Randhawa, G. J. (2001). Molecular characterization of germplasm. In B. S. Dhillon, K. S. Varaprasad, M. Singh, S. Archak, U. Srivastava, & G. D. Sharma (Eds.), *National Bureau of Plant Genetic Resources: A compendium of achievements* (pp. 166–182). New Delhi: National Bureau of Plant Genetic Resources.

Kawalkar, T. G., Bisht, I. S., Mahajan, R. K., Patel, D. P., Gupta, P. N., & Chandel, K. P. S. (1996). *Catalogue on green gram [Vigna radiata L. Wilczek] germplasm*. New Delhi, India: National Bureau of Plant Genetic Resources. p. 130.

Konarev, A., Tomooka, N., & Vaughan, D. A. (2002). Proteinase inhibitor polymorphism in the genus *Vigna* subgenus *ceratotropis* and its biosystematic implications. *Euphytica*, *123*, 165–177.

Konno, S., & Narikawa, T. (1978). Recent studies and problems on breeding and cultivation of azuki bean in Japan: *Proceedings Ist international mungbean symposium*. Shanhua, Taiwan: Asian Vegetable Research and Development Centre. pp. 236–239.

Kumar, D., & Rodge, A. B. (2012). Status, scope and strategies of arid legumes research in India – A review. *Journal of Food Legumes, 25*, 255–272.

Kumar, S. V., Tan, S. G., Quah, S. C., & Yusoff, K. (2002). Isolation of microsatellite markers in mungbean, *Vigna radiata. Molecular Ecology Notes, 2*, 96–98.

Lambrides, C. J., Lawn, R. J., Godwin, I. D., Manners, J., & Imrie, B. C. (2000). Two genetic linkage maps of mungbean using RFLP and RAPD markers. *Australian Journal of Agricultural Research, 51*, 415–425.

Lawn, R. J. (1995). The Asian Vigna species. In J. Smartt & N. W. Simmonds (Eds.), *The evolution of crop plants* (pp. 321–326) (2nd ed.). Harlow, UK: Longman.

Lee, Y. S., Choi, K. J., Choi, J. K., Park, T. D., Nam, C., Kim, B. H., et al. (1998). A new multi-disasters restant and high yielding mungbean variety 'Keumseongnogdu'. *RDAJ Crop Science, (II), 40*, 121–125.

Lukoki, L., Maréchal, R., & Otoul, E. (1980). Les ancetres sauvages des haricots cultives: *Vigna radiata* (L.) Wilczed et V. Mungo (l.) Hepper. *Bulletin du Jardin Botanique, 50*, 385–391.

Lumpkin, T. A., & McClary, D. C. (1994). *Adsuki bean: Botany production and uses*. UK: Commonwealth Agriculture Bureau International. p. 255.

Machado, M., Tai, W., & Baker, L. R. (1982). Cytogenetic analysis of the interspecific hybrid *Vigna radiata* × *V. umbellata. Journal of Heredity, 73*, 205–208.

Maekawa, F. (1995). Topo-morphological and taxonomical studies in phaseoliae, leguminosae. *Japanase Journal of Botany, 15*, 103–106.

Mahajan, R. K., Sapra, R. L., Srivastava, U., Singh, M., & Sharma, G. D. (2000). *Minimal descriptors for characterization and evaluation of agri-horticultural crops (Part I)*. New Delhi, India: National Bureau of Plant Genetic Resources. p. 230.

Malik, S. S., Srivastava, U., Tomar, J. B., Bhandari, D. C., Pandey, A., Hore, D. K., et al. (2001). Plant exploration and germplasm collection. In B. S. Dhillon, K. S. Varaprasad, M. Singh, S. Archak, U. Srivastava, & G. D. Sharma (Eds.), *National bureau of plant genetic resources: A compendium of achievement* (pp. 31–68). New Delhi, India: National Bureau of Plant Genetic Resources.

Maréchal, R., Mascherpa, J. M., & Stainer, F. (1978). Etude taxonomique d'un groupe complexe d' especes des genres Phaseolus et Vigna (Papilionaceae) sur La base de donne'es morphologiques et polliniques, traitees par I' analyse informatique. *Memoires des Conservatoire et Jardin de Geneve Boissiera (Geneva), 28*, 1–273.

Menancio-Hautea, D., Fatokun, C. A., Kumar, L., Danesh, D., & Young, N. D. (1993). Comparative genome analysis of mungbean (*Vigna radiata* L. Wilczek) and cowpea (*V. unguiculata* L. Walpers) using RFLP mapping data. *Theoretical and Applied Genetics, 86*, 797–810.

Miyagi, M., Humphry, M., Ma, Z. Y., Lambrides, C. J., Bateson, M., & Liu, C. J. (2004). Construction of bacterial artificial chromosome libraries and their application in developing PCR-based markers closely linked to a major locus conditioning bruchid resistance in mungbean (*Vigna radiata* L. Wilczek). *Theoretical and Applied Genetics, 110*, 151–156.

Negi, K. S., Pant, K. C., Muneem, K. C., Agarwal, R. C., Sharma, B. D., Hore, D. K., et al. (1998). *Catalogue on rice bean [Vigna umbelleta (Thunb.) ohwi and ohashi] germplasm*. Bhowali, Nainital, Uttar Pradesh, India: National Bureau of plant genetic resources, regional station. p. 73.

Paroda, R. S., & Thomas, T. A. (1988). Genetic resources of mungbean (*Vigna radiata* (L.) Wilczek) in India. In S. Shanmugasundaram & B. T. McLean (Eds.), *Proc 2nd int. symp mungbean* (pp. 19–28). Shanhua, Taiwan: AVRDC.

Prakit, S., Seehalak, W., & Srinives, P. (2009). Development, characterization and cross-species amplification of mungbean (*Vigna radiata*) genic microsatellite markers. *Conservation Genetics*, *10*, 1939–1943.

Purseglove, J. W. (1974). *Tropical crops, dicotyledons*. Harlow, UK: Longman. pp. 539–555.

Rachie, K. O., & Roberts, L. M. (1974). Grain legumes of the lowland tropics. *Advances in Agronomy*, *26*, 1–132.

Rai, S. D., & Patil, B. D. (1979). Legume forage- you will like to grow cowpea, clover, Lucerne and rice bean. *Intensive Agriculture*, *17*, 31–32.

Rao, A. V., Venkateswarlu, B., & Henry, A. (1984). Genetic variation in nodulation and nitrogenase activity in guar and moth. *Indian Journal of Genetics*, *44*, 425–428.

Rushid, K. A., Smartt, J., & Haq, N. (1987). Hybridization in the genus *Vigna*. In S. Shanmungasundaran & B. T. McLean (Eds.), *Mungbean. Proceedings of international symposium* (pp. 205–214). Taipei, Taiwan: AVRDC Press.

Sarma, B. K., Singh, M., Gupta, H. S., Singh, G., & Srivastava, L. S. (1995). *Studies in rice bean germplasm*: *Research Bulletin No. 34*. Barapani, Meghalaya, India: ICAR Research Complex for NEH Region.

Sastrapradja, S., & Sutarno, H. (1977). *Vigna umbellata* (L.) DC. In *Annals bogoriensis,* VI, Indonesia, pp. 155–167.

Sato, T. (1995). *Basic study of biotechnology in Azuki bean (Vigna angularis* Ohwi and Ohashi). Report of Hokkaido Prefectural Agric. Exp. Stations, No. 87 (in Japanese with English summary).

Satyan, B. A., Mahishi, D. M., & Shivashankar, G. (1982). Meiosis in the hybrid between green gram and rice bean. *Indian Journal of Genetics*, *42*, 356–359.

Sen, N. K., & Ghosh, A. K. (1961). Genetic studies in green gram. *Indian Journal of Genetics*, *19*, 210–227.

Singh, D. P. (1994). Breeding for resistance to diseases in mungbean: Problems and prospects. In A. N. Asthana & D. H. Kim (Eds.), *Recent advances in pulses research* (pp. 152–164). Kanpur, India: Indian Society of Pulses Research (IIPR).

Singh, H. B., Joshi, B. S., Chandel, K. P. S., Pant, K. C., & Saxena, R. K. (1974). Genetic diversity in some Asian phaseolus species and its conservation Proceedings 2nd SABRAO congress. *Indian Journal of Genetics*, *34*, 52–57.

Singh, H. B., & Thomas, T. A. (1970). Moth bean *Phaseolus acondifolius* Jacq: *Pulse Crops of India*. New Delhi: Indian Council of Agricultural Research. pp.156–158.

Singh, M., Bisht, I. S., Sardana, S., Gautam, N. K., Husain, Z., Gupta, S., et al. (2006). Asiatic *Vigna*. In B. S. Dhillon, S. Saxena, A. Agrawal, & R. K. Tyagi (Eds.), *Plant genetic resources: Foodgrain crops* (pp. 275–301). New Delhi: Narosa Publishing House Pvt. Ltd..

Singh, M., Kumar, G., Abraham, Z., Phogat, B. S., Sharma, B. D., & Patel, D. P. (2001). Germplasm evaluation. In B. S. Dhillon, K. S. Varaprasad, Mahendra Singh, Sunil Archak Umesh Srivastava, & G. D. Sharma (Eds.), *National Bureau of Plant Genetic Resources: A compendium of achievements* (pp. 116–165). New Delhi: NBPGR.

Singh, R. V., Chand, D., Brahmi, P., Verma, N., Tyagi, V., & Singh, S. P. (2001). Germplasm exchange. In B. S. Dhillon, K. S. Varaprasad, M. Singh, S. Archak Umesh Srivastava, & G. D. Sharma (Eds.), *National Bureau of Plant Genetic Resources: A compendium of achievements* (pp. 69–89). New Delhi: NBPGR.

Smartt, J. (1990). Grain legume: Evolution and genetic resources: *The old world pulses: Vigna species*. Cambridge, UK: Cambridge University Press.

Souframanien, J., & Gopalakrishna, T. (2006). ISSR and SCAR markers linked to the mungbean yellow mosaic virus (MYMV) resistance gene in black gram [*Vigna mungo* (L.) Hepper]. *Plant Breeding*, *125*, 619–622.

Swag, J. G., Chung, J. W., Chung, H. K., Lee, J. H., Ma, K. H., & Dixit, A. (2006). Characterization of new microsatellite markers in mung bean, *Vigna radiata* (L.). *Molecular Ecology Notes*, *6*, 1132–1134.

Takahashi, Y. (1917). Experimental results of azuki characterization and breeding in Japan. *Bulletin of Hokkaido Agricultural Experimentation*, *7*, 1–181.

Talekar, N. S. (1994). Sources of resistance to major insect pests of mungbean in Asia. In A. N. Asthana & D. H. Kim (Eds.), *Recent advances in mungbean research* (pp. 40–49). Kanpur, India: Indian Society of Pulses Research (IIPR).

Tateishi, Y. (1985). A revision of the azuki bean group, *the subgenus ceratotropis of the genus vigna (Leguminoseae). Ph.D. Thesis*. Japan: Tohoku University, p. 292.

Tateishi, Y. (1996). Systematics of the species of Vigna subgenus ceratotropis. In P. Srinives, C. Kitbamroong, & S. Miyazaki (Eds.), *Mungbean germplasm: Collection, evaluation and utilisation for breeding programs* (pp. 9–24). Tsukuba, Japan: Japan International Research Center for Agricultural Sciences.

Tay, D. C. S., Huang, V. K., & Chen, C. Y. (1989). *Germplasm catalogue of mungbean (Vigna radiata) and other Vigna species*. Shanhua, Taiwan: Asia Vegetable Research and Development Centre (AVRDC). pp. 316–318, Pub. No. 89.

Tickoo, J. L., Gajraj, M. R., & Manji, C. (1994). Plant type in mung bean (*Vigna radiata* L. Wilczek). In A. N. Asthana & D. H. Kim (Eds.), *Recent advances in mungbean research* (pp. 197–213). Kanpur, India: Indian Society of Pulses Research (IIPR).

Tomooka, N., Egawa, Y., & Kaga, A. (2000). Biosystematics and genetic resources of the genus *Vigna* subspecies *ceratotropis*. In D. Vaughen, N. Tomooka, & A. Kaga (Eds.), *The 7th MAFF international workshop on genetic resources. Part 1: wild legumes* (pp. 337–362). Japan: Ministry of Agriculture, Forestry and Fisheries and National Institute of Agrobiological Resources.

Tomooka, N., Lairungreang, C., & Egawa, Y. (1996). Taxonomic position of wild *Vigna* species collected in Thailand based on RAPD analysis. In P. Srinives, C. Kitamroong, & S. Miazaki (Eds.), *Moongbean germplasm: Collection, evaluation and utilization for breeding programs* (pp. 31–40). Tsukuba, Japan: Japan International Research Centre for Agricultural Sciences.

Tomooka, N., Lairungreang, C., Nakeeraks, P., Egawa, Y., & Thavarasook, C. (1991). *Mung bean and the genetic resources*. Japan: TARC.

Tomooka, N., Lairungreang, C., Nakeeraks, P., Egawa, Y., & Thararasook, C. (1992). Development of bruchid resistant mungbean line using wild mungbean germplasm in Thailand. *Plant Breeding*, *109*, 60–66.

Tomooka, N., Vaughan, D. A., & Moss, H. (2002). *The Asian vigna: Genus vigna subgenus ceratotropis genetic resources*. The Netherlands: Kluwer Academic Publishers. p. 277.

Tuba, A. K., Gupta, S., & Datta, S. (2010). Mapping of mungbean yellow mosaic india virus (MYMIV) and powdery mildew resistant gene in black gram [*Vigna mungo* (L.) Hepper]. *Electronic Journal of Plant Breeding*, *1*, 1148–1152.

Vavilov, N. I. (1926). *Studies on the origin of cultivated plants*. Leningrad: Institute of Applied Botany and Plant Breeding.

Verdcourt, B. (1970). Studies in the leguminosae–papilionoideae for the flora of tropical East Africa. IV. *Kew Bulletin*, *24*, 507–569.

Young, N. D., Danesh, D., Menancio-Hautea, D., & Kumar, L. (1993). Mapping oligogenic resistance to powdery mildew in mungbean with RFLPs. *Theoretical and Applied Genetics*, *87*, 243–249.

Young, N. D., Kumar, L., Menancio-Hautea, D., Danesh, D., Talekar, N. S., Shanmungasundrum, S., et al. (1992). RFLP mapping of a major bruchid resistance gene in mungbean [*Vigna radiata* (L.) Wilczek]. *Theoretical and Applied Genetics*, *84*, 839–844.

11 Grass Pea

Shiv Kumar[1], *Priyanka Gupta*[1], *Surendra Barpete*[1], *A. Sarker*[2], *Ahmed Amri*[1], *P.N. Mathur*[3] *and Michael Baum*[1]

[1]International Center for Agricultural Research in the Dry Areas (ICARDA), Aleppo, Syria, [2]South Asia and China Regional Program (SACRP) of ICARDA, New Delhi, India, [3]Bioversity International Office for South Asia, New Delhi, India

11.1 Introduction

Grass pea (*Lathyrus sativus* L.) is one of the hardiest but most underutilized crops for adaptation to fragile agro-ecosystems, because of its ability to survive under extreme climatic conditions such as drought, water stagnation and heat stress. It is an annual cool-season legume crop of economic and ecological significance in South Asia and sub-Saharan Africa, and to a limited extent in Central and West Asia, North Africa (CWANA), southern Europe and South America. It is grown mainly for eating purposes in India, Bangladesh, Nepal, Pakistan and Ethiopia, and for feed and fodder purposes in other countries (Campbell, 1997; Kumar, Bejiga, Ahmed, Nakkoul, & Sarker, 2011; Siddique, Loss, Herwig, & Wilson, 1996). Grass pea grains are a good protein supplement (24–31%) to the cereal-based diet of poor people in areas of its major production (Aletor, Abd-El-Moneim, & Goodchild, 1994). Globally, the area under grass pea cultivation is estimated at 1.50 million ha, with annual production of 1.20 million tonnes (Kumar, Bejiga et al., 2011). The crop has not attained much progress, due to the limited research on genetic and genomic resources available for grass pea in the gene banks of world. The knowledge that the excessive consumption of grass pea can lead to a neurological disorder in humans has further discouraged adaptive research on this orphaned crop. Therefore, conservation and sustainable use of genetic resources are of paramount importance for grass pea improvement. In this review, we discuss the present status of the genetic and genomic resources of *Lathyrus* and their importance in crop improvement.

11.2 Origin, Distribution, Diversity and Taxonomy

Grass pea is believed to have originated and become domesticated in the Mediterranean region and later spread to other continents. Vavilov (1951) described

Genetic and Genomic Resources of Grain Legume Improvement. DOI: http://dx.doi.org/10.1016/B978-0-12-397935-3.00011-6

Central Asia and Abyssinia as the centres of origin. However, archaeological evidence revealed that its cultivation began in the Balkan Peninsula in the early Neolithic period. Kislev (1989) believed it originated in Southwest and Central Asia, later extending into the eastern Mediterranean region. Charred seeds of *Lathyrus* species were unearthed during excavation in Israel; it is believed that the seeds were carried to the Levant from the Aegean, in the Bronze Age, by traders or with Philistine immigrants (Mahler-Slasky & Kislev, 2010). However, the natural distribution of *L. sativus* is completely obscured by cultivation, making it difficult to precisely locate its centre of origin. The lack of morphological differences between wild and domesticated populations presumably arose from the dual-purpose use of *L. sativus* (grain and forage) in those areas to which it is native. The small-seeded accessions and subaccessions are primitive types with hard seeds, while selection for forage use has resulted in landraces with broad leaves, pods and seeds but low seed yield in the Mediterranean region. Grass pea underwent further diversification and domestication in the Near East and North African region. Diversity of *Lathyrus* species is found in Europe, Asia and North America, and extended to South America and East Africa, but the main centre of diversity remains primarily in the Mediterranean and Irano-Turanian regions (Kupicha, 1981). It is adapted to temperate regions but can also be found at high altitudes in tropical Africa. The genus contains many restricted endemic species present in all continents except Australia and Antarctica (Kupicha, 1981). The ecogeographic distribution of all but a few *Lathyrus* species is poorly understood, particularly those in the section *Notolathyrus* that are endemic to South America. There is a need for a detailed ecogeographic study of the whole genus if it is to be effectively and efficiently conserved and utilized for grass pea genetic improvement. The most widely cultivated species for human consumption is *L. sativus*. Other species which are grown for forage and/or grain purposes are *L. cicera*, *L. ochrus*, *L. clymenum*, *L. tingitanus*, *L. latifolius* and *L. sylvestris* (IPGRI, 2000). However, *L. cicera* is cultivated in Greece, Cyprus, Iran, Iraq, Jordan, Spain and Syria, and *L. ochrus* in Cyprus, Greece, Syria and Turkey (Saxena, Abd El Moneim, & Raninam, 1993). Some other species, like *L. hirsutus* and *L. clymenum*, are cultivated as minor forage or fodder crops in southern United States and Greece (Sarker, Abd El Moneim, & Maxted, 2001). Some species, such as *L. odoratus*, *L. latifolius* and *L. sylvestris*, are grown as ornamental crops.

The genus *Lathyrus*, along with *Vicia*, *Lens*, *Pisum*, and *Vavilovia*, belongs to the tribe *Vicieae* of the subfamily Papilionoideae. The precise generic boundaries between these genera have been much debated, but the oroboid species are considered to form a bridge between *Lathyrus* and *Vicia* (Kupicha, 1981). There are about 187 species in the genus *Lathyrus* (Allkin, Goyder, Bisby, & White, 1983, 1986). The taxonomic classification proposed by Kupicha (1983) dividing these species into 13 sections has been accepted, but the phylogenetic relationships among sections and species require further detailed investigation involving morphological, biochemical, cytogenetic and molecular markers (Table 11.1). Based on morphology and taxonomy, *Lathyrus* species are classified into five groups: *Clymenum*, *Aphaca*, *Nissolia*, *Cicerula* and *Lathyrus*. The first four groups are composed of annual species, whereas the remaining species, mostly perennials, are assigned to progressively smaller, more numerous and better-defined sections (Asmussen & Liston,

Table 11.1 Classification and Distribution of the Genus *Lathyrus*

Section	Species	Important Species	Geographical Distribution
Orobus	54		Europe, West and East Asia, Northwest Africa, North and Central America
Lathyrostylis	20		Central and Southern Europe, West Asia, Northwest Africa
Orobon	1		Anatolia, Caucasia, Crimea, Iran
Lathyrus	33	*L. annuus, L. cicera, L. sativus, L. sylvestris, L. tingitanus, L. tuberosus, L. gorgoni, L. hirsutus, L. latifolius, L. odoratus, L. rotundifolius, L. blepharicarpus*	Europe, Canaries, West and Central Asia, North Africa
Pratensis	6	*L. pratensis*	Europe, West and Central Asia, North Africa
Aphaca	2	*L. aphaca*	Europe, West and Central Asia, North Africa
Clymenum	3	*L. clymenum, L. ochrus*	Mediterranean
Orobastrum	1		Mediterranean, Crimea, Caucasia
Viciopsis	1		Southern Europe, Eastern Anatolia, North Africa
Linearicarpus	7		Europe, West and Central Asia, North and East Africa
Nissolia	1		Europe, West and Central Asia, Northwest Africa
Neurolobus	1		West Crete
Notolathyrus	23		Temperate South America, Southeast USA

Source: Kupicha (1983).

1998; Bässler, 1966, 1973, 1981; Czefranova, 1971; Kenicer, Kajita, Pennington, & Murata, 2005; Kupicha, 1983). Based on morphological characters, Asmussen and Liston (1998) summarized the evolution of taxonomic identification of genus *Lathyrus*. On the basis of crossability relationships, *Lathyrus* species have been grouped into primary, secondary and tertiary gene pools (Jackson & Yunus, 1984; Kearney, 1993; Kearney & Smartt, 1995; Yunus & Jackson, 1991). The primary gene pool of *Lathyrus* is limited to cultivars, landraces and escapes from cultivation, while the secondary gene pool includes *L. chrysanthus, L. gorgoni, L. marmoratus, L. pseudocicera, L. amphicarpus, L. blepharicarpus, L. chloranthus, L. cicera, L. hierosolymitanus* and *L. hirsutus*. The remaining species are included in the tertiary gene pool, which can only be exploited for crop improvement purpose with the help of bridging species and tissue culture techniques. The progenitor of *L. sativus* remains unknown, but several Mediterranean candidate species, namely *L. cicera*, *L. marmoratus*, *L. blepharicarpus* and *L. pseudocicera*, qualify as candidates since they resemble the cultigens morphologically. However, *L. cicera* is the most probable progenitor of *L. sativus* as it is morphologically and cytogenetically closest to the cultivated species (Jackson & Yunus, 1984; Hopf, 1986).

11.3 Cytotaxonomy and Genomic Evolution

Most species in the genus *Lathyrus* are true diploids ($2n=2x=14$ chromosomes) with some degree of variation in karyotype (Campbell, 1997; Ozcan, Hayirlioglu, & Inceer, 2006; Rees & Narayan, 1997; Yunus, 1990). There are a few polyploid species among the perennials including hexaploid (*L. palustris*, $2n=6x=42$ chromosomes) and tetraploid (*L. venosus*, $2n=4x=28$ chromosomes) (Darlington & Wylie, 1995; Narayan & Durrant, 1983). Natural and induced autopolyploids have also been reported in *L. sativus*, *L. odoratus*, *L. pratensis* and *L. veosus* (Khawaja, Sybenga, & Ellis, 1997). Polyploid and aneuploid plants reported in *Lathyrus* species showed the same basic chromosome number (Broich, 1989; Khawaja, 1988; Murray, Bennett, & Hammett, 1992). This reveals that a conserved basic chromosome number remains a common phenomenon in *Lathyrus* with polyploidy as rare exception (Kalmt & Wittmann, 2000; Seijo & Fernandez, 2001). Within the species, variation has been reported in chromosome size, centromere location and the number, size and location of secondary constrictions, in spite of the identical number of chromosomes (Barpete, Parmar, Sharma, & Kumar, 2012; Battistin & Fernandez, 1994; Broich, 1989; Fouzard & Tandon, 1975). Variation in chromosome size is often the result of amplification or deletion of a chromatin segment during species diversification. Intra- and interspecific variations in chromosome size indicate marked variation in the amount of DNA affecting the complement size; a high percentage of DNA is moderately repetitive (Rees & Narayan, 1997). The nuclear 2C DNA amount is reported to be in the range of 13.8–15.6pg in *L. sativus* (Ali, Meister, & Schubert, 2000; Nandini, Murray, O'Brien, & Hammett, 1997). There are reports of variation in DNA content involving euchromatin and heterochromatin, as well as repetitive and nonrepetitive DNA sequences (Battistin, Biondo, & May, 1999). Despite this stability in chromosome number, large variations in chromosome size have played an important role in the genomic evolution of *Lathyrus* species, which are associated with a fourfold variation in 2C nuclear DNA amount (Narayan & Rees, 1976).

11.4 Phylogenetic Relationships and Genetic Diversity

Several methods have been used to study the phylogenic relationships among different *Lathyrus* species including morphological traits (Shehadeh, 2011), crossability

(Yunus, 1990), karyotype analysis (Battistin & Fernandez, 1994; Murray et al., 1992; Schifino-Wittman, Lan, & Simioni, 1994), chromosome banding and *in situ* hybridization (Murray et al., 1992; Unal, Wallace, & Callow, 1995) and molecular markers (Badr, ElShazly, ElRabey, & Watson, 2002; Ceccarelli, Sarri, Polizzi, Andreozzi, & Cionini, 2010; Croft, Pang, & Taylor, 1999). Yunus (1990) established phylogenetic relationships among *Lathyrus* species by crossability studies. Ali et al. (2000) found that karyotype features reflect well the phylogenetic relationships among *Lathyrus* species belonging to different sections proposed by Kupicha (1983). Isozyme patterns on 18 accessions of five *Lathyrus* species allowed an unexpected grouping between *L. pubescens* and *L. sativus* (Schifino-Wittmann, 2001). By using the storage protein gene sequences, de Miera, Ramos, and Pe´rez de la Vega (2008) showed that *L. sativus*, *L. annuus*, *L. cicera* and *L. tingitanus*, all belonging to section *Lathyrus*, formed a monophyletic group, while *L. latifolius* of the same section is included in the group formed by *L. clymenum* and *L. ochrus* of the section *Clymenum*. Asmussen and Liston (1998) conducted a detailed investigation of *Lathyrus* to date which allowed a review of the classification proposed by Kupicha (1983). Kenicer et al. (2005) used nuclear ribosomal and chloroplast DNA to study the systematics and biogeography of 53 *Lathyrus* species. The results supported generally the recent classification based on morphological characters and resolved the clades between *Lathyrus* and *Lathyrostylis* sections, but questioned the monophyly of the section *Orobus sensu* (Kupicha, 1983). These studies have also brought some suggestions of the geographic origin of different species. Ceccarelli et al. (2010) used satellite DNA to show the close phylogenetic relationship between *L. sylvestris* and *L. latifolius*, confirming the results of Asmussen and Liston (1998) using chloroplast DNA study.

Further, molecular approaches have been increasingly applied to plant systematics and phylogenetics to elucidate relationships between allied taxa (Soltis, Soltis, & Doyle, 1992). Molecular diversity analysis supported a close phylogenetic proximity between *L. sativus* and *L. cicera* based previously on morphological and hybridization studies (Jackson & Yunus, 1984; Kupicha, 1983; Yunus & Jackson, 1991). Chtourou-Ghorbel, Lauga, Combes, and Marrakchi (2001) concluded that random amplified polymorphism DNA markers (RAPDs) are equivalent to restriction fragment length polymorphisms (RFLPs) in assessing the genetic diversity of *Lathyrus* species belonging to the sections *Lathyrus* and *Clymenum*. Recently, six amplified fragment length polymorphism (AFLP) markers along with 47 morphological characters were used to clarify the taxonomic and phylogenetic relationships within and between the sections and the species of the genus *Lathyrus*, subjecting 184 accessions belonging to 9 predefined sections and 144 originating from the Mediterranean basin and Caucasus, Central and West Asia regions (Shehadeh, 2011). The results showed that the sections *Aphaca*, *Clymenum*, *Lathyrostylis* and a large part of the *Lathyrus* section could be differentiated either by using morphological characters or AFLP markers. Genetic diversity of numerous *Lathyrus* species has been assessed with DNA markers in addition to morphological analyses (Belaid, Chtourou-Ghorbel, Marrakchi, & Trifi-Farah, 2006; Chtourou-Ghorbel et al., 2001; Croft et al., 1999; Lioi et al., 2011; Shehadeh, 2011). Different levels of diversity have been detected in different species using isozymes (Ben-Brahim, Salho, Chtorou, Combes, & Marrakcho, 2002; Kiyoshi, Toshiyuki, & Blumenreich, 1985), RFLPs (Chtourou-Ghorbel et al., 2001), RAPDs (Barik, Acharya, Mukherjee, & Chand, 2007; Croft et al., 1999), chloroplast DNA restriction sites (Asmussen & Liston, 1998) and AFLPs (Badr et al., 2002). Chowdhury and Slinkard (2000) studied genetic diversity in 348 accessions and subaccessions of *L. sativus* from 10 geographical regions using polymorphism for 20 isozymes. They observed the closest genetic distance between populations from the Near East and North Africa. Populations from South Asia and Sudan–Ethiopia, though geographically widely separated, exhibited a closer genetic distance from each other than from other regions.

11.5 Erosion of Genetic Diversity from the Traditional Areas

Genetic diversity of *Lathyrus* has experienced serious genetic erosion, largely as a result of intensification of agriculture, overgrazing, decline of permanent pastures and disappearance of sclerophyll evergreen trees, as well as maquis and garrigue shrub vegetation in the Mediterranean region (Maxted & Bisby, 1986, 1987). Many weedy *Lathyrus* species are associated with traditional farming systems that are also disappearing rapidly throughout the region. Most of the dry lands of the CWANA region are also subject to the adverse effects of climate change, which is amplifying the biodiversity loss. Turkey used to have the richest diversity area of *Lathyrus* and was reported to cultivate *L. sativus*, *L. cicera*, *L. clymenum*, *L. hirsutus* and *L. ochrus* (Cetin, 2006; Davis, 1970; Genc & Sahin, 2001; Tosun, 1974). The absence of *L. cicera* and *L. ochrus* among the encountered species in the recent exploration in Turkey clearly indicates that the cultivated *Lathyrus* materials had been exposed to genetic erosion during the last 50 years (Basaran, Asci, Mut, Acar, & Ayan, 2011). In South Asia, the generic diversity of *Lathyrus* has suffered a great deal from the government policy of ban on its sale, causing serious erosion of landraces from the region. There has been a growing interest among germplasm curators for *in situ* and *ex situ* conservation of plant genetic resources. *In situ* conservation, whether in a natural reserves or on farms, has so far not been adopted for *Lathyrus* species, except for an initial attempt in Turkey in the Kaz Dag, Amanos and Ceylan Pínner region (Ertug & Tan, 1997). Maxted (1995) proposed the establishment of sites for reserves for *Vicieae* species in Syria and Turkey, but these ideas have not yet been initiated. There is an urgent need to make encouraging steps to establish reserves both for the wild species of *Lathyrus* and on-farm projects to conserve the ancient landraces of cultivated species in the region. The GEF-ICARDA regional project on 'conservation and sustainable use of dryland agro-biodiversity' concluded that natural habitats in most of the monitoring areas surveyed in Jordan, Lebanon, Palestine and Syria are under severe threat by overgrazing and habitat destruction (Amri et al., 2005), resulting in the recommendation of areas for *in situ* conservation of wild relatives of cereals and legumes including *Lathyrus*. Many annual *Lathyrus* species are weedy species of disturbed land, making them very vulnerable to changes in human activity such as changes in agricultural practice, increased or decreased stocking levels and application of herbicides. With the limited adoption of new cultivars, it is believed that landraces of grass pea are still widely grown by farmers under harsh conditions in spite of the drastic reduction in area under its cultivation. Traditional cultivation of *L. cicera* is disappearing rapidly in the Mediterranean basin, but one area where cultivation is maintained is in the Djebel Al-Arab of southern Syria, which needs on-farm conservation in place. Conservation initiatives for the wild *Lathyrus* species need to be expedited before potentially valuable sources of genetic variability are permanently lost (Gurung & Pang, 2011).

11.6 Status of Germplasm Resources Conservation

Sporadic attempts were made in the past to conserve the genetic diversity of the genus *Lathyrus* using *ex situ* and *in situ* methods. Recently, conservation of *Lathyrus* genetic resources has attracted more attention because of their future role under the climate change scenario. The Global Crop Diversity Trust (GCDT) in collaboration with ICARDA has developed a long-term conservation strategy for the major food legumes including *Lathyrus* (GCDT, 2009). In the regional strategies, *Lathyrus* was given lower priority compared to the major crop species such as cereals. In South Asia it ranked 22nd of the top 24 highest priority crop species and in Ethiopia 19th of the 21 highest priority crop species. In the rest of the world it ranks as of only negligible value.

Past explorations have led to large *ex situ* collections of *Lathyrus* germplasm in different national and international gene banks (Arora, Mathur, Riley, & Adham, 1996; Mathur, Alercia, & Jain, 2005; Panos, 1940, 1957; Panos, Sotiriadis, & Fikas, 1961; Zalkind, 1933, 1937). Recent compilation of the *Lathyrus* germplasm collections in different countries indicated 463 accessions in Algeria, 1001 in Australia, 2432 in Bangladesh, 31 in Cyprus, 96 in Ethiopia, 4387 in France, 445 in Germany, 307 in Hungary, 2580 in India, 36 in Jordan, 149 in Nepal, 130 in Pakistan, 1240 in Russia, 307 in Spain and 529 in USA (Mathur et al., 2005). India has 2720 accessions of grass pea in the national gene bank at New Delhi and 2604 accessions of active collections at Indira Gandhi Krishi Vishwavidyalaya, Raipur (Pandey et al., 2008). Currently, there are 586 accessions in the grass pea collection maintained at the Institute of Biodiversity Conservation in Ethiopia (Girma & Korbu, 2012). Of these, 560 accessions are maintained under long-term storage. The grass pea collection in the Ethiopian gene bank contains predominantly 45% accessions from the Shewa region (Shiferaw, Pe, Porceddu, & Ponnaiah, 2011). Thus, it would be useful to increase representative samples from other regions to capture the maximum diversity. The *Lathyrus* database produced as a result of the *Lathyrus* global conservation strategy contains around 23,000 accessions with main collections held by University of Pau in France, ICARDA, NBPGR and Genetic Resources Center in Bangladesh (Tables 11.2 and 11.3). Global collection at ICARDA represents 45 species from 45 countries. This collection is unique because 45% and 54% of the accessions are wild

Table 11.2 Major *Ex Situ Lathyrus* Collections in the World

Country	Total Accessions	Wild Relatives	Landraces	Breeding Materials
ICARDA	3327	45%	54%	0.1%
France	4477	n.a.	n.a.	n.a.
India	2619	3%	85%	12%
Bangladesh	1841	–	100%	–
Chile	1424	n.a.	n.a.	n.a.
Australia	986	28%	39%	19%
Russia	848	43%	30%	18%
Canada	840	10%	90%	–
USA	669	n.a.	n.a.	n.a.
Ethiopia	588	2%	75%	25%
Germany	568	40%	n.a.	n.a.
Spain	543		n.a.	n.a.
Algeria	437		n.a.	n.a.
Hungary	394	1%	22%	n.a.
Spain	377	14%	86%	–
Bulgaria	368	1.6%	80%.	n.a.
Turkey	363	94%	n.a.	n.a.
Nepal	164	–	100%	–
Armenia	157	98%.	1%	1%
Pakistan	130	n.a.	n.a.	n.a.
Portugal	199	5%.	30%	n.a.
China	80	n.a.	n.a.	n.a.
Azerbaijan	66	47%	33%.	20%.
Czech Republic	52	75%	–	25%
Greece	47	–	2%	98%
Slovakia	47	–	87%	13%
Cyprus	31	–	100%	–
Poland	10	–	–	100%

Source: Shehadeh (2011).

relatives and landraces, respectively, mainly of *L. sativus*, followed by *L. cicera* and *L. ochrus*. Furthermore, it is necessary to study the genetic diversity of the available collections in order to understand their full utilization potential and possible gaps (Maxted, Guarino, & Shehadeh, 2003). ICARDA has characterized more than 60% accessions for main descriptors (Robertson & Abd-El-Moneim, 1997).

A comprehensive global database of *Lathyrus* species originating from the Mediterranean basin, Caucasus, Central and West Asian regions has recently been developed using accessions of the major gene banks and information from eight herbaria in Europe. This global *Lathyrus* database was used to conduct gap analysis to guide future collection missions and *in situ* conservation efforts for 37 priority species. The results showed the highest concentration of priority species in the countries of the Fertile Crescent, France, Italy and Greece. The region extending from south

Table 11.3 Present Status of *Ex Situ Lathyrus* Collections in Major Gene Banks

Species	Institute					
	AARI	**ATFC**	**ICARDA**	**IPK**	**IBEAS**	**W-6**
L. annuus	44	6	68	2	0	3
L. chrysanthus	1	1	3	0	0	0
L. cicera	90	141	182	63	785	31
L. clymenum	1	10	2	25	0	20
L. gorgoni	27	6	60	2	0	1
L. hierosolymitanus	22	13	104	1	0	4
L. hirsutus	2	9	17	8	0	16
L. Latifolius	0	1	1	13	326	10
L. marmoratus	4	0	33	0	0	0
L. ochrus	1	85	136	46	0	23
L. odoratus	2	3	3	6	0	23
L. pseudocicera	8	1	65	1	0	0
L. sativus	17	572	1627	170	2382	222
Other *Lathyrus* sp.	300	172	698	108	984	111
Total	519 (32)	1020 (42)	3001(44)	445	4477 (6)	464 (23)

Numbers in brackets indicate the number of other *Lathyrus* species conserved.
AARI, Aegean Agricultural Research Institute, Menemen, Turkey; ATFC, Australian Temperate Field Crop collection, Horsham, Australia; IBEAS, IBEAS, Université de Pau et des Pays de l'Adour, Pau, France; W-6, Western Regional Plant Introduction Station, Pullman, Washington, USA; ICARDA, International Center for Agricultural Research in the Dry Areas, Syria; IPK, Institut fur Pflanzengenetik und Kulturpflanzenforschung (IPK), Getersleben, Germany.
Source: Shehadeh (2011).

central Turkey, through the western Mediterranean mountains of Syria to the northern Bekaa valley in Lebanon, and precisely the area around the Lebanese/Syrian border near the Tel Kalakh region in Homs, was identified as the hot spot for establishing genetic reserves. The gap analysis for *ex situ* conservation shows that only 6 of the 37 priority species are adequately sampled, showing a need for more collection missions in the areas and for collecting closely related wild species of *Lathyrus*. Six priority species have no *ex situ* collections, requiring targeted collection missions. An ecogeographic survey revealed that conservation efforts need to be focused on *L. sativus*, *L. cicera* and *L. ochrus* and other species over the whole of their native distribution (GCDT, 2009; Hawtin, 2007). The Near East and North Africa exhibited the greater genetic diversity. Thus, countries of this region should be explored further for the additional genetic variability of grass pea (Table 11.4).

11.7 Germplasm Evaluation

Evaluation of *Lathyrus* germplasm has been undertaken sporadically for different traits to identify useful donors for important parameters including low ODAP content, appropriate phenology and high biomass including yield-related traits (Campbell et al., 1994; Grela, Rybinski, Klebaniuk, & Mantras, 2010; Hanbury,

Table 11.4 Possible Gaps in Global *Lathyrus Ex Situ* Conservation

Country	*L. sativus*	*L. cicera*	*L. ochrus*
Egypt	+	+	
Iraq	+	+	
Iran	+	+	
Tunisia		+	+
Greece			+
Turkey			+
Russia	Black Sea Coast and Volga-Kama region		
Iraq	Kurdish area		
Bangladesh	Syleth area (high altitude)		
India	Northeast and Eastern parts		
Ethiopia	High altitude areas, recently opened by roads		
Afghanistan	Northeast and Central part		
Spain	Almeria (Andalucía) and Murcia		

Source: GCDT (2009).

Sarker, Siddique, & Perry, 1995; Kaul, Islam, & Humid, 1986; Pandey, Chitale, Sharma, & Geda, 1997; Pandey, Kashyap, Geda, & Tripathi, 1996; Pandey et al., 1995; Pandey et al., 2008; Sharma, Kashyap, Chitale, & Pandey, 1997). A total of 1082 accessions belonging to 30 species were evaluated for 21 descriptors and agronomic traits at ICARDA (Robertson & Abd-El-Moneim, 1997). The results have shown a wide range of variability for various traits (Table 11.5). For ODAP content, studies have shown a wide range of variation within the existing germplasm, ranging from 0.02% to 2.59% (Table 11.6). Hanbury, Siddique, Galwey, and Cocks (1999) reported a range of 0.04–0.76% for ODAP content in a set of 503 accessions procured from ICARDA. Pandey et al. (1997) reported a range of 0.128–0.872% for ODAP content among 1187 accessions. A detailed catalogue of grass pea germplasm comprising characterization and evaluation information on 63 traits for 1963 accessions has recently been published in India (Pandey et al., 2008). A wide range of variability was observed for all the traits of interest, such as crop duration, plant height, pods per plant, seeds per pod, seed weight, biomass score, seed yield and ODAP content (0.067–0.712%). Some of the accessions having <0.1% ODAP are IPLY9, Prateek, AKL 19, BioL202, BioL203, Ratan, No. 2203 and No. 2208. Kumar, Bejiga et al. (2011) also screened 1128 accessions of *L. sativus* and found a wide range (0.150–0.952%) for ODAP content. Only two accessions, IG118563 (0.150%) and IG64888 (0.198%), had low ODAP content. Multi-location evaluation of grass pea germplasm at ICARDA between 1999 and 2006 indicated the maximum variability for ODAP content in Ethiopian germplasm (Table 11.7). Grass pea germplasm from Ethiopia and the Indian subcontinent is generally high in ODAP (0.7–2.4%) as compared to 0.02–1.2% in germplasm from the Near East (Abd-El-Moneim, Van-Dorrestein, Baum, & Mulugeta, 2000). A recent study by Gutierrez-Marcos, Vaquero, de Miera, and Vences (2006) on 2987 individuals belonging to 110

Table 11.5 Variability for Agro-morphological Traits in Major *Lathyrus* Species Evaluated at ICARDA, Aleppo, Syria

Trait	*L. cicera*			*L. ochrus*			*L. sativus*		
	Mean	**Min**	**Max**	**Mean**	**Min**	**Max**	**Mean**	**Min**	**Max**
Days to 50 % flowering	123.9	115	136	120	115	145	126	119	142
Days to 90% maturity	163.9	156	181	157	149	184	173.8	145	189
Days to 90% podding	128.3	122	148	124	118	154	137.5	122	154
Plant height (cm)	35.4	24.1	49.8	34.7	23	48	41.1	5	60
Height to first flower (cm)	8.1	2.4	13.2	13	7	19	9.2	3	17
Seeds per pod	3.8	2.3	9.6	4.6	3.32	5.7	3.1	1.48	6.5
Harvest index (%)	33.8	12.7	52	36.2	12.7	48.6	19.5	1.9	54.7
1000-seed weight (g)	83.1	13.9	116.7	121.3	57.2	156.3	86.8	34.5	225.9
Seed yield (kg/ha)	1120	117	2030	815	105	1454	445	29	1406
Biomass yield (kg/ha)	3101	635	4972	2221	726	3741	2167	516	5200
Straw yield (kg/ha)	2578	488	3067	1406	564	2499	1722	440	3861
ODAP content (%)	0.160	0.030	0.220	1.400	0.460	2.500	1.300	0.020	2.400

Table 11.6 Genetic Variation for ODAP Content in Grass Pea Germplasm

Country/ Institution	Number of Accessions	ODAP (%) in Seeds		References
		Minimum	Maximum	
Bangladesh	–	0.450	1.400	Kaul et al. (1986)
Bangladesh	116	0.040	0.780	Sarwar, Malek, Sarker, and Hassan (1996)
China	73	0.075	0.993	Campbell et al. (1994)
Ethiopia	150	0.149	0.916	Tadesse and Bekele (2003)
India (IARI)	576	0.100	2.590	Nagarajan and Gopalan (1968)
India (IARI)	1500	0.150	0.300	Jeswani, Lal, and Shivprakash (1970)
India (IARI)	643	0.100	0.780	Somayajulu, Barat, Prakash, Mishra, and Shrivastava (1975)
India (IARI)	1000	0.200	2.000	Leakey (1979)
India (IGKV)	1187	0.128	0.872	Pandey et al. (1995, 1996)
India (IGKV)	1963	0.067	0.712	Pandey et al. (2008)
ICARDA	81	0.020	0.740	Robertson and Abd-El-Moneim (1997)
ICARDA	1128	0.150	0.952	Kumar, Bejiga et al. (2011)
Australia	503	0.040	0.760	Hanbury et al. (1999)
Chile	76	0.180	0.520	Tay, Valenzuela, and Venegas (1999)

different global samples revealed considerable genetic diversity in grass pea collections throughout the world.

Wild crop gene pool is a rich reservoir of rare alleles. Therefore, efforts have been made to evaluate wild relatives to identify zero ODAP genetic resources (Jackson & Yunus, 1984). Assessment of ODAP in wild relatives indicated that none of the species is free from ODAP (Aletor et al., 1994; Hanbury et al., 1999; Siddique et al., 1996). However, on average, the ODAP concentration in *L. cicera* is lower compared to *L. sativus* (Table 11.8). Hanbury et al. (1999) observed the lowest ODAP in *L. cicera* (0.18%) followed by *L. sativus* (0.39%) and *L. ochrus* (1.01%). Aletor et al. (1994) reported four to five times lower ODAP content in *L. cicera* (0.13%) than in *L. ochrus* (0.56%) and *L. sativus* (0.49%). Similarly, Abd-El-Moneim et al. (2000) reported ranges of 0.02–2.40% in *L. sativus*, 0.03–0.22% in *L. cicera* and 0.46–2.50% in *L. ochrus*. Eichinger, Rothnie, Delaere, and Tate (2000) screened *Lathyrus* germplasm using capillary electrophoresis and found that *L. cicera* is consistently low in ODAP as compared to *L. sativus* and *L. ochrus*. Evaluation of 142 accessions of *L. cicera* at ICARDA during 2009 showed a range of 0.073–0.513% for ODAP content, which is much lower than the cultivated species. Therefore, *L.*

Table 11.7 Geographical Distribution of ODAP Content in Grass Pea Germplasm Evaluated at ICARDA During 1999–2006

Country of Origin	Accessions Evaluated	ODAP Content (%)			Environments
		Minimum	Maximum	Mean	
Bangladesh	317	0.376	0.699	0.482	18
Ethiopia	98	0.067	0.848	0.341	21
Nepal	47	0.403	0.531	0.487	13
Pakistan	62	0.336	0.517	0.466	13
Europe[a]	115	0.198	0.908	0.458	2

[a]Also includes accessions from Central Asia.

Table 11.8 Variation in ODAP Content in Different *Lathyrus* Species

Species	ODAP (%) Content			No. of Accessions	Country	Reference
	Minimum	Maximum	Mean			
L. sativus	0.15	0.87	n.a.	–	China	Yu (1995)
L. cicera	0.07	0.10	n.a.	–	China	Yu (1995)
L. sativus	0.04	0.76	0.39	407	Australia	Hanbury et al. (1999)
L. cicera	0.08	0.34	0.18	96	Australia	Hanbury et al. (1999)
L. ochrus	0.64	1.35	1.01	32	Australia	Hanbury et al. (1999)
L. sativus	0.33	0.57	0.49	36	Syria	Aletor et al. (1994) and El-Haramein, Abd-El Moneim, and Nakkoul (1998)
L. cicera	0.09	0.16	0.13	16	Syria	Aletor et al. (1994) and El-Haramein et al. (1998)
L. ochrus	0.48	0.63	0.56	16	Syria	Aletor et al. (1994) and El-Haramein et al. (1998)
L. sativus	0.15	0.95	0.47	1128	Syria	Kumar, Bejiga et al. (2011)
L. cicera	0.07	0.51	0.30	141	Syria	Kumar, Bejiga et al. (2011)

cicera accessions hold promise as a source of low ODAP content in grass pea breeding programmes.

11.8 Use of Germplasm in Crop Improvement

Significant efforts have been directed towards genetic improvement of grass pea in India, Canada, Bangladesh, Ethiopia and Nepal during late 1970s and at ICARDA since 1989. Breeding efforts are mostly focused on three species, *L. sativus*, *L. cicera* and *L. ochrus*, and to a lesser extent *L. clymenum*, with an aim to improve grain yield, biomass, resistance to biotic and abiotic stresses and most importantly to reduce the neurotoxin from its seeds. A conventional breeding approach has resulted in development of high-yielding low ODAP varieties (Table 11.9). In India, Pusa 24, Prateek and Mahateora, with low ODAP and high yield, were developed through intraspecific hybridization. In Bangladesh, low ODAP and high-yielding varieties BARI Khesari 1 and BARI Khesari 2 were developed for commercial cultivation. At ICARDA, several breeding lines with <0.1% ODAP concentration were bred, which have led to the release of BARI Khesari 3 in Bangladesh, Wasie in Ethiopia and Ali Bar in Kazakhstan. In Canada, a low ODAP (0.03%) line, LS 8246 was released for fodder and feed purposes. In Australia, two varieties, Ceora and Chalus, were released for diversification of the wheat-based system. Mutation breeding has also been occasionally employed to create additional genetic variability in order to develop zero/low ODAP varieties (Talukdar, 2009). Two varieties, namely Poltavskaya in the former USSR and Bina Khesari 1 in Bangladesh, were developed through mutation breeding using Ethyl methane sulphonate (EMS) (0.01%) and gamma rays (250 Gy), respectively. Somaclonal variation can also contribute to the development of varieties with low ODAP (Mehta, Ali, & Barna, 1994; Mehta

Table 11.9 Improved Varieties of Grass Pea Released for Cultivation in Different Countries

Country	Improved Varieties
Australia	Ceora, Chalus
Bangladesh	Bari Khesari 1, Bari Khesari 2, Bari Khesari 3, Bina Khesari 1
Bulgaria	Strandja
Canada	LS 82046
Chile	Luanco-INIA
Ethiopia	Wasie
Kazakhstan	Ali Bar
India	Pusa 24, Prateek, Ratan, Mahateora
Nepal	CLIMA 2 pink, 19A, 20B, Bari Khesari 2
Pakistan	Italian
Poland	Derek, Krab
Turkey	Gurbuz 2001
Russia	Poltavskaya

& Santha, 1996; Santha & Mehta, 2001). Ratan is released as a variety in India from selection in the somaclonal variation. More efforts are needed to exploit the genetic diversity existing within species of grass pea gene pools.

11.9 Limitations in Germplasm Use

The problem of genetic resources is not only the size but also the lack of systematic characterization, evaluation and existence of duplicates that hinder their effective use in breeding programmes. The core set has been proposed to overcome the problem of limited use of genetic resources (Frankel & Brown, 1984). Core sets of *Lathyrus* species were identified to develop manageable subsets that capture most of the variation from the original dataset of 2674 accessions belonging to 31 *Lathyrus* species. Among modifications, development of mini-core sets has been proposed to address the concern of limitation in germplasm use, in particular the use in relation to the trait of interest (Upadhyaya, Bramel, Ortiz, & Singh, 2002). However, mini-core sets may not be needed at present for grass pea, as the global collection is estimated at only 3360 accessions. For adaptive traits, core and mini-core collections may not capture the needed diversity (Gepts, 2006; Pessoa-Filho, Rangel, & Ferreira, 2010). As an alternative to the core, the Focused Identification of Germplasm Strategy (FIGS) approach is developed, which is a trait-based approach with high probability of identification of desired genetic material. The FIGS approach has been applied in *Lathyrus* at ICARDA to derive a heat- and drought-tolerant subset based on maximum temperature and aridity index. These subsets with manageable size and higher probability of finding the desired traits will allow linking conservation with utilization of genetic resources and reduce the pressure to frequently regenerate species with cross-pollination, as is the case with grass pea.

11.10 Germplasm Enhancement Through Wide Crosses

Over the years, ICARDA has collected and conserved 1555 accessions of 45 wild *Lathyrus* species from 45 countries in its global germplasm repository (Kumar, Bejiga et al., 2011). These species may play an important role in the genetic improvement of the cultivated species. For example, a toxin-free gene has been identified in *L. tingitanus*, which can be used to develop toxin-free grass pea varieties (Zhou & Arora, 1996). *Lathyrus* species such as *L. ochrus* and *L. clymenum* (Sillero, Cubero, Fernández-Aparicio, & Rubiales, 2005) and *L. cicera* (Fernández-Aparicio, Flores, & Rubiales, 2009; Fernández-Aparicio & Rubiales, 2010) are identified as possessing resistance to *Orobanche* which is not available within the cultivated germplasm. *L. cicera* is also a good source for earliness and cold tolerance. However, alien gene transfer has hardly been attempted in grass pea in spite of the success of interspecific hybridization between *L. sativus* and two wild *Lathyrus* species (*L. cicera* and *L. amphicarpus*) with viable seeds (Addis & Narayan, 2000;

Davies, 1957, 1958; Khawaja, 1988; Yunus, 1990). Yunus (1990) crossed 11 wild species with *L. sativus* and found viable seeds only with *L. cicera* and *L. amphicarpus*. Other species formed pods but did not produce fully developed viable seeds (Kearney, 1993; Yamamoto, Fujiware, & Blumenreich, 1989; Yunus, 1990). Some other successful interspecific hybrids reported in the genus *Lathyrus* were *L. annuus* with *L. hierosolymilanus* (Hammett, Murray, Markham, & Hallett, 1994; Hammett, Murray, Markham, Hallett, & Osterloh, 1996; Yamamoto et al., 1989); *L. articulatus* with *L. clymenus* and *L. ochrus* (Davies, 1958; Trankovskij, 1962); *L. cicera* with *L. blepharicarpus*, *L. gorgoni*, *L. marmoratus* and *L. pseudocicera* (Kearney, 1993; Yamamoto et al., 1989); *L. gorgoni* with *L. pseudocicera* (Kearney, 1993; Yamamoto et al., 1989); *L. hirsutus* with *L. odoratus* (Davies, 1958; Khawaja, 1988; Trankovskij, 1962; Yamamoto et al., 1989); *L. marmoratus* with *L. blepharicarpus* (Kearney, 1993; Yamamoto et al., 1989); *L. odoratus* with *L. belinenesis* (Hammett et al., 1994, 1996); *L. rotundifolius* with *L. tuberosus* (Marsden-Jones, 1919) and *L. sylvestris* with *L. latifolius* (Davies, 1957). From the information available on crossing, fertility and chromosome behaviour of the hybrids, it may be concluded that breeding strategies involving alien genetic transfer for the improvement of grass pea are possible through the readily crossable species *L. cicera* and *L. amphicarpus* and any other gene transfer technology involving other species will have to be assisted by biotechnology tools (Ochatt, Durieu, Jacas, & Pontecaille, 2001).

11.11 Grass Pea Genomic Resources

Considerable progress has been made in recent years in developing genomic resources in food and model legumes (Kumar, Pratap et al., 2011). However, in grass pea only a few reports on genomic resource development are available, apparently because of the large genome size and poorly characterized germplasm set (Lioi et al., 2011; Ponnaiah, Shiferaw, Pe, & Porceddu, 2011; Shiferaw et al., 2011). Molecular markers, such as inter-simple sequence repeat (ISSR), RAPDs, sequence tagged site (STS), RFLP and AFLP, have been developed and used to examine the genetic variation and phylogenetic relationships within the genus *Lathyrus* (Badr et al., 2002; Barik et al., 2007; Belaid et al., 2006; Chtourou-Ghorbel et al., 2001; Croft et al., 1999; Skiba, Ford, & Pang, 2003; Tavoletti & Iommarini, 2007). In *Lathyrus*, only 15 SSR primers were reported earlier (Lioi et al., 2011; Shiferaw et al., 2011). Recently, 300 expressed sequence tag-simple sequence repeats (EST–SSR) primer pairs were identified and loci characterized for size polymorphism among 24 grass pea accessions (Sun et al., 2012). Among them 117 SSR loci were monomorphic and 44 SSR loci were polymorphic. These novel markers will be useful and convenient to study the gene mapping and molecular breeding in grass pea. In terms of plant resources for functional genomic studies, various mapping populations including recombinant inbred lines (RILs), Near isogenic lines (NILs) and Targeting induced local lesions in genomes (TILLING) populations are critically needed for trait–marker association and gene inactivation/deletion studies.

11.12 Conclusions

The success of genetic resources in improving the crop lies in the manner in which we harness the wealth of allelic diversity provided by nature and currently warehoused in gene banks. Until now, only modest success has been made through conventional breeding methods in utilizing these resources. Recent developments in biotechnology indicate that there is a tremendous opportunity to realize the potential variability conserved in various gene banks of the world. The genus *Lathyrus* is well placed to help face the challenges posed by climate change because of the genetic resources available for crop improvement. Coordinated efforts to collect and conserve *Lathyrus* crop species have been initiated in the last 10–15 years and have gained momentum with the development of a grass pea conservation strategy as part of the Global Crop Diversity Trust and Bioversity International. Furthermore, it is necessary to study the genetic diversity of the available collections systematically in order to understand their full potential. Proper documentation of all passport, characterization and evaluation information needs to be improved through the development of *Lathyrus* catalogues to avoid duplicates and to ensure the easy use of genetic resources. In the last few years there has been a growing emphasis on the characterization of germplasm collections by molecular markers, which has served to enhance the use of germplasm collections in crop improvement via plant breeding. This also aids the management of collections themselves, through an improved understanding of the relationships between accessions and underlying patterns of diversity. Issues like whether or not genetic variation is being lost with progressive domestication or how the variation is distributed among populations can also be addressed by genetic diversity analysis. Further research is needed to expand on the use of molecular markers in species identification. Much more efforts are needed to augment the genetic resources of both cultivated and wild lines for the genetic improvement through conventional as well as by contemporary approaches. A key goal is to solve the problem of ODAP content in grass pea.

References

Abd-El-Moneim, A. M., Van-Dorrestein, B., Baum, M., & Mulugeta, W. (2000). Improving the nutritional quality and yield potential of grasspea (*Lathyrus sativus* L.). *Food and Nutrition Bulletin*, *21*, 493–496.

Addis, G., & Narayan, R. K. J. (2000). Inter-specific hybridization of *Lathyrus sativus* (Guaya) with wild *Lathyrus* species and embryo rescue. *African Crop Science Journal*, *8*(2), 129–136.

Aletor, V. A., Abd-El-Moneim, A. M., & Goodchild, A. V. (1994). Evaluation of the seeds of selected lines of three *Lathyrus* spp. for β-*N*-oxalyl amino-L-alanine (BOAA), tannins, trypsin inhibitor activity and certain *in vitro* characteristics. *Journal of the Science of Food and Agriculture*, *65*, 143–151.

Ali, H. B. M., Meister, A., & Schubert, I. (2000). DNA content, rDNA loci and DAPI bands reflect the phylogenetic distance between *Lathyrus* species. *Genome*, *43*, 1027–1032.

Allkin, R., Goyder, D. J., Bisby, F. A., & White, R. J. (1983). List of species and subspecies in the vicieae. *Vicieae Database Project*, *1*, 4–11.

Allkin, R., Goyder, D. J., Bisby, F. A., & White, R. J. (1986). Names and synonyms of species and subspecies in the Vicieae. *Vicieae Database Project*, *7*, 1–75.

Amri, A., Monzer, M., Al-Oqla, A., Atawneh, N., Shehahdeh, A., & Konopka, J. (2005). Status and threats to crop wild relatives in selected natural habitats in West Asia region: *Proceedings of the international conference on promoting community-driven conservation and sustainable use of dryland agrobiodiversity*. (pp. 13–27). Aleppo, Syria: ICARDA.

Arora, R. K., Mathur, P. N., Riley, K. W., & Adham, Y. (1996). Lathyrus genetic resources in Asia: *Proceedings of a regional workshop, Indira Gandhi Agricultural University, Raipur, India, 27–29 December 1995*. New Delhi, India: IPGRI Office for South Asia.

Asmussen, C. B., & Liston, A. (1998). Chloroplast DNA characters, phylogeny, and classification of *Lathyrus* (Fabaceae). *American Journal of Botany*, *85*, 387–401.

Badr, A., ElShazly, H., ElRabey, H., & Watson, L. E. (2002). Systematic relationships in *Lathyrus* sect. Lathyrus (Fabaceae) based on amplified fragment length polymorphism (AFLP) data. *Canadian Journal of Botany*, *80*, 962–969.

Barik, D. P., Acharya, L., Mukherjee, A. K., & Chand, P. K. (2007). Analysis of genetic diversity among selected grasspea (*Lathyrus sativus* L.) genotypes using RAPD markers. *Zeitschrift für Naturforschung*, *62c*, 869–874.

Barpete, S., Parmar, D., Sharma, N. C., & Kumar, S. (2012). Karyotype analysis in grasspea (*Lathyrus sativus* L.). *Journal of Food Legumes*, *25*, 14–17.

Basaran, U., Asci, O. O., Mut, H., Acar, Z., & Ayan, I. (2011). Some quality traits and neurotoxin β-*N*-oxalyl-L-α,β-diaminopropionic acid (β-ODAP) contents of *Lathyrus* sp. cultivated in Turkey. *African Journal of Biotechnology*, *10*, 4072–4080.

Bässler, M. (1966). Die Stellung des Subgenus *Orobus* (L.) Baker in der Gattung *Lathyrus* L. und seine systematische Gliederung. *Feddes Repertorium*, *72*, 69–97.

Bässler, M. (1973). Revision der eurasiatischen arten von *Lathyrus* L. Sect. *Orobus* (L.) Gren. et. Godr. *Feddes Repertorium*, *84*, 329–447.

Bässler, M. (1981). Revision von *Lathyrus* L. Sect. *Lathyrostylis* (Griseb.) Bässler (Fabaceae). *Feddes Repertorium*, *92*, 179–254.

Battistin, A., Biondo, E., & May, C. L. G. (1999). Chromosomal characterization of three native and one cultivated species of *Lathyrus* L. in southern Brazil. *Journal of Genetics and Molecular Biology*, *22*, 557–563.

Battistin, A., & Fernandez, A. (1994). Karyotype of four species of South America natives and one cultivated species of *Lathyrus* L.. *Caryologia*, *47*, 325–330.

Belaid, Y., Chtourou-Ghorbel, N., Marrakchi, M., & Trifi-Farah, N. (2006). Genetic diversity within and between populations of *Lathyrus* genus (Fabaceae) revealed by ISSR markers. *Genetic Resources & Crop Evolution*, *53*, 1413–1418.

Ben-Brahim, N., Salho, A., Chtorou, N., Combes, D., & Marrakcho, M. (2002). Isozymic polymorphism and phylogeny of 10 *Lathyrus* species. *Genetic Resources and Crop Evolution*, *49*, 427–436.

Broich, S. L. (1989). Chromosome number of North American *Lathyrus* (Fabaceae). *Madrona*, *36*, 41–48.

Campbell, C. G. (1997). Grasspea (*Lathyrus sativus* L.): *Promoting the conservation and use of underutilized and neglected crops, 18*. Rome, Italy: Institute of Plant Genetics and Crop Plant Research, Gatersleben/International Plant Genetic Resources Institute.

Campbell, C. G., Mehra, R. B., Agrawal, S. K., Chen, Y. Z., Abd-El-Moneim, A. M., Khawaja, H. I. T., et al. (1994). Current status and future strategy in breeding grass pea (*Lathyrus sativus*). *Euphytica*, *73*, 167–175.

Ceccarelli, M., Sarri, V., Polizzi, E., Andreozzi, G., & Cionini, P. G. (2010). Characterization, evolution and chromosomal distribution of two satellite DNA sequence families in *Lathyrus* species. *Cytogenetic & Genome Research, 128*, 236–244.

Cetin, T. (2006). Karyotype analysis of some *Lathyrus* L. species. *M.Sc. Thesis*. Isparta: Sueyman Demirel University, Institute of Natural and Applied Sciences, Department of Biyologi.

Chowdhury, M. A., & Slinkard, A. E. (2000). Genetic diversity in grasspea (*Lathyrus sativus* L.). *Genetic Resources and Crop Evolution, 47*, 163–169.

Chtourou-Ghorbel, N., Lauga, B., Combes, D., & Marrakchi, M. (2001). Comparative genetic diversity studies in genus *Lathyrus* using RFLP and RAPD markers. *Lathyrus Lathyrism Newsletter, 2*, 62–68.

Croft, A. M., Pang, E. C. K., & Taylor, P. W. J. (1999). Molecular analysis of *Lathyrus sativus* L. (grasspea) and related Lathyrus species. *Euphytica, 107*, 167–176.

Czefranova, Z. (1971). Conspectus systematis generis *Lathyrus* L. *Novititas Systematic Plant Vascular (Leningrad), 8*, 191–201.

Darlington, C. D., & Wylie, A. P. (1995). *Chromosome atlas of flowering plants*. London: George Allen and Unwin. pp. 155–156.

Davies, A. J. S. (1957). Successful crossing in the genus *Lathyrus* through stylar amputation. *Nature, 180*, 612.

Davies, A.J.S. (1958). A cytogenetic study in the genus *Lathyrus*. *Unpublished Ph.D. Thesis*. UK: University of Manchester.

Davis, P.H. (1970). *Flora of Turkey and East Aegean islands* (pp. 328–369). Edinburgh, UK.

de Miera, L. E. S., Ramos, J., & Pe´rez de la Vega, M. (2008). A comparative study of convicilin storage protein gene sequences in species of the tribe Vicieae. *Genome, 51*, 511–523.

Eichinger, P. C. H., Rothnie, N. E., Delaere, I., & Tate, M. E. (2000). New technologies for toxin analyses in food legumes. In R. Knight (Ed.), *Linking research and marketing opportunities for pulses in the 21st century* (pp. 685–692). Dordrecht: Kluwer Academic Publishers.

El-Haramein, F. J., Abd-El Moneim, A. M., & Nakkoul, H. (1998). Prediction of the neurotoxin beta-*N*-oxylyl-amino-L-alanine in Lathyrus species, using near infrared reflectance spectroscopy. *Journal of Near Infrared Spectroscopy, 6*, A93–A96.

Ertug, F. A., & Tan, A. (1997). *In situ* conservation of genetic diversity in Turkey. In N. Maxted, B. V. Ford-Lloyd, & J. G. Hawkes (Eds.), *Plant genetic conservation: The in situ approach* (pp. 254–262). London: Chapman & Hall.

Fernández-Aparicio, M., Flores, F., & Rubiales, D. (2009). Field response of *Lathyrus cicera* germplasm to crenate broomrape (*Orobanche crenata*). *Field Crops Research, 113*, 187–358.

Fernández-Aparicio, M., & Rubiales, D. (2010). Characterisation of resistance to crenate broomrape (*Orobanche crenata* Forsk.). *Lathyrus cicera L. Euphytica, 173*, 77–84.

Fouzard, A., & Tandon, S. L. (1975). Cytotaxonomic investigations in the genus. *Lathyrus. Nucleus, 18*, 24–33.

Frankel, O. H., & Brown, A. H. D. (1984). *Current plant genetic resources -- A critical appraisal in Genetics: New frontiers* (Vol. IV). New Delhi, India: Oxford and IBH Publ. Co..

GCDT, (2009). *Strategy for the ex situ Conservation of Lathyrus (grass pea), with special reference to Lathyrus sativus, L. cicera, and L. ochrus*. Rome, Italy: Global Crop Diversity Trust (GCDT). <http://www.croptrust.org/documents/cropstrategies/Grasspea.pdf/>.

Genc, H., & Sahin, A. (2001). Cytotaxonomic researches in some *Lathyrus* species grown in West Mediterranean and South Aegean regions III. *Journal of SDU Natural Applied Science, 5(1)*, 98–112.

Gepts, P. (2006). Plant genetic resources conservation and utilization: The accomplishments and future of a societal insurance policy. *Crop Science*, *46*, 2278–2292.

Girma, D., & Korbu, L. (2012). Genetic improvement of grasspea (*Lathyrus sativus*) in Ethiopia: An unfulfilled promise. *Plant Breeding*, *131*, 231–236. doi:10.1111/j.1439-0523.2011.01935.x.

Grela, E. R., Rybinski, W., Klebaniuk, R., & Mantras, J. (2010). Morphological characteristics of some accessions of grasspea (*Lathyrus sativus* L.) grown in Europe and nutritional traits of their seeds. *Genetic Resources & Crop Evolution*, *57*, 693–701.

Gurung, A. M., & Pang, E. C. K. (2011). Lathyrus. In C. Kole (Ed.), *Wild crop relatives: Genomic and breeding resources-legume crop and forage* (pp. 117–126). Berlin, Heidelberg: Springer.

Gutierrez-Marcos, J. F., Vaquero, F., de Miera, L. E. S., & Vences, F. J. (2006). High genetic diversity in a worldwide collection of *Lathyrus sativus* L. revealed by isozymatic analysis. *Plant Genetic Resources*, *4(3)*, 159–171.

Hammett, K. R. W., Murray, B. G., Markham, K. R., & Hallett, I. C. (1994). Interspecific hybridization between *Lathyrus odoratus* and *L. belinensis*. *International Journal of Plant Science*, *155*, 763–771.

Hammett, K. R. W., Murray, B. G., Markham, K. R., Hallett, I. C., & Osterloh, I. (1996). New interspecific hybrids in *Lathyrus* (Leguminosae): *Lathyrus annuus* × *L. hierosolymitanus*. *Botanical Journal of Linnean Society*, *122*, 89–101.

Hanbury, C. D., Sarker, A., Siddique, K. H. M., & Perry, M. W. (1995). *Evaluation of Lathyrus germplasm in a Mediterranean type environment in south-western Australia.* Perth: CLIMA. Occasional Publication No. 8.

Hanbury, C. D., Siddique, K. H. M., Galwey, N. W., & Cocks, P. S. (1999). Genotype-environment interaction for seed yield and ODAP concentration of *Lathyrus sativus* L. and *L. cicera* L. in Mediterranean-type environments. *Euphytica*, *110*, 45–60.

Hawtin, G. (2007). *Strategy for the ex situ conservation of Lathyrus (grasspea), with special reference to Lathyrus sativus, L. cicera, L. ochrus*. Rome, Italy: International Plant Genetic Resources Institute. (p. 36).

Hopf, M. (1986). Archaeological evidence of the spread and use of some members of the Leguminosae family. In C. Barigozzi (Ed.), *The original and domestication of cultivated plants* (pp. 35–60). Oxford: Elsevier.

IPGRI, (2000). *Descriptors for Lathyrus ssp*. Rome, Italy: International Plant Genetic Resources Institute (IPGRI).

Jackson, M. T., & Yunus, A. G. (1984). Variation in the grasspea (*L. sativus* L.) and wild species. *Euphytica*, *33*, 549–559.

Jeswani, L. M., Lal, B. M., & Shivprakash, (1970). Studies on the development of low neurotoxin (B-*N*-oxayl amino alanine lines in Khesari (*L. sativus* L.). *Current Science*, *39*, 518–519.

Kalmt, A., & Wittmann, M. T. (2000). Karyotype morphology and evolution in some *Lathyrus* (Fabaceae) species of southern Brazil. *Journal of Genetics and Molecular Biology*, *23*, 463–467.

Kaul, A. K., Islam, M. Q., & Humid, A. (1986). Screening of *Lathyrus* germplasms of Bangladesh for BOAA content and some agronomic characters. In A. K. Kaul & D. Combes (Eds.), *Lathyrus and lathyrism* (pp. 130–141). New York: Third World Medical Research Foundation.

Kearney, J.P. (1993). Wild *Lathyrus* species as genetic resources for improvement of grasspea (*L. Sativus*). *Unpublished Ph.D. Thesis*. UK: University of Southampton.

Kearney, J. P., & Smartt, J. (1995). The grasspea *Lathyrus sativus* (Leguminosae-Papilionoideae). In J. Smartt & N. W. Simmonds (Eds.), *Evolution of crop plants* (pp. 266–270) (2nd ed.). London: Longman.

Kenicer, G., Kajita, T., Pennington, R., & Murata, J. (2005). Systematics and biogeography of *Lathyrus* (Leguminosae) based on internal transcribed spacer and cpDNA sequence data. *American Journal of Botany*, *92*, 1199–1209.

Khawaja, H. I. T. (1988). A new inter-specific *Lathyrus* hybrid to introduce the yellow flower color into sweat pea. *Euphytica*, *37*, 69–75.

Khawaja, H. I. T., Sybenga, J., & Ellis, J. R. (1997). Chromosome pairing and chiasma formation in autopolyploids of different *Lathyrus* species. *Genome*, *40*, 937–944.

Kislev, M. E. (1989). Origin of the cultivation of *Lathyrus sativus* and *L. cicera* (Fabaceae). *Economic Botany*, *43*, 262–270.

Kiyoshi, Y., Toshiyuki, F., & Blumenreich, I. D. (1985).. In A. K. Kaul & D. Combes (Eds.), *Isozymic variation and interspecific crossability in annual species of genus Lathyrus L* (pp. 118–129). Pau, France: Lathyrus and Lathyrism.

Kumar, J., Pratap, A., Solanki, R. K., Gupta, D. S., Goyal, A., Chaturvedi, S. K., et al. (2011). Genomic resources for improving food legume crops. *Journal of Agricultural Science (Cambridge)*, *150*, 289–318. doi:10.1017/S0021859611000554.

Kumar, S., Bejiga, G., Ahmed, S., Nakkoul, H., & Sarker, A. (2011). Genetic improvement of grasspea for low neurotoxin (ODAP) content. *Food & Chemical Toxicology*, *49*, 589–600.

Kupicha, F. K. (1981). Vicieae. In R. M. Polhill & P. M. Raven (Eds.), *Advances in legume systematics* (pp. 377–381). Kew, UK: Royal Botanic Gardens.

Kupicha, F. K. (1983). The infrageneric structure of *Lathyrus*. Notes Royal Botanical Garden. *Edinburg*, *41*, 209–244.

Leakey, C. (1979). Khesari dhal. The poisonous pea. *Appropriate Technology*, *6*, 15–16.

Lioi, L., Sparvoli, F., Sonnante, G., Laghetti, G., Lupo, F., & Zaccardelli, M. (2011). Characterization of Italian grasspea (*Lathyrus sativus* L.) germplasm using agronomic traits, biochemical and molecular markers. *Genetic Resources & Crop Evolution*, *58*, 425–437.

Mahler-Slasky, Y., & Kislev, M. E. (2010). Lathyrus consumption in late bronze and Iron Age sites in Israel: An Aegean affinity. *Journal of Archaeological Science*, *37*, 2477–2485.

Marsden-Jones, M. (1919). Hybrids of *Lathyrus*. *Journal of the Royal Horticulture Society*, *45*, 92–93.

Mathur, P. N., Alercia, A., & Jain, C. (2005). *Lathyrus germplasm collections directory*. New Delhi, India: International Plant Genetic Resource Institute, IPGRI Office of South Asia. (p. 58).

Maxted, N. (1995). An eco-geographic study of *Vicia* subgenus *Vicia Systematic and eco-geographic studies in crop genepools* (Vol. 8). Rome, Italy: IPGRI.

Maxted, N., & Bisby, F. A. (1986). *IBPGR final report-wild forage legume collection in Syria*. Rome, Italy: IBPGR.

Maxted, N., & Bisby, F. A. (1987). *IBPGR final report-wild forage legume collection in Turkey/Phase I-South-west Turkey*. Rome, Italy: IBPGR.

Maxted, N., Guarino, L., & Shehadeh, A. (2003). *In situ* techniques for efficient genetic conservation and use: A case study for *Lathyrus*. In P. L. Forsline, C. Fideghelli, H. Knüpffer, A. Meerow, J. Nienhus, K. Richards, A. Stoner, E. Thorn, A. F. C. Tombolato, & D. Williams, (Eds.), *XXVI international horticultural congress: Plant genetic resources, the fabric of horticulture's future. Acta horticulturae (ISHS)*, *623*, 41–60.

Mehta, S. L., & Santha, I. M. (1996). Plant biotechnology for development of non-toxic strains of *Lathyrus sativus*. In R. K. Arora, P. N. Mathur, K. W. Riley, & Y. Adham (Eds.), *Lathyrus genetic resources in Asia* (pp. 129–138). Rome, Italy: International Plant Genetic Resources Institute.

Mehta, S. L., Ali, K., & Barna, K. S. (1994). Somaclonal variation in a food legume – *Lathyrus sativus*. *Indian Journal of Plant Biochemistry and Biotechnology*, *3*, 73–77.

Murray, B. G., Bennett, M. D., & Hammett, K. R. W. (1992). Secondary constrictions and NORs of *Lathyrus* investigated by silver staining and *in situ* hybridization. *Heredity*, *68*, 473–478.

Nagarajan, V., & Gopalan, C. (1968). Variation in the neurotoxin (b-*N*-oxalyl amino alanine) content in *Lathyrus sativus* samples from Madhya Pradesh. *Indian Journal of Medical Research*, *56*, 95–99.

Nandini, A. V., Murray, B. G., O'Brien, I. E. W., & Hammett, K. R. W. (1997). Intra-and interspecific variation in genome size in *Lathyrus* (Leguminosae). *Botanical Journal of Linnean Society*, *125*, 359–366.

Narayan, R. K. J., & Durrant, A. (1983). DNA distribution in chromosomes of *Lathyrus* species. *Genetica*, *61*, 47–53.

Narayan, R. K. J., & Rees, H. (1976). Nuclear DNA variation in *Lathyrus*. *Chromosoma*, *54*, 141–154.

Ochatt, S. J., Durieu, P., Jacas, L., & Pontecaille, C. (2001). Protoplast, cell and tissue cultures for the biotechnological breeding of grasspea (*Lathyrus sativus* L.). *Lathyrus Lathyrism Newsletter*, *2*, 35–38.

Ozcan, M., Hayirlioglu, A. S., & Inceer, H. (2006). Karyotype analysis of some *Lathyrus L.* taxa (Fabaceae) from north-eastern Anatolia. *Acta Botanica Gallica*, *153*, 375–385.

Pandey, R. L., Chitale, M. W., Sharma, R. N., & Geda, A. K. (1997). Evaluation and characterization of germplasm of grasspea (*Lathyrus sativus*). *Journal of Medicinal and Aromatic Plant Sciences*, *19*, 14–16.

Pandey, R. L., Chitale, M. W., Sharma, R. N., Kashyap, O. P., Agrawal, S. K., Geda, A. K., et al. (1995). *Catalogue on grasspea (Lathyrus sativus) germplasm*. Raipur, India: Indira Gandhi Krishi Vishwavidyalaya. (p. 60).

Pandey, R. L., Kashyap, O. P., Geda, A. K., & Tripathi, S. K. (1996). Germplasm evaluation for grain yield and neurotoxin content in grasspea (*Lathyrus sativus* L.). *Indian Journal of Pulses Research*, *9*, 169–172.

Pandey, R. L., Kashyap, O. P., Sharma, R. N., Nanda, H. C., Geda, A. K., & Nair, S. (2008). *Catalogue on grasspea (Lathyrus sativus L.) germplasm*. Raipur, India: Indira Gandhi Krishi Vishwavidyalaya. (p. 213).

Panos, D. A. (1940). Pflanzenzuchterische arbeiten in Griechenland im Jahre 1938/39 (Plant breeding in Greece, 1938 to 1939). *Der Zuechter*, *12*, 295–298.

Panos, D. A. (1957). Introduction and breeding of food and fodder crops in Greece. *Euphytica*, *6*, 255–258.

Panos, D. A., Sotiriadis, S., & Fikas, E. (1961). Grassland's progress in Greece. *Der Zuechter*, *31*(2), 37–47.

Pessoa-Filho, M., Rangel, P. H. N., & Ferreira, M. E. (2010). Extracting samples of high diversity from thematic collections of large gene banks using a genetic-distance based approach. *BMC Plant Biology*, *10*, 127.

Ponnaiah, M., Shiferaw, E., Pe, M. E., & Porceddu, E. (2011). Development and application of EST-SSRs for diversity analysis in Ethiopian grasspea. *Plant Genetic Resources*, *9*, 276–280.

Rees, H., & Narayan, R. K. J. (1997). Evolutionary DNA variation in *Lathyrus*. *Chromosomes Today*, *6*, 131–139.

Robertson, L. D., & Abd-El-Moneim, A. M. (1997). Status of *Lathyrus* germplasm held at ICARDA and its use in breeding programs. In P. N. Mathur, V. R. Rao, & R. K. Arora, (Eds.), *Lathyrus genetic resources network, Proceedings of a IPGRI-ICARDA-ICAR regional working group meeting*, 8–10 December 1997, New Delhi, India, pp. 30–41.

Santha, I. M., & Mehta, S. L. (2001). Development of low ODAP somaclones of *Lathyrus sativus*. *Lathyrus Lathyrism Newsletter*, *2*, 42.

Sarker, A., Abd El Moneim, A. M., & Maxted, N. (2001). Grasspea and chicklings (*Lathyrus* L.). In N. Maxted & S. J. Bennett (Eds.), *Plant genetic resources of legumes in the Mediterranean* (pp. 159–180). Dordrecht: Kluwer.

Sarwar, C. D. M., Malek, M. A., Sarker, A., & Hassan, M. S. (1996). Genetic resources of grasspea (*Lathyrus sativus* L.) in Bangladesh. In R. K. Arora, P. N. Mathur, K. W. Riley, & Y. Adham (Eds.), *Lathyrus genetic resources in Asia* (pp. 13–18). Rome, Italy: International Plant Genetic Resources Institute.

Saxena, M. C., Abd El Moneim, A. M., & Raninam, M. (1993). Vetches (*Vicia* spp.) and chicklings (*Lathyrus* spp.) in the farming systems in West Asia and North Africa and improvement of these crops at ICARDA. In J. R. Garlinge & M. W. Perry (Eds.), *Potential for vicia and Lathyrus species as new grain and fodder legumes for Southern Australia* (pp. 2–9). Perth: CLIMA.

Schifino-Wittmann, M. T. (2001). Determinação da quantidade de DNA nuclear em plantas. *Ciência Rural*, *31*, 897–902.

Schifino-Wittman, M. T., Lan, A. H., & Simioni, C. (1994). The genera *Vicia* and *Lathyrus* in Rio Grande do Sul (Southern Brazil): Cytogenetic of native, naturalized and exotic species. *Brazilian Journal of Genetics*, *17*, 313–319.

Seijo, J. G., & Fernandez, A. (2001). Cytogenetic analysis in *Lathyrus japonicus* Wild. (Leguminosae). *Caryologia*, *66*, 173–179.

Sharma, R. N., Kashyap, O. P., Chitale, M. W., & Pandey, R. L. (1997). Genetic analysis for seed attributes over the years in grasspea (*Lathyrus sativus* L.). *Indian Journal of Genetics and Plant Breeding*, *57*(2), 154–157.

Shehadeh, A.A. (2011). Ecogeographic, genetic and genomic studies of the genus *Lathyrus* L. Thesis submitted to University of Birmingham, UK.

Shiferaw, E., Pe, M. R., Porceddu, E., & Ponnaiah, M. (2011). Exploring the genetic diversity of Ethiopian grasspea (*Lathyrus sativus* L.) using EST-SSR markers. *Molecular Breeding*, *30*(*2*), 789–797. doi:10.1007/s11032-011-9662-y.

Siddique, K. H. M., Loss, S. P., Herwig, S. P., & Wilson, J. M. (1996). Growth, yield and neurotoxin (ODAP) concentration of three *Lathyrus* species in Mediterranean type environments of Western Australia. *Australian Journal of Experimental Agriculture*, *36*, 209–218.

Sillero, J. C., Cubero, J. I., Fernández-Aparicio, M., & Rubiales, D. (2005). Search for resistance to crenate broomrape (*Orobanche crenata*) in Lathyrus. *Lathyrus Lathyrism Newsletter*, *4*, 7–9.

Skiba, B., Ford, R., & Pang, E. C. K. (2003). Amplification and detection of polymorphic sequence tagged sites in *Lathyrus sativus*. *Plant Molecular Biology Reporter*, *21*, 391–404.

Soltis, P. S., Soltis, D. E., & Doyle, J. J. (1992). *Molecular systematics of plants*. New York: Chapman and Hall.

Somayajulu, P. I. N., Barat, G. K., Prakash, S., Mishra, B. K., & Shrivastava, V. C. (1975). Breeding of low toxin containing varieties of *L. sativus*. *Proceedings of Nutrition Society of India*, *19*, 35–39.

Sun, X. L., Yang, T., Guan, J. P., Ma, M., Jiang, J. Y., Cao, R., et al. (2012). Development of 161 novel EST-SSR markers from *Lathyrus sativus* (Fabaceae). *American Journal of Botany*, *99*, 379–390. doi:10.3732/ajb.1100346.

Tadesse, W., & Bekele, E. (2003). Variation and association of morphological and biochemical characters in grass pea (*Lathyrus sativus* L.). *Euphytica*, *130*, 315–324.

Talukdar, D. (2009). Dwarf mutations in grasspea (*Lathyrus sativus* L.): origin, morphology, inheritance and linkage studies. *Journal of Genetics*, *88*(*2*), 165–175.

Tavoletti, S., & Iommarini, L. (2007). Molecular marker analysis of genetic variation characterizing a grasspea (*Lathyrus sativus* L.) collection from central Italy. *Plant Breeding*, *126*, 607–611.

Tay, J., Valenzuela, A., & Venegas, F. (1999). Collecting and evaluating Chilean germplasm of grass pea (*Lathyrus sativus* L.). *FABIS Newsletter*, *42*, 3–5.

Tosun, F. (1974). *Forage legume and grasses*. Turkey: Ataturk University Press. (pp. 180–181) No. 242.

Trankovskij, D. A. (1962). Interspecific hybridization in the genus. *Lathyrus*. *Bulletin of Moscow: Nature and Biology Series*, *67*, 140–141.

Unal, F., Wallace, A. J., & Callow, R. S. (1995). Diverse heterochromatin in *Lathyrus*. *Caryologia*, *48*, 47–63.

Upadhyaya, H. D., Bramel, P. J., Ortiz, R., & Singh, S. (2002). Developing a mini core of peanut for utilization of genetic resources. *Crop Science*, *42*, 2150–2156.

Vavilov, N. I. (1951). The origin, variation, immunity and breeding of cultivated plants. *Chronica Botany*, *13*, 13–47.

Yamamoto, K., Fujiware, T., & Blumenreich, L. (1989). Isozymic variation and interspecific crossability in annual species of the genus *Lathyrus* L.. In A. K. Kaul & D. Combes (Eds.), *Lathyrus and lathyrism* (pp. 118–121). New York: Third World Medical Research Foundation.

Yu, J. Z. (1995). Cultivation of grass pea and selection of lower toxin varieties and species of Lathyrus in China. In H. K. M. Yusuf & F. Lambein (Eds.), *Lathyrus sativus and human Lathyrism: Progress and prospects* (pp. 145–151). Dhaka, Bangladesh: University of Dhaka.

Yunus, A.G. (1990). Biosystematics of *Lathyrus* section *Lathyrus* with special reference to the grasspea, *L. sativus* L. *Unpublished Ph.D. Thesis*. UK: University of Birmingham.

Yunus, A. G., & Jackson, M. T. (1991). The gene pools of the grasspea (*Lathyrus sativus* L.). *Plant Breeding*, *106*, 319–328.

Zalkind, F. L. (1933). *Lathyrus* spp. *Rastenievodstvo SSR*, *1*(2), 363–366.

Zalkind, F. L. (1937). *Lathyrus* L.. In E. V. Wulff & N. I. Vavilov (Eds.), *Flora of cultivated plants. Vol. 4: Grain Leguminosae* (pp. 171–227). Moscow & Leningrad: Lenin Academy of Agricultural Sciences Institute of Plant Industry, State Agricultural Publishing Company.

Zhou, M., & Arora, R. K. (1996). Conservation and use of underutilized crops in Asia. In R. K. Arora, P. N. Mathur, K. W. Riley, & Y. Adham (Eds.), *Lathyrus genetic resources in Asia* (pp. 91–96). Rome, Italy: International Plant Genetic Resources Institute.

12 Horsegram

R.K. Chahota[1], *T.R. Sharma*[1], *S.K. Sharma*[1], *Naresh Kumar*[1] *and J.C. Rana*[2]

[1]Department of Agricultural Biotechnology, CSK Himachal Pradesh Agricultural University, Palampur, Himachal Pradesh, India, [2]NBPGR, Regional Research Station, Phagli, Shimla, India

12.1 Introduction

Horsegram (*Macrotyloma uniflorum* (Lam.) Verdcourt (Syn., *Dolichos uniflorus* Lam., *Dolichos biflorus* auct. non L.)) is a pulse and fodder crop native to Southeast Asia and tropical Africa, but the centre of origin of cultivated species is considered to be southern India (Vavilov, 1951; Zohary, 1970). The name *Macrotyloma* is derived from the Greek words *makros* meaning large, *tylos* meaning knob and *loma* meaning margin, in reference to knobby statures on the pods (Blumenthal & Staples, 1993). It is a true diploid having chromosome number $2n=2x=20$. It is cultivated in India, Myanmar, Nepal, Malaysia, Mauritius and Sri Lanka for food purposes and in Australia and Africa primarily for fodder purposes (Asha et al., 2006). The limited use of dry seeds of horsegram is due to its poor cooking quality. However, it is consumed as soups and sprouts in many parts of India (Sudha, Mushtari Begum, Shambulingappa, & Babu, 1995). Owing to its medicinal importance and its capability to thrive under drought-like conditions, the US National Academy of Sciences has identified this legume as a potential food source for the future (National Academy of Sciences, 1978). India is the only country cultivating horsegram on a large acreage, where it is used as human food. However, horsegram is a versatile crop and can be grown from near sea level to 1800 m. It is highly suitable for rainfed and marginal agriculture but does not tolerate frost and waterlogging. It is a drought-tolerant plant and can be grown with rainfall as low as 380 mm. Leaf diseases and root rot are major production constraints in high rainfall areas. Being a leguminous crop, it adds nitrogen to the soils where it grows, thus improving the soil fertility. The protein content in cultivated horsegram is reported to be 16.9–30.4% (Patel, Dabas, Sapra, & Mandal, 1995). It also has high lysine content, an essential amino acid (Gopalan, Ramashastri, & Balasubramanyan, 1989). Horsegram is also rich in phosphorus, iron and vitamins such as carotene, thiamine, riboflavin, niacin and vitamin C (Sodani, Paliwal, & Jain, 2004). It is known to contain many medicinal and therapeutic benefits, although many of them are yet to be proven

Genetic and Genomic Resources of Grain Legume Improvement. DOI: http://dx.doi.org/10.1016/B978-0-12-397935-3.00012-8

scientifically. It can be an ayurvedic medicine, used to treat edema, piles, renal stones, and so on. It has polyphenols that have high antioxidant properties, molybdenum that regulates calcium intake and iron that helps in transporting oxygen to cells and forms part of haemoglobin in blood (Murthy, Devaraj, Anitha, & Tejavathi, 2012; Ramesh, Rehman, Prabhakar, Vijay Avin, & Aditya, 2011). Horsegram is rich source of Haemagglutinin, which is an agent or substance responsible for red blood cells and agglutinate. Chaitanya et al. (2010) proved that the seeds of *M. uniflorum* are endowed with significant antiurolithiatic activity. Certain tests have proven that lipids extracted from horsegram are known to heal rats with peptic ulcers (Jayaraj, Tovey, Lewin, & Clark, 2000). With the continuously expanding need for suitable cultivar development, there is an urgent need for systematic collection, evaluation and utilization of genetic resources for both the present and posterity.

12.2 Origin, Distribution, Diversity and Taxonomy

The origin of horsegram is not clearly mentioned in the literature. Though wild members of *M. uniflorum* exist in both Africa and India (Verdcourt, 1971), its centre of origin as cultivated plant is regarded as India (Purseglove, 1974; Smartt, 1985; Vavilov, 1951; Zohary, 1970). Arora and Chandel (1972) have been more specific in arguing that the primary centre of origin and use of *M. uniflorum* var. *uniflorum* as a cultivated plant is southwestern India. Mehra and Magoon (1974), on the other hand, suggest that *M. uniflorum* has both African and Indian gene centres. The other varieties, var. *stenocarpum* and var. *verucosum*, are basically of African origin, although a wild or long naturalized form is found in northeastern Australia (Bailey, 1900). The region of maximum genetic diversity is considered to be in the Old World tropics, especially the southern part of India and the Himalayas (Zeven and de Wet, 1982). But some studies consider it as a plant native to African countries. It was probably domesticated in India, where its cultivation is known since prehistoric times and it is still an important cultivated crop. Nowadays horsegram is cultivated as a low-grade pulse crop in many Southeast Asian countries, such as India, Bangladesh, Myanmar, Sri Lanka and Bhutan. It is also grown as a forage and green manure in many tropical countries, especially in Australia and Africa, but it is unclear to what extent it is currently grown. The wild relatives of horsegram are reported mainly in Australia, Papua New Guinea, Africa and India. There is no report that horsegram is cultivated as a pulse crop, in central, eastern and southern Africa where it occurs wild. (Blumenthal, O'Rourke, Hidler & Williams, 1989).

Horsegram is a slender, twining annual herb with cylindrical tomentose stems. As a pure crop it cannot stand due to its weak stem and forms a dense mat of 30–60 cm height, but in association with cereals as a mixed crop it may climb on the companion species to a height of 60–110 cm. It has trifoliate leaves, 7–10 mm long persistent stipules and 3–7 cm long petiole. Leaflets are ovate, rounded at base, acute or slightly acuminate, commonly 3.5–7.5 cm long, 2–4 cm broad, length and breadth ratio of 1.5–2.5. Flowers are short, sessile or subsessile10–12 mm long, two- to four-flowered axillary racemes, greenish yellow with a vinous spot on the standard. Calyx is tomentose with 2–3 mm long tube, and the lobes are lanceolate setaceous, 3–8 mm long. Standard is oblong, slightly emarginate at the summit, 9–10.5 mm long, 7–8 mm broad, with two linear appendages about 5 mm long, wings 8–9.5 mm long as long as the keel. Ovary is appressed with dense white hairs, style attenuate and stigma surrounded by a ring of short dense hairs. Pods are stipitate, slightly curved, tomentose, 4.5–6 cm long and about 6 mm broad. Seeds are usually six or seven per pod, 6–8 mm long, 4–5 mm broad, pale fawn, sometimes with faint mottles or with small scattered black spots and hilum placed centrally (Purseglove, 1974).

Initially horsegram was included in the genus *Dolichos* by Linnaeus but Verdcourt (1980) reorganized the different species formerly assigned to *Dolichos* and assigned the genus *Macrotyloma* to horsegram. The style, standard and pollen characteristics distinguish *Macrotyloma* from *Dolichos* (Verdcourt, 1970). Most of the wild species of the genus are restricted to Africa but some wild species have also been reported in Asia and Australia. *M. uniflorum* is the only cultivated species grown in the Indian subcontinent. The horsegram plant belongs to the kingdom Plantae, subkingdom Tracheobionta, division Magnoliophyta and class Magnoliopsida. The genus *Macrotyloma* (Wight & Arn.) Verdc. – *Macrotyloma* of family Fabaceae – consists of about 25 wild species having the chromosome numbers $2n=2x=20$ and $2n=2x=22$ (Allen, O.N. & Allen, E.K. 1981; Lackey, 1981).

Within *M. uniflorum*, four varieties have been distinguished:

1. var. *uniflorum*: pods 6–8 mm wide; wild in southern Asia and Namibia, widely cultivated in the tropics as a cover and forage crop.
2. var. *stenocarpum* (Brenan) Verdc.: pods 4–5.5 mm wide; shortly stiped and with more or less smooth margins, leaflets pubescent; occurring in central, eastern and southern Africa and in India, up to 1700 m altitude in grassland, bushland and thicket, often on sandy soils and in disturbed locations; cultivated in Australia and California (the United States).
3. var. *verrucosum* Verdc.: pods 4–5.5 mm wide; distinctly stiped and with obscurely to markedly warted margins, leaflets pubescent; occurring in eastern and southern Africa up to 550 m altitude in grassland and thicket.
4. var. *benadirianum* (Chiov.) Verdc.: pods 4–5.5 mm wide; shortly stiped and with slightly warted margins, leaflets densely velvety; occurring in East Africa (Somalia, Kenya) at sea level on sand dunes and thin soils on coral rag.

The geographical distribution of different species is provided in Table 12.1. It effectively nodulates with nitrogen-fixing bacteria of the *Bradyrhizobium* group (Brink, 2006).

12.3 Erosion of Genetic Diversity from the Traditional Areas

The quest for increasing food production and the ensuing success achieved in major crops has increased the thrust and expectations to repeat the success in other minor crops. Variability refers to heterogeneity of alleles and genotypes with their attendant morphotypes and phenotypes. Genetic erosion implies that disappearance of genetic variability in a population is altered so that the net change in diversity is negative. Considerable genetic erosion started in the early 1960s due to changes in

Table 12.1 Geographical Distribution of *Macrotyloma* Species

S. No.	Species Name	Area of Distribution
1	*Macrotyloma africanum* (Wilczek) Verdc.	Africa
2	*Macrotyloma axillare* (E.Mey.) Verdc.	Africa and Australia
3	*Macrotyloma bieense* (Torre) Verdc.	Africa
4	*Macrotyloma biflorum* (Schum. & Thonn.) Hepper	Africa
5	*Macrotyloma brevicaule* (Baker) Verdc.	Africa
6	*Macrotyloma ciliatum* (Willd.) Verdc.	Asia and Africa
7	*Macrotyloma coddii* Verdc.	Africa
8	*Macrotyloma daltonii* (Webb) Verdc.	Africa
9	*Macrotyloma decipiens* Verdc.	Africa
10	*Macrotyloma densiflorum* (Baker) Verdc.	Africa
11	*Macrotyloma dewildemanianum* (Wilczek) Verdc.	Africa
12	*Macrotyloma ellipticum* (R.E.Fr.) Verdc.	Africa
13	*Macrotyloma fimbriatum* (Harms) Verdc.	Africa
14	*Macrotyloma geocarpum* (Harms) Marechal & Baudet	Africa
15	*Macrotyloma hockii* (De Wild.) Verdc.	Africa
16	*Macrotyloma kasaiense* (R. Wilczek) Verdc.	Africa
17	*Macrotyloma maranguense* (Taub.) Verdc.	Africa
18	*Macrotyloma oliganthum* (Brenan) Verdc.	Africa
19	*Macrotyloma prostratum* Verdc.	Africa
20	*Macrotyloma rupestre* (Baker) Verdc.	Africa
21	*Macrotyloma schweinfurthii* Verdc.	Africa
22	*Macrotyloma stenophyllum* (Harms) Verdc.	Africa
23	*Macrotyloma stipulosum* (Baker) Verdc.	Africa
24	*Macrotyloma tenuiflorum* (Micheli) Verdc.	Africa
25	*Macrotyloma uniflorum* (Lam.) Verdc.	Asia, Africa and Australia

cropping pattern and induction of new crops in the Indian farming system. However, population growth, urbanization, developmental pressures on the land resources, deforestation, changes in land use patterns and natural disasters are contributing to considerable habitat fragmentation and destruction of the crops and their wild relatives. Horsegram is a neglected crop cultivated by poor and marginal farmers in tribal localities and drought-prone areas of India (Jansen, 1989). There are no concerted efforts for varietal developments reported from any part of the world, barring some isolated efforts in a few research institutions in India. Therefore, genetic erosion is not attributable in this case to the diffusion of high-yielding varieties to replace the landraces. Rather, the main cause of genetic is the cultivation of commercial crops in the horsegram–growing areas.

12.4 Status of Germplasm Resources Conservation

Horsegram is an important pulse crop of Indian sub-continent; therefore, the efforts to conserve the germplasm at global level are also lacking. Therefore, most of the conservation work was undertaken by Indian Institutes. The Germplasm Resources Information Network (GRIN) of the US Department of Agriculture (USDA) has conserved only 35 accessions of horsegram in its gene bank. Protabase, responsible for germplasm conservation for African countries, has 21 accessions at the National Gene Bank of Kenya, Crop Plant Genetic Resources Centre, Kenya Agricultural Research Institute (KARI), Kikuyu, Kenya. The Australian Tropical Crops and Forages Genetic Resources Centre, Biloela, Queensland has 38 accessions of horsegram germplasm (Brink, 2006). Only the National Bureau of Plant Genetic Resources (NBPGR) in New Delhi has a systematic collection of this important legume. The efforts to collect and conserve the horsegram germplasm started way back in the 1970s with the inception of the PL480 scheme (a scheme under collaboration between Indian Council of Agricultural Research (ICAR) and the USDA project on food security in Haiti, using Public Law 480), and since then germplasm has been collected from almost all the horsegram–growing areas. Under different exploration and collection programmes, a total of 1627 accessions of horsegram have been collected and maintained at different satellite stations of NBPGR.

12.5 Germplasm Evaluation and Maintenance

Horsegram is being treated as a orphan crop therefore much attention has not been paid to the systematic evaluation of germplasm, except maintaining it in the gene banks. There are only a limited number of accessions conserved in the gene banks worldwide. *Ex situ* conservation by different countries is given in Table 12.2. In India, a total of 1627 accessions of horsegram are conserved in the national gene bank, and out of these 1161 accessions were characterized during 1999–2004. Latha (2006) made some observations while studying on agro-morphological traits in Indian *Dolichos* germplasm that yield and yield component traits in general showed that all promising lines with higher seed yield are of long duration type. The seed yield per plant ranged from 0.22 to 7.31 g in short duration type, from 0.27 to 7.07 g in medium duration and from 0.21 to 11.86 g in long duration type. Rana (2010) also observed variability in qualitative characteristics and revealed that growth habit ranged from semi-erect to vine types, leafiness between sparse and abundant, leaf pubescence from puberulant to densely pubescent and stem

Table 12.2 *Ex Situ* Conservation at Different Gene Banks of the World

S. No.	Country	Name of the Organization	Accessions
1.	India	National Bureau of Plant Genetic Resources, New Delhi	1627
2	The United States	Germplasm Resources Information Network of US Department of Agriculture	35
3	Australia	Tropical Crops and Forages Genetic Resources Centre, Biloela, Queensland	38
4	Kenya	National Gene Bank of Kenya, Crop Plant Genetic Resources Centre, KARI, Kikuyu	21

Table 12.3 Evaluation of Germplasm by Different Institutes for Agro-Morphological Traits

S. No.	Name of the Institute	Number of Germplasm Accessions Evaluated	Year of Evaluation	References
1	National Bureau of Plant Genetic Resources, New Delhi, India	1426	1984–1990	Patel et al. (1995)
2	NBPGR, New Delhi, India	22	2005 and 2006	Latha, (2006)
3	Parvartiya Krishi Anusdhan Kendra Almora, India	10	2007	Mahajan et al. (2007)
4	Himachal Pradesh Agricultural University, Palampur, India	63	2005	Chahota et al. (2005)
5	Commerce and Science College Jalna, India	22	2011	Kulkarni (2010)

colour between green and purple. However, range in variability was maximum in pod and seed colour. Mature pod colour varied from straw, tan, cream, light brown, brown, dark brown to brownish black. The plant height ranged from 17 to 145 cm and primary branches per plant varied from 1.0 to 9.8 in number. Other yield component traits such as pods per plant (4–148), pod length (3.07–6.17 cm), 100-seed weight (0.92–4.10 g) and biological yield (0.21–11.86 g) revealed variability. NBPGR has published a catalogue with details of 11 economically important traits of 1426 accessions. During the *Kharif* (autumn) of 1984–1990, about 506 accessions at New Delhi and 920 accessions at NBPGR satellite research station Akola were characterized and documented on the basis of evaluation data for various qualitative and quantitative traits (Patel et al., Dabas, Sapra & Mandal, 1995). The Vivekanand Parvartiya Krishi Anusandhan Sansthan (VPKAS), Almora, has evaluated 10 lines for agro-morphological traits (Mahajan et al., 2007). Chahota, Sharma, Dhiman, and Kishore (2005) evaluated 63 horsegram accessions procured from NBPGR, Phagli, Shimla for 12 agro-morphologic characters at CSK Himachal Pradesh Agricultural University, Palampur. Kulkarni and Mogle (2011) and Kulkarni (2010) evaluated 22 germplasm lines for different agronomic traits and identified five high-yielding genotypes. Sudha et al. (1995) and Subba Rao and Sampath (1979) evaluated horsegram lines for various nutritional and anti-nutritional factors (Table 12.3). An attempt was made by Prakash, Channayya Hiremath, Devarnavdgi & Salimath (2010) to assess the genetic divergence among 100 lines collected from different parts of Karnataka, using Mahalanobis D^2 statistics. D^2 is the distance between the different clusters having lines. In addition, considerable numbers of studies have been conducted on various aspects of the crop by several researchers (Dobhal & Rana, 1994a, 1994b; Jayan & Maya, 2001; Joshi, Chikkadevaiah, & Shashidhas, 1994; Lad, Chavan, & Dumbre, 1999; Patil, Deshmukh, & Singh, 1994; Savithriamma, Shambulingappa, & Rao, 1990; Nagaraja, Nehru, & Manjunath, 1999; Sharma, 1995; Tripathi, 1999).

12.6 Use of Germplasm in Crop Improvement

New plant resources need to be exploited in order to meet the growing needs of human society, which incidentally has depended on only a small portion of plant wealth. Accordingly, many of underutilized plants have the potential for improving agriculture in various ways and have great potential for exploitation in view of the value of their economic products (Bhag & Joshi, 1991). Although a lot of germplasm has been collected from different parts of the world and conserved in the national gene banks of different countries, very little effort has been made to improve this plant as a commercial crop. The lack of efforts both at institutional and governmental levels has undermined the importance of this crop. The evaluation and documentation of germplasm have not been updated in many countries, so the utilization of germplasm could not be taken up by the concerned breeders. In India, there are about 1800 accessions of horsegram germplasm, of which only 912 lines have been evaluated and documented. The genetic improvement of horsegram has been undertaken at just a few institutions in India, but no improvement programme is in place at the global level.

In India, the cultivars released for cultivation are region specific and do not hold promise for commercial agriculture, as the plant types contain many weedy traits, such as twining and indeterminate growth habit, asynchronous and delayed maturity and photosensitivity. Sufficient diversity is available for different traits as revealed by germplasm evaluation data, but effort are lacking to develop ideal cultivars or to introgress desirable traits scattered in different genotypes. Hybridization studies conducted between photosensitive and day neutral varieties with black and brown coloured seeds revealed that photoperiod response is a qualitative trait that is controlled by at least two genes. In case of inheritance of seed colour, the black seed colour was observed to be dominant over brown. Two genes in polymeric gene action were found to control seed colour (Sreenivasan, 2003). Most of the horsegram varieties released for cultivation in different states in India originated from the local germplasm following their effective and specific evaluation. The varieties developed in different states (Table 12.4) include BR 5, BR 10 and Madhu from Bihar; HPK-2 and HPK-4 from Himachal Pradesh; PDM 1 and VZM 1 from Andhra Pradesh;

Table 12.4 Improved Varieties Released by Different States in India for Cultivation

S. No.	Variety	Place of Release
1	BR 5, BR 10 and Madhu	Bihar
2	HPK-4 and VLG 1	Himachal Pradesh
3	PDM 1 and VZM 1	Andhra Pradesh
4	K82 and Birsa Kulthi	Jharkhand
5	S27, S8, S39 and S1264	Orissa
6	Co-1, 35-5-122 and 35-5-123	Tamil Nadu
7	Hebbal Hurali 2, PHG 9 and KBH 1	Karnataka
8	Maru Kulthi, KS 2, AK 21 and AK 42	Rajasthan
9	VLG 1	Uttarakhand

K82 and Birsa Kulthi from Jharkhand; S27, S8, S39 and S1264 for Orissa; Co-1, 35-5-122 and 35-5-123 from Tamil Nadu; Hebbal Hurali 2, PHG 9 and KBH 1 from Karnataka; Maru Kulthi, KS 2, AK 21 and AK 42 from Rajasthan and VLG 1 from Uttarakhand. Some of the improved varieties developed through single plant selection from the bulk collected included Co-1. No 35-5-122 and 123. Hebbal Hurali 1 and 2 were developed by the single plant selection (Kumar, 2005).

The nonavailability of important traits in the germplasm has encouraged many workers to induce desirable traits by using gamma radiation and chemical mutants. Gupta, Sharma, and Rathore (1994) induced variability for seed yield per plant, biological yield per plant, pods per plant, pod length, seeds per pods and 100-seed weight. Jamadagani and Birari (1996) developed three photo-insensitive mutants by irradiating a photosensitive variety Dapali-1 with 20 kR. Ramakanth, Setharama, and Patil (1979) attempted to induce mutation following treatment with five doses of gamma rays. Chahota (2009) treated HPKC-2, a promising line, with a 25-kR dose of gamma radiation and succeeded in inducing important agronomic traits in horsegram. Wild forms of horsegram have also been reported in the Western Ghats, especially in the wildlife sanctuaries. *Macrotyloma ciliatum* (Willd.) Verdc. is found in Tamil Nadu (Mathew, 1983; Nair & Henry, 1983) and Andhra Pradesh (Pullaiah & Chennaiah, 1997). *Macrotyloma sar-garhwalensis* is a wild relative of horsegram found in the Central Himalayas of India (Gaur & Dangwal, 1997). It is a non-twining annual herb with a high protein content of 38.35%, which can be utilized in the breeding programmes for the improvement of protein content (Negi, Yadav, Mandal & Bhandari, 2002). *Macrotyloma axillare* and *M. africanum*, the two other species of this genus, have also shown potential as forage plants.

12.7 Germplasm Enhancement Through Wide Crosses

Horsegram is cultivated as a pulse crop only in the Indian subcontinent, whereas in rest of the world, it is cultivated as a feed and fodder crop for animals. In pastures and grasslands broadcasting of seeds is done to improve the grass quality. The major bottleneck in the improvement of this crop is the lack of variability at the morphological as well as molecular level. Therefore, wide hybridization could be a useful tool to create additional variability for broadening its base. Though the genus *Macrotyloma* consists of more than 25 species, there is no report regarding the evaluation of these wild species for desirable traits. Morris (2008) compared *M. uniflorum* with *M. axillare* and described a set of descriptors to differentiate these species. Evaluation of few wild species of *Macrotyloma* has been undertaken at the CSK, Himachal Pradesh Agricultural University, Palampur, India, to initiate a systematic hybridization programme involving cultivated and wild species to transfer desirable traits from *M. axillare* and *Macrotyloma sar-garhwalensis* to cultivated background. *M. axillare* has many desirable traits such as high number of pods per plant, high seed yield per plant and tolerance to cold and drought conditions (Staples, 1966, 1982). The cultivated species of *M. uniflorum* is infected by a number of diseases, particularly in high rainfall areas, such as Anthracnose, *Cercospora* leaf spot, *Fusarium* wilt, rust, *Pellicularia* root rot and *Aschochyta* blight. Though the *M. axillare* is reported to have resistance against many diseases hybridization between *M. uniflorum* and *M. axilare* resulted in juvenile flowering in the first year of F_1 plant, hence prolonging the breeding process.

12.8 Horsegram Genomic Resources

The horsegram plant is considered unsuitable for commercial cultivation due to the presence of many undesirable traits, such as longer days to maturity accompanied by asynchrony, photosensitivity and indeterminate growth habit. However, some work on the development of suitable ideotype is being conducted at CSK, Himachal Pradesh Agricultural University, Palampur since 1995. Various breeding techniques are being used to improve the plant type. Furthermore, it was felt that before embarking on a breeding programme, the information on genetics of different traits of interest is also an important aspect to combine important characters in the well-adapted genetic backgrounds. The lack of genomic information in *M. uniflorum* in particular and *Macrotyloma* genus in general is another hurdle for its systematic breeding.

The legume family has been divided into three subfamilies, namely Casalpinieae, Mimosoideae and Papilionoideae. Most of the economically important legumes are members of the monophylotic subfamily paplionoideae, which is further divided into four clades; clade *phaseoloids* have important warm-season legumes such as *Glycine*, *Phaseolus*, *Vigna*, *Cajanus* and *Macrotyloma* species (Doyle & Luckow, 2003; Gept et al., 2005). There is complete genomic information available for the two model legumes, *Medicago truncatum* and *Arabidopsis*, but that may not be very useful in horsegram due to its distance from the warm-season grain legumes, as they are in another clade. The recently sequenced *Cajanus cajan* genome can act as the model plant for these orphaned warm-season legume crops. Therefore, sequence information available in *C. cajan* can be crucial in understanding comparative genomics of horsegram. Marker resources can also be used for constructing linkage maps and identifying genomic regions linked to traits of agronomic value. Such cross-species genetic information may be very important for 'orphan crops' such as horsegram that have limited or no genomic resources available. Intron-targeted amplified polymorphism (ITAP) markers among various legumes have a very high degree of transferability rate and have been used to prepare linkage maps of *Lupinus albus* (Phan, Ellwood, Adhikari, Nelson, & Oliver, 2006). Similarly a consensus genetic map of cowpea has been developed from the genetic information available in *Glycine* and *Phaseolus* species (Wellington et al., 2009). Some preliminary work in this direction has been initiated at CSK, Himachal Pradesh Agricultural University, Palampur to study the transferability of genomic Simple Sequnce Repeats (SSR) markers of related legume species to prepare a framework genetic linkage map of horsegram. This map will help to initiate a scientific breeding programme or marker-assisted selection to develop improved plant types of horsegram.

12.9 Conclusions

Horsegram is an important pulse crop of the Indian subcontinent; therefore, collection and systematic evaluation work on the germplasm are confined to India only. A total of 1721 accessions of horsegram are being conserved in different gene banks around the world. Of these collections, about 95% are conserved in the NBPGR, New Delhi, and its regional research station. Regional Reserch Station of NBPGR, Thrissur, Kerala, has been designated as the active site for the conservation and evaluation of horsegram germplasm amassed in Indian gene banks. All these accessions need proper characterization and evaluation to enable their exploitation in a horsegram breeding programme. Molecular markers provide precise information on genetic diversity and help in more rapid breeding gain when it used in Markers Assisted Selection (MAS). But unfortunately, in spite of its medicinal importance and drought tolerance, the potential of this crop has not been realized by the government, nor at the institutional levels. Very few researchers have explored its phenotypic and biochemical diversity, while diversity at the DNA level is totally lacking and no molecular markers has been developed in this crop to date.

References

Allen, O. N., & Allen, E. K. (1981). *The Leguminosae. A source book of characteristics, uses and nodulation.* Madison, WI: The University of Wisconsin Press.

Arora, R. K., & Chandel, K. P. S. (1972). Botanical source areas of wild herbage legumes in India. *Tropical Grasslands, 6*, 213–221.

Asha, K. I., Latha, M., Abraham, Z., Jayan, P. K., Nair, M. C., & Mishra, S. K. (2006). Genetic resources. In D. Kumar (Ed.), *Horsegram in India (pp. 11–28).* Jodhpur, India: Scientific Publisher.

Bailey, F. M. (1900). *The Queensland flora. Part II.* Brisbane: Government Printer.

Bhag, M., & Joshi, V. (1991). Under-utilised plant resources. In R. S. Paroda & R. K. Arora (Eds.), *Plant Genetic Resources-conservation and management* (p. 211). New Delhi, India: Malhotra Publishing House.

Blumenthal, M. J., O'Rourke, P. K., Hilder, T. B., & Williams, R. J. (1989). Classification of the Australian collection of the legume *Macrotyloma. Australian Journal of Agricultural Research, 40*, 591–604.

Blumenthal, M. J., & Staples, I. B. (1993). Origin, evaluation and use of *Macrotyloma* as forage – a review. *Tropical Grassland, 27*, 16–29.

Brink, M. (2006). *Macrotyloma uniflorum* (Lam.) Verdc. In M. Brink & G. Belay (Eds.), *PROTA1: Cereals and pulses.* Wageningen, the Netherlands: PROTA.

Chahota, R.K. (2009). Induction of early flowering and other important traits in horse gram (*Macrotyloma uniflorum*) using gamma irradiation. Final project report submitted to Bhabha Atomic Research Centre, Trombay, India.

Chahota, R. K., Sharma, T. R., Dhiman, K. C., & Kishore, N. (2005). Characterization and evaluation of horse gram (*Macrotyloma uniflorum* Roxb.) germplasm from Himachal Pradesh. *Indian Journal of Plant Genetic Resources, 18*, 221–223.

Chaitanya, D. A. K., Santosh Kumar, M, Reddy, A. M., Mukherjee, N. S. V., Sumanth, M. H., & Ramesh, D. A. (2010). Anti urolithiatic activity of *Macrotyloma uniflorum* seed extract on ethylene glycol induced urolithiasis in albino rats. *Journalof Innovative Trends in Pharmaceutical Sciences*, *1*(5), 216–226.

Dobhal, V. K., & Rana, J. C. (1994a). Multivariate analysis in horsegram. *Legume Research*, *17*(3-4), 157–161.

Dobhal, V. K., & Rana, J. C. (1994b). Genetic analysis for yield and its components in horsegram *(Macrotyloma uniflorum)*. *Legume Research*, *17*(3-4), 179–182.

Doyle, J. J., & Luckow, M. A. (2003). The rest of the iceberg. Legume diversity and evolution in a phylogenetic context. *Plant Physiology (Rockville)*, *131*, 900–910.

Gaur, R. D., & Dangwal, L. R. (1997). A new species of macrotyloma from Garhwal Himalaya, U.P., India. *Bombay Natural Historical Society*, *94*, 381–383.

Gept, P., Beavis, W. D., Brummer, E. C., Shoemaker, R. C., Stalker, H. T., Weeden, N. F., et al. (2005). Legumes as a model plant family. Genomics for food and feed report of the cross-legumes advances through genomics conference. *Plant Physiology*, *137*, 1228–1235.

Gopalan, C., Ramashastri, B. V., & Balasubramanyan, S. C. (1989). *Nutritive value of Indian foods*. Hyderabad, India: National Institute of Nutrition, ICMR.

Gupta, V. P., Sharma, S. K., & Rathore, P. K. (1994). Induced mutagenic response of gamma irradiation on horsegram (*Macrotyloma uniflorum*). *Indian Journal of Agricultural Sciences*, *64*, 160–164.

Jamadagani, B. M., & Birari, S. P. (1996). Modification in photoperiodic response of flowering by gamma-rays induced mutation in horsegram. *Journal of Maharashtra Agricultural University*, *21*, 384–386.

Jansen, P. C. M. (1989). *Macrotyloma uniflorum* (Lam.) Verdc.: *Plant resources of South-East Asia, pulses*. (pp. 53–54). Wageningen, the Netherlands: Pudoc.

Jayan, P. K., & Maya, C. N. (2001). Studies on germplasm collections of horsegram. *Indian Journal of Plant Genetic Resources*, *14*, 443–447.

Jayaraj, A. P., Tovey, F. I., Lewin, M. R., & Clark, G. C. (2000). Duodenal ulcer prevalence: experimental evidence for possible role of dietary lipids. *Journal of Gastroenterology and Hepatology*, *15*, 610–616.

Joshi, S. S., Chikkadevaiah, & Shashidhas, H. E. (1994). Classification of horse gram germplasm. *Crop Research*, *7*(3), 375–391.

Kulkarni, G. B. (2010). Evaluation of genetic diversity of horse gram (*Macrotyloma uniflorum*) germplasm through phenotypic trait analysis. *Green Farming*, *1*(6), 563–565.

Kulkarni, G. B., & Mogle, U. P. (2011). Characterization and evaluation of horse gram (*Macrotyloma uniflorum* (Lam.) Verdcourt) genotypes. *Science Research Reporter*, *1*(3), 122–125.

Kumar, D. (2005). Status and direction of arid legumes research in India. *Indian Journal of Agriculture Sciences*, *75*, 375–391.

Lackey, J. A. (1981). Phaseolus. In R. M. Polhill & P. H. Raven (Eds.), *Advances in legume systematic Part I* (pp. 301–327). Kew, London: Royal Botanic Garden.

Lad, B. D., Chavan, S. A., & Dumbre, A. D. (1999). Genetic variability and correlation studies in horsegram. *Journal of Maharashtra Agricultural University*, *23*, 268–271.

Latha, M. (2006). Germplasm characterization and evaluation of horticultural crops. Annual report of NBPGR report-2006, New Delhi.

Mahajan, V., Shukla, S. K., Gupta, N. S., Majumdera, D., Tiwari, V., & Piiasad, S. V. S. (2007). Identifying phenotypically stable genotypes and developing strategy far yield improvement in horsegram (*Macrotyloma uniflorum*) for mid-altitudes of Northwestern Himalayas. *Indian Journal of Agricultural Sciences*, *3*, 19–22.

Mathew, K. M. (1983). *The flora of Tamil Nadu Carnatic. Part one Polypetalae*. Tiruchirapalli, India: The Rapinat Herbarium, St. Joseph's College.

Mehra, K.L. & Magoon, M.L., (1974). Gene centres of tropical and sub-tropical pastures legumes and their significance in plant introduction. *Proceedings of the XII International Grassland Congress*, (pp. 251–256) Moscow.

Morris, J. B. (2008). *Macrotyloma axillare* and *M. uniflorum*: descriptor analysis, anthocyanin indexes, and potential uses. *Genetic Resources Crop Evolution, 55*, 5–8.

Murthy, S. M., Devaraj, V. R., Anitha, P., & Tejavathi, D. H. (2012). Studies on the activities of antioxidant enzymes under induced drought stress *in vivo* and *in vitro* plants of *Macrotyloma uniflorum* (Lam.) Verdc. *Recent Research in Science and Technology, 4*, 34–37.

Nagaraja, N., Nehru, S. D., & Manjunath, A. (1999). Plant types for high yield in horsegram as evidenced by path coefficients and selection indices. *Karnataka Journal of Agricultural Sciences, 12*(1-4), 32–37.

Nair, N. C., & Henry, A. N. (1983). *Flora of Tamil Nadu. India Series* I: *Analysis*: (Vol. 1). (pp. 1–184). Coimbatore, India: Botanical Survey of India Southern Circle.

National Academy of Sciences, (1978). *Mothbean in tropical legumes : Resources for the future*. Washington, DC: National Academy of Sciences.

Negi, K. S., Yadav, S., Mandal, S., & Bhandari, D. C. (2002). Registration of wild horsegram germplasm (INGR No. 02007; IC2I2722). *Indian Journal of Plant Genetic Resources, 15*(3), 300–301.

Patel, D. P., Dabas, B. S., Sapra, R. S., & Mandal, S. (1995). *Evaluation of horsegram (Macrotyloma uniflorum) (Lam.) germplasm*. New Delhi, India: National Bureau of Plant Genetic Resources Publication.

Patil, J. V., Deshmukh, R. B., & Singh, S. (1994). Selection of genotypes in horsegram based on genetic diversity. *Journal of Maharashtra Agricultural University, 19*, 412–414.

Phan, H. T. T., Ellwood, S. R., Adhikari, K., Nelson, M., & Oliver, R. P. (2006). The first genetic and comparative map of white Lupin (*Lupinus albus* L.) identification of QTLs for anthracnose resistance and flowering time, and a locus for alkaloid content. *DNA Research, 14*, 59–70.

Prakash, B. G., Channayya Hiremath, P., Devarnavdgi, S. B., & Salimath, P. M. (2010). Assessment of genetic diversity among germplasm lines of horsegram (*Macrotyloma uniflorum*) at Bijapur. *Electronic Journal of Plant Breeding, 1*, 414–419.

Pullaiah, T., & Chennaiah, E. (1997). *Flora of Andhra Pradesh (India), Rananculace-Alangiaceae: Vol. 1*. Jodhpur, India: Scientific Publishers.

Purseglove, J. W. (1974). *Dolichos uniflorus. Tropical crops: Dicotyledons*. (pp. 263–264). London and New York: Longman.

Ramakanth, R. S., Setharama, A., & Patil, N. M. (1979). Induced mutations in dolichos lablab. *Mutation Breeding Newsletter, 14*, 1–2.

Ramesh, C. K., Rehman, A., Prabhakar, B. T., Vijay Avin, B. R., & Aditya, S. J. (2011). Antioxidant potential in sprouts vs. seeds of *Vigna radiata* and *Macrotyloma uniflorum. Journal of Applied Pharmaceutical Sciences, 1*, 99–103.

Rana, J.C., (2010). Evaluation of horsegram germplasm. *Annual progress report of the NBPGR*, Shimla, India.

Savithriamma, D. L., Shambulingappa, K. G., & Rao, M. G. (1990). Evaluation of horsegram germplasm. *Current Research, 19*(9), 150–153.

Sharma, R. K. (1995). Nature of variation and association among grain yield and yield components in horsegram. *Crop Improvement, 22*(1), 73–76.

Smartt, J. (1985). Evolution of grain legumes. II. Old and new world pulses of lesser economic importance. *Experimental Agriculture, 21*, 1–18.

Sodani, S. N., Paliwal, R. V., & Jain, L. K. (2004). Phenotypic stability for seed yield in rain-fed horse gram (*Macrotyloma uniflorum* (Lam.) Verdc.): *Proceedings of the national symposium on arid legumes for sustainable agriculture and trade*. Jodhpur, India: Central Arid Zone Research Institute. November 5–7.

Sreenivasan, E. (2003). Hybridization studies in horse gram. In A. Henry, D. Kumar, & N. B. Singh (Eds.), *Advances in arid legumes research* (pp. 115–121). Jodhpur, India: Scientific Publishers.

Staples, I. B. (1966). A promising new legume. *Queensland Agricultural Journal*, *92*, 388–392.

Staples, I.B., (1982). Intra and interspecific crosses in *Macrotyloma axillare*. *Proceedings of the Second Australian Agronomy Conference*, p. 340, Wagga.

Subba Rao, A., & Sampath, S. R. (1979). Chemical composition and nutritive value of horsegram (*Dolichos biflorus*). *Mysore Journal of Agricultural Sciences*, *13*, 198–205.

Sudha, N., Mushtari Begum, J., Shambulingappa, K. G., & Babu, C. K. (1995). Nutrients and some anti-nutrients in horsegram (*Macrotyloma uniflorum* (Lam.) Verdc.) *Food and Nutrition Bulletin*, *16*, 81–83.

Tripathi, A. K. (1999). Variability in horsegram. *Annals of Agricultural Research*, *20*, 382–383.

Vavilov, N. I. (1951). Phytogeographic basis of plant breeding. *Chronica Botanica*, *13*, 13–54.

Verdcourt, B. (1970). Studies in the *Leguminosae–Papilionoideae* for the 'Flora of Tropical East Africa', III. *Kew Bulletin*, *24*, 379–447.

Verdcourt, B. (1971). Phaseoleae. In J. C. Gillet, R. M. Polhill, & V. Verdcourt (Eds.), *Flora of east tropical Africa. Leguminosae–subfamily Papilionoideae* (*Vol. 2*, pp. 581–594). Crown Agents.

Verdcourt, B. (1980). The classification of *Dolichos* L. emend. Verdc., *Lablab* Adans., *Phaseolus* L., *Vigna* Savi and their allies. In R. J. Summerfield & A. H. Bunting (Eds.), *Advances in legume science* (pp. 45–48). London, Kew Royal Botanic Gardens.

Wellington, M., Diop, N. N., Bhat, P. R., Fenton, R. D., Wanamaker, S., Porrorff, M., et al. (2009). A consensus genetic map of cowpea (*Vigna unguiculata* (L.) Walp.) and synteny based on EST-derived SNPs. *PNAS*, *27*, 18159–18164.

Zeven, A. C., & de Wet, J. M. J. (1982). *Dictionary of cultivated plants and their regions of diversity*. Wageningen, the Netherlands: Centre for Agricultural Publishing and Documentation.

Zohary, D. (1970). Centers of diversity and centers of origin. In O. H. Frankel & F. Bennett (Eds.), *Genetic resources in plants* (pp. 33–42). Oxford: Blackwell Scientific Publications.

Printed in the United States
By Bookmasters